Times of Feast,
Times of Famine

A HISTORY OF CLIMATE

SINCE THE YEAR 1000

Emmanuel Le Roy Ladurie

TIMES OF FEAST, TIMES OF FAMINE
A History of Climate Since the Year 1000

TRANSLATED BY BARBARA BRAY

THE NOONDAY PRESS

FARRAR, STRAUS AND GIROUX

NEW YORK

CONTENTS

FIGURES

PLATES

APPENDICES

MANUSCRIPT SOURCES

Departmental archives of Haute-Savoie: Series G: 10 G 204 to 329: collegiate church of Sallanches, papers of the Chamonix priory (very abundant throughout). I examined in particular 10 G 226 to 265 (accounts) and 307 to 329 (legal proceedings). Also 1 G 98 to 130 (pastoral visitations by the bishops of Geneva, to which see the priory of Chamonix belongs). Lastly, I consulted, in the collection of early surveys, the *old map* of Chamonix (1730).

Communal archives of Chamonix (Haute-Savoie): In particular the series HH, and most of all CC, which contain records of the inquiries into disasters caused by the glaciers in the seventeenth century.

For manuscript sources for wine harvest dates and series relating to meteorological events, see Appendices 11 and 12, and *Les paysans de Languedoc,* I, Chapter 1, and II, Appendices 1, 2, and 3.

Bibliothèque nationale, cabinet des Estampes (Print Room), dossiers on the États sardes and Savoie, cotés Vb 1 and Vb 2 (pictorial records of the glaciers).

Archives nationales: LL 1210, green book of the abbey of Saint-Denis (fourteenth–fifteenth century).

ACKNOWLEDGMENTS

This book could not have been written without the constant and disinterested help of those who have followed and assisted me in my researches, sometimes for more than ten years: in particular, Fernand Braudel, professor at the Collège de France; Ernest Labrousse and Pierre Pédelaborde, professors at the Sorbonne; and Maurice Arléry of the Météorologie nationale. I should also like to thank my friends Jacques Bertin and Janine Récurat, as well as Jacques Langevin, who gave me valuable help with the maps. I am indebted to my wife for most of the photographs.

ABBREVIATIONS

AC	Archives communales
AC Cham.	Archives communales of Chamonix
ACM	Archives communales of Montpellier
ADHS	Archives départementales of Haute-Savoie
A.I.H.S.	Association internationale d'hydrologie scientifique
A.N.Y.A.S.	*Annals of the New York Academy of Science*
B.A.G.F.	*Bulletin de l'Association des Géographes français*
Fonds Chobaut	Fonds Chobaut preserved in the Calvet Museum at Avignon; dossiers on meteorological disasters and on wine harvests
Geog. Ann.	*Geografiska Annaler*
Geog. Rev.	*Geographical Review*
J. Geol.	*Journal of Geology*
J. Glac.	*Journal of Glaciology*
Mét.	*La Météorologie*
M.G.A.	Meteorological and Geo-astrophysical Abstracts
M.L.R.L.	Madeleine Le Roy Ladurie
Q.J.R.M.S.	*Quarterly Journal of the Royal Meteorological Society*
R.G.D.	*Revue de géomorphologie dynamique*
Z.D.O.A.	*Zeitschrift des deutschen une oesterreichen Alpenvereins*
Z.f. Glk.	*Zeitschrift für Gletscherkunde*

Times of Feast,
Times of Famine

A HISTORY OF CLIMATE

SINCE THE YEAR 1000

INTRODUCTION

Historians of climate can be ranged for the most part in one of two categories, according to their intellectual origins. Some, like Bryson, Fritts, Lamb, Mitchell, and Von Rudloff, are experts in the natural sciences: biologists, or more often meteorologists, they quite naturally want to complement the knowledge and explanation of present-day climate, which both limits and delimits their colleagues' researches, with another dimension stretching back into the past. Honor where honor is due: these scholars have been able to adduce new and far-reaching ideas on the physical causes of, and major variations in, climatic change.

But the contribution of a second group of researchers is equally irreplaceable. This group consists of geographers, archaeologists, and professional historians whose specialization in economics or demography brings them in contact with old archives, either documentary or archaeological, concerning climatic events. Thus Gustav Utterström, starting from the analysis of agricultural conditions in eighteenth-century Sweden,[1] has been able to give an account of climatic fluctuation in modern Scandinavia, showing that the exceptionally mild winters in Sweden between 1721 and 1735 had a beneficial influence on the sowing of grain, pasturing of cattle, employment, public health, and longevity. He is also able to show this momentary mildness of the Nordic winter, as one of the factors in the great leap forward in the population of Sweden in the 1720s and 1730s. In England, dealing with a much earlier period, the medievalist John Titow, using the manorial accounts of the see of Winchester, has been able to draw up an outline of the incidence of bad weather which is of cardinal importance for our knowledge of climate in the West during the thirteenth century.

Like Utterström and Titow, I belong to this second group, historians who burrow among archives. It was in fact the history of agriculture that led me, by a logical and even inevitable transition, to the history of climate. Twelve or so years ago I was studying, through archives and registers, the history of certain groups of French peasants in the sixteenth and seventeenth centuries.[2] As is usually the case, the documents were extremely informative about the chronology of climate: meteorological references to severe winters and wet summers accompanied all the records of poor harvests, famines, shortages, and occasionally years of abundance. The country dwellers of traditional societies were constantly at the mercy of climatic benediction or calamity. But, fascinating as these descriptive documents were, in themselves they did not enable one to throw much light on contemporary meteorology. The handwritten comments on climate from some parish register, or the worm-eaten and illegible records of some lawyer, were too accidental and irregular to provide material for really organized knowledge. Of course they had their uses, but one wondered whether the absence, at that period, of continuous information and systematic observation of temperature and precipitation, such as we have for the nineteenth and twentieth centuries, was not irremediable. The historian of seventeenth-century climate needed in fact to be able to apply a quantitative method comparable in rigor if not in accuracy and variety to the methods used by present-day meteorologists in the study of twentieth-century climate.

The elements for such an approach, and for the quantification of the history of climate, did exist: they were even available for periods long before thermometers and barometers were perfected and brought into general use. In America, for example, dendrochronologists have used tree rings to create elaborate techniques for the study of climate. For this they have laid under contribution the incredible "sequoia historian" and very old pines and firs: their growth rings yield, year by year, sometimes for over a thousand years, a complete pluviometrical record.

In Europe too the dendrochronologists have achieved excellent results. But except perhaps in Germany there are no continuous tree series comparable in duration or in abundance of original data

to those which have been established for Arizona. Fortunately, for
Europe and for those parts of Asia once the sites of great civiliza-
tions, there is a substitute. For almost a century researchers into
ancient climates, in France, Germany, and Japan,[3] have made use
of the phenological method. Phenology is the study of the dates at
which certain phenomena occur in plants about which we possess
records; for example, flowering and ripening in vines and cherry
trees. In 1955 M. Garnier published an article on this subject in
La Météorologie which though modest in form was cardinal in
content.[4] In it he rediscovered the forgotten virtues of Angot's
work on wine harvest dates.[5] He showed that by using very long
series of these one could check, or even roughly reconstruct, temperature curves: thanks to the information hidden in the wine
harvests one could establish more firmly the long chronicle of hot
and cool summers, mild or chilly springs. A collection of masterly
series just as rigorous as those of the dendrochronologists was
available to western historians, who for a long time had unjustly
despised phenological sources.

The work of Angot and Garnier was the starting point for my
own initial research between 1955 and 1960. I was trying to find
series of wine harvest dates previous to 1750 or 1800. I found a
considerable number in municipal records, ecclesiastical accounts,
and legal and police archives of the seventeenth and eighteenth
centuries. In this way I was able to supplement André Angot's
great record for the south of France. The hunt ended in an important discovery when in 1959 I came upon the unexpected
treasure that every historian meets two or three times in his
life. In Avignon, in the Musée Calvet, I literally stumbled on the
huge pile of harvest dates collected by Hyacinthe Chobaut, together with much other data, in the course of a lifetime of scholarship. This self-effacing Avignon archivist was one of the true pioneers of the scientific history of European climate: his figures are
to be found before my own in the tables at the end of this book.

But wine harvest dates alone would not have been enough to
resurrect the meteorology of a bygone age. Another factor presented itself, not exhaustive but equally indispensable. It did not
relate directly to a strictly quantitative history of climate, but

threw invaluable light on it in the long term, in terms of centuries or spans of centuries. This new factor was historical glaciology of the documentary period.

Nearly thirty years ago, in an article called "Glaciers,"[6] Matthes, following Mougin and Kinzl, told the fascinating story of the Chamonix hamlets buried at the end of the sixteenth century by the advance of Alpine glaciers. Their inhabitants were the terrified and indirect observers of the subtle climatic variation of modern times. First through Matthes, then through Mougin, Richter, Kinzl, Monterin, and many others, I gradually became acquainted with the original and often neglected bibliography of articles on the modern history of Alpine glaciers, in the sixteenth, seventeenth, and eighteenth centuries.[7]

Most of these articles were piously, even coyly buried in the esoteric and dusty pages of agricultural revues, bulletins of learned societies, or the unobtainable yearbook of some Teutonic Alpine club of the 1880s. But the difficult task of hunting them out was well worth while, and led me on to important new sources. Their authors referred me to even older experts on glacial history, and I gradually learned to trace back these bibliographies to their sources just as one traces back the remotest tributaries and most complicated ramifications of a river. As I went along I disregarded parasite branches, unprofitable or redundant authors, plagiarists and those who merely repeated others, seeking always for the fresh water of documents and texts. At last I came to the sources themselves, in the old archives of Chamonix which are preserved at Annecy. I examined the ancient plans and maps; read the reports of such early visitors to the glaciers as Sebastian Münster; and, following the example of my predecessors, both historians and glaciologists, I compared all this data with the results of observation on the spot.

This last phase led me to the Chamonix valley itself and to Grindelwald, Courmayeur, Rhonegletscher, and Vernagtferner: all the places where one could compare present-day glacial fronts with those which appear before 1800 in published or unpublished engravings, maps and texts.

So forests, wine harvests and glaciers were my starting points. But my search widened out on the way to include other things besides its original objects. For a historiography of recent climate had to take into account the work of the most various and recent of specialists, from the professional meteorologists who reveal the climatology of the nineteenth and twentieth centuries to the pioneers of pollen research, invaluable to anyone interested in the climatic variations of the early Middle Ages. The following chapters refer as occasion arises to these various spheres of research.

As the search through the archives and the exploration through books gradually combined, new horizons opened before me. My subject, and that of this book, was no longer climate merely in its human or ecological aspects, but climate as a subject of historical study in itself.

It was an attempt worth making. My researches among unpublished documents, among harvest dates and archives relating to glaciers, revealed a strange landscape which was already magnificently pioneered, but which could still have further light thrown on certain aspects of it. It was a landscape that appeared almost unmoving. Yet slow fluctuations became perceptible when one observed them over several centuries. No doubt these fluctuations were slight and therefore comparatively unimportant in relation to human evolution proper. But the documentary historian, supplementing his privileged access to original sources with the results of the work of other specialists, might well make a modest attempt to establish an accurate chronology of these fluctuations. This book makes no claim to be an exhaustive history of world climate, taken as a whole over the last thousand years. It would be absurd to attempt so immense a task. What the reader will find here is essentially a methodological introduction to the problems presented by the historiography of recent climate in Europe and America, from the point of view of a documentary historian. The period covered includes those centuries when observation of climate came within the province of the historian: that is, the sixteenth, seventeenth, and eighteenth centuries. But for purposes of comparison, and in order to throw light on the past from the present, there will be frequent references to data accumulated by professional meteorologists in the nineteenth and twentieth centuries. And when unpublished documents or the results of parallel research warrant it, I shall also deal with certain aspects of the climatic fluctuations of the Middle Ages.

CHAPTER I

THE HISTORICAL STUDY
OF CLIMATE

Climate is a function of time. It varies; it is subject to fluctuation; it has a history. While it is true that the idea of climate is a summary description of meteorological conditions over a series of years,[1] it must be added at once that such descriptions never exactly match from one period to another. Even the most extensive figures we have, those covering periods of a hundred years, do not give an exact picture of the climate of any given region.[2] All this is well known to meteorologists, geographers, glaciologists, geologists, and palynologists,[3] who come across evidence of it every day in their researches. And yet historians themselves, although they are directly concerned, have not yet really addressed themselves to the historiography of climate as a subject of special research: such few attempts as have been made have often ended in failure. Why is this?

The explanation lies, in the first place, in the attitude adopted by the first historians of climate, who instead of studying climatic change in itself and for its own sake launched themselves into something quite different and highly dangerous, namely the climatic interpretation of human history. Elsworth Huntington, for example,[4] did not really make an unprejudiced study of climatic fluctuation in Asia, but tried from the outset to account for the Mongol migrations in terms of climate. Similarly Le Danois, studying ocean climate, was primarily interested in movements of fish and fishing grounds, and in fluctuations in Paris fashions as an index of reactions to changing weather conditions.

Ignazio Olagüe "explains" the history of various Mediterranean countries in terms of fluctuations in rainfall.[5] There are plenty of other examples of this exaggeratedly anthropocentric approach, but let one more suffice.

In 1955, in the *Scandinavian Economic History Review*, the Swedish historian Gustav Utterström published an important, interesting, and well-informed study on "Climatic Fluctuations and Population Problems in Early Modern History."[6] This article brings together almost all the available data on the influence of climate in medieval and modern history, and represents one of the furthest points ever reached by the traditional method. It may therefore be useful to analyze it at some length before describing some other techniques.

The author sets out to prove that there were long periods of climatic deterioration which had disastrous economic effects on Europe. He deals mainly with two periods: the fourteenth and fifteenth centuries, and the seventeenth.

According to him there was a general cooling of the climate in the fourteenth and fifteenth centuries. Utterström supports this assertion with many but somewhat motley facts. The first symptom he adduces is the fact that between 1300 and 1350 fishing took the place of cereal-growing as the chief economic activity in Iceland. One might have thought this change to be as open to an economic interpretation as to a climatic one. But the climatic interpretation is backed up by an appeal to the chronology of glaciers and ice sheets (we shall come back to this chronology): the advance of the glaciers, supposed to have begun "after 1200," went on in Iceland during the fourteenth and fifteenth centuries, "continued" in the sixteenth, and reached its maximum in the seventeenth and eighteenth. This extension of the ice sheets is supposed to be both confirmed and dated by the destruction in the fourteenth century of Norman colonies in Greenland. The destruction must have come about in a very subtle fashion, since the Normans are supposed to have been victims both of the spreading of the ice cap and of the resulting mass "descent" of the Eskimos, pursuing seals and icebergs to the south.

The decline of vinegrowing in England in the fourteenth century, after it had been at its height in the thirteenth, is also put forward as the result of a climatic revolution rather than, as had been supposed, a mere sign of economic change. The heyday of the English vineyard in the twelfth and thirteenth centuries had already led to the conclusion that the English summer must have been warmer then than it is now. True, German vineyards did not decline at the same rate as the English after 1300–1350, but apart from certain short periods their good years were only "occasional" in the fourteenth and fifteenth centuries, and this was taken as another sign of a general deterioration in climatic conditions.

According to Utterström, the end of the fifteenth century, after 1460, and the first half of the sixteenth century enjoyed a much more clement climate than the period immediately before, but about 1560 another cool and unpleasant period began, which lasted right through to the seventeenth century. By way of proof we are told that Swedish grain harvests "dwindled" between 1554 and 1640; but one would very much like to know how anyone has been able to measure variations in yield in Scandinavia in the sixteenth and seventeenth centuries. But let us pass on to the suggestion that the southwest Baltic and the Thames, which had not frozen between 1460 and 1550, had much more rigorous winters during the second half of the sixteenth century and the first half of the seventeenth. In England, cherry trees spread north at the beginning of the sixteenth century, but during the reign of Queen Elizabeth the weather is supposed to have become "cooler" again. Lastly, the glaciers began to advance again at the end of the sixteenth and during the seventeenth century. This was the "little ice age." This advance, said to be the greatest since the Ice Age, is supposed to have been most marked in the Alps and in Iceland. There were various episodes, but no real retreat of the ice until about 1890.

The author supports his thesis by quoting the economically catastrophic years that occurred in Scandinavia during the period concerned: 1596–1603; the years immediately before and after 1630; 1649–1652; 1675–1677; and the 1690s.

He also presents the flow of Baltic grain toward the Mediterranean after 1590, and the depopulation of Spain in the seventeenth century, as clear signs of a reversal of climate. Another symptom of a "change of climate" is the fall in the number of sheep in Spain after 1560, and especially after 1600.

In short, the whole "crisis of the seventeenth century," of such great historical importance, is supposed to be of climatic origin, and unexplainable in terms of a mere internal analysis of contemporary European society and economics.

The article provides a rich harvest of facts and data of all kinds. But a lot of these facts are not a priori climatic: the decline of vinegrowing or sheep-rearing, for instance; the spread of wheat or cherry trees; and still more the changes in the cereal trade. In the present state of our knowledge all these things can be explained equally well, or better, in terms of purely economic considerations.[7] When, on the other hand, the author quotes years of climatic hardship and agricultural deficit, he is adducing real meteorological evidence. But in order to prove his point he would need to demonstrate, rigorously and with statistics, that these bad years all resulted from more or less comparable meteorological conditions; and then that they occurred with noticeable frequency during the long period in question, or at least that they were noticeably less frequent during the preceding or following periods. Until it has been shown that there is a significant difference between two periods, the bad years have to be regarded not as a long series but merely as episodes in short-term meteorological fluctuation. That being the case, the author is not justified in using them to support his argument about long-term fluctuations in climate. It is as if a historian or an economist were to try to demonstrate a lasting rise in prices by taking a few periodic peaks of a graph while taking no account of its general curve. By the same token, a few outstandingly cold winters here and there during the fifteenth century are not enough to prove that the century as a whole was a cold one.

Utterström's glaciological data are those which are the most indicative of some long-term climatic movement. But the chronology of these movements is too vague and their influence on human history too uncertain to warrant such ambitious conclusions as he draws from them. What should we think of a historian who tried, even partly, to explain economic progress in Europe since 1850 by the warming up revealed in the retreat of Alpine and other glaciers since that date? Utterström is doing much the same thing when he tries to establish a close connection between the advance of the glaciers and the economic crises in Europe during the fourteenth, fifteenth, and seventeenth centuries.

But anthropocentrism is not the only weakness to be found in the various attempts at historical interpretation of climate. Other researchers, some historians, some not, have gone cycle-mad. Douglass, the eminent pioneer of American dendrochronology, lost years and devoted incredible statistical refinement to searching his tree rings for an eleven-year sunspot cycle.[8] Jevons junior and senior and Henry Moore did likewise with the maize trade, unemployment and the price of pigs in Chicago[9]; and Beveridge himself was influenced by their work. Brückner adopted a thirty-five-year cycle for temperature, wine harvest dates and glacier termini.[10] As for "astroclimatic" cycles of forty thousand years, divided into "subcycles," they were still working up havoc till quite recently.[11] Sometimes such speculations lead to astonishing predictions about the future: one cycle-hunter, basing his calculations on the periodicity he has discovered, is bold enough to forecast what the level of the Seine will be in the year 2000.[12]

Such researches are to the real history of climate what the philosopher's stone was to oxygen. Yet they have been carried on ceaselessly by generations of researchers. But now they have received a deathblow. Without absolutely denying the possibility, in theory, of regular periodicity, climatologists no longer really believe in cycles of unvarying duration or in an "éternel retour" of climate. They are much more interested in the idea of fluctuation, which undoubtedly occurs, but over varying periods.

Serious historians did not wait for the verdict of scientific meteorology. They shrugged their shoulders at the daring reconstructions of the climatic romancers. One economic historian spoke for most of his colleagues when he said, "I don't trust climatic explanations."[13] His mistrust was quite justified: for each "explanation" of this kind it is easy to adduce a purely human and immediate consideration which is both adequate and intelligible. Demography, lack of currency, scarcity, and low productivity account for "crises" in certain sectors in certain periods of the seventeenth century. As for the so-called "crisis of the seventeenth century" as a whole, historians still dispute among themselves[14] whether it really existed as such, and whether it really affected a whole century. It would be rather premature to seek a climatic explanation for a phenomenon whose existence is not even certain! Economic motives or the caprices of taste are often enough to explain other changes, such as those in fishing grounds and even more in fashion.

To turn toward the Mediterranean, the relative decline of Spain is not due to a fall in humidity but to the social structure, a totalitarian religion, the monetary fluctuations of the Renaissance and Baroque periods, a system of values not adapted to capitalism, and geographical conditions inadequate to modern economic needs. As for the series of disasters in the fourteenth and fifteenth centuries, the occurrence of which no one calls in question, epidemics of pulmonary or bubonic plague, together with many other factors, had more to do with them than any hypothetical wave of cold or wet.

To tell the truth, controversy has been and still is so active on this last problem of the catastrophes of the late Middle Ages (1348–1450) that it may be useful here to sketch the beginnings of an ultimate climatic study of the "great crisis" of the fourteenth and fifteenth centuries. Among its many possible causes, one that is frequently invoked, at least as a hypothesis, is the recurrence of wet years[15] during the century that begins with the year 1310, to take a good round figure. These wet years, which destroyed or rotted wheat and grape harvests, are supposed to have helped plunge agriculture in the West into depression and sometimes even famine during that very long period.

Such a hypothesis calls, in the first place, for an accurate examination of the volume of the harvests during the period concerned, and for a comparison with the periods before and after.

But while such an examination, conducted for example on the great wheat-producing area of northern France, may confirm the theory of a deterioration in climate, it does not necessarily confirm that of a deterioration in harvests sufficient to have disastrous consequences on human life. It is true that during the very wet decade beginning with 1310 the floods of 1315[16] inundated harvests, vineyards, and seeds in the ground with immediate consequences that were terrible enough. The 1316 wheat harvest was very poor; food was scarce; ingenious bakers mixed what little flour they had with the droppings of pigeons and pigs. Prices rose; eggs were sold in Limoges at a penny apiece; in France, Flanders, Germany, and England the poor died in millions from hunger and epidemics. Cannibalism was recorded in Britain and Livonia. In southern France, in 1316, half-religious, half-popular processions of snails were inaugurated, presumably to ward off the rain supposed to be their familiar element.[17] At Tournai, in gloomy Wallonia, the inhabitants were even reduced to drinking the local wines because the French vineyards that usually supplied them had had such a bad harvest: the grapes had been washed away by the rain.

All these dreadful or picturesque details, however, only concern the short term—the decade of 1310,[18] in which the famine of 1316 particularly stands out. But in the long term, on the scale which interests the historians of the fourteenth century, it seems that agricultural production very soon got over the disasters of 1315. We may confine ourselves to a few important examples. The receipts in wheat, wine and hay of the abbey of Saint-Denis near Paris, which represents a wide sampling of estates and vineyards,[19] are as steady as a gyroscope between 1284 and 1342. It would be very difficult to discern the slightest decline in the harvests there; of a long crisis during those sixty years there can be no question. It is true that brief climatic or other incidents occasionally affect the graph of the garnerings of Saint-Denis, but they have no lasting effect on the trend, which is in the long term even and horizontal over the whole period from 1284 to 1342. During this time the hypothetical "deterioration of climate," if there was one, certainly had no negative consequence for the rich, stable agriculture of the Paris area. It was the entirely nonmeteorological series of disasters brought about by the plague, the English and the brigands after 1346–1348 which make the hitherto impassive wine and wheat graphs for Saint-Denis go hurtling downward.[20]

The same situation is found in Flanders and northern France. If the hypothesis of an unusually wet fourteenth century and a corresponding decline in harvests were correct, one would expect to find, in this area which even normally is covered all the year in heavy rain clouds, yields of wheat which if not poor were at least declining. But no such thing. In Picardy, Artois and Flanders, on the cereal-growing estates of Thierry d'Hireçon (an astonishing character who was one of Guillaume de Nogaret's bodyguards, then devoted many years to agriculture, finally becoming a monk and dying bishop of Arras), and on the Douai estates of the abbey of Notre-Dame-des-Prés, the harvests around 1319 to 1340 were good and abundant. Whether the year was good or bad, Thierry and the abbey of Notre-Dame harvested an average of eight to ten grains for every one sown: this is entirely comparable to the results obtained in the same region during the good years of the eighteenth century. It is only after 1340, when first military operations, then epidemics, ravaged the country, that the rural prosperity of Artois and Hainaut was halted for about thirty years. As for excessive rain, according to excellent documentary evidence it appears to have played no part as a factor in this later decline.[21]

It is also far from certain that the decline in winegrowing in the north,[22] so clearly marked in the fourteenth century, can be ascribed to a deterioration in climate. Let us take as an example a marginal vineyard which no longer exists: the one that used to lie to the north of Paris. From 1350 to 1400 and after, it was in full decline. Was this the fault of excessively severe winters which killed the plants, or of extra frequent spring frosts, or of wet summers which prevented the fruit from ripening? Not at all. Contemporary texts give the real answer. In 1399 the abbot of Saint-Denis sent a party of monks who were experts in matters of wine to inquire into the lamentable situation of the monastery's vineyards. They found them lying uncultivated and abandoned. How on earth could this have come about?

At first sight the answer handed down to posterity in the *Livre vert* (Green Book) of Saint-Denis[23] is very optimistic. "Your vines," the investigators tell the abbot, "in principle give an excellent yield. Well looked after, they could all produce four pipes to the acre." This means a minimum of twenty hectoliters per hectare, and probably much more, thirty hectoliters at least, if, as seems highly probable,[24] the monks calculated according to the Paris *arpent* or acre (34 ares) and the *queue* or pipe of Saint-Denis (386.5 liters). Such yields are magnificent for that period, and indicate clearly that the Paris vineyards, northerly though they were, were no more subject to unfavorable climatic stress in the fourteenth century than at other times.

But that being the case, why the crisis, why were the vineyards in such bad shape? The monks' considered answer to these questions again rules out climate. They told the abbot that the vines, if cultivated normally, would produce four pipes of wine to the acre. As a pipe sold for 32 sous, each acre would thus bring in 128 sous. But unfortunately profits were eaten up by expenses. The mere cultivation of the vines, without counting the always high cost of vintage, absorbed 140 sous per acre—more than the proceeds of the sale of the wine. They concluded that the best thing was to abandon winegrowing altogether as no longer profitable.

I have dwelt on this report by the monks of Saint-Denis because it seems to me to throw a very clear light on the whole matter. If the marginal northern vineyards receded in the postplague years of the fourteenth century (after 1348), it was not because the climate did not provide the vines with their fair share of warmth. It was because, as a result of the depopulation caused by epidemic and war, labor had become too dear. The counterproof to this is easy: a century and a half later, around 1560, according to figures taken from winegrowers' accounts, the yield of the Paris vineyards had not increased but, since real salaries had become much lower than in 1400, profits were considerable. So whether it rained, blew or froze, the Paris winegrowers, right at the beginning of the Little Ice Age, still went about their work with a will. With their profits they paid for their sons' studies and for their grown-up daughters to learn to play the lute and spinet and even to write.[25] So once again a close examination of unpublished documents rules out the climatic explanation, too generally and too ambitiously applied. A thorough study of meteorological conditions in the first half of the fourteenth century[26] shows, moreover, that the five decades from 1300 to 1350, taken as a whole, were not exceptionally wet, and therefore not exceptionally unpropitious to wheat- or wine-growing, in the areas near the Channel. In the scale established by H. H. Lamb the ten-yearly index for the English summer between 1300 and 1340 fluctuates around 6.5, 6.5 again, and 6.7. In other words, it is comparable to the indices for the driest decades in the eleventh, twelfth, and thirteenth centuries. It is only during the absolutely exceptional decade from 1310 to 1319 that this index temporarily soars to 15.0. This was caused by a very temporary wave of heavy rainfall, which as we have seen was a contributory factor in the great famines of 1315–1316. It does not indicate a lasting trend of wet summers clouding all the first half of the English fourteenth century.

On the contrary, it was after 1350 that England had a series of decades (1350–1360, 1360–1370, and 1370–1380) characterized by wetter summers. But in fact in the generation from 1348 to 1380 the effects of the plague, which were not operative in the previous half century, had become so crushing that the climatic factor becomes only secondary. It has been said that the plague of 1348 carried off 25 percent of England's population; that of 1360, 22.7 percent; that of 1369, 13.1 percent; that of 1375, 12.7 percent. Even if these figures, which are those of Russell,[27] are contested in detail, and historians have not denied themselves that pleasure, there is no doubt of the enormous human losses caused by those four outbreaks of plague. In comparison with these terrible blows, what could be the effect of a few exceptionally heavy falls of rain, or a few spoiled harvests? It is quite clear that in the West, after 1348, the chief factor responsible for the stagnation, the depopulation, and the subsequent economic crisis was not the climate. It was definitely, among other factors, Yersin's bacillus; and in the second place, on the Continent, war, brigandage, and the huge wave of criminality and gangsterism which spread over France at the time of the Hundred Years' War, and in comparison with which the Chicago of the 1920s was a haven of peace.

But to get back to the main issue: the anthropocentric approach which consists in taking a vast human crisis like that of the late

Middle Ages and trying to give it a climatic explanation is not a helpful one. The naïve anthropocentrism of the first historians of climate sometimes even took the form of a vicious circle. Huntington explained the Mongol migrations by the fluctuations in rainfall and barometric pressure in the arid zones of central Asia.[28] Brooks[29] carried on the good work by basing a graph of rainfall in central Asia on the migrations of the Mongols! The first extrapolated from the barometer to the Mongols, and the second, with even less justification, from the Mongols to the barometer.[30] What better example of a serpent biting its own tail?

Such "methods" have provoked in many scholars and historians a completely negative reaction and sometimes the temptation of an easy triumph. Some simply deny that there have been any "recent" variations in climate, i.e., any that have taken place in historical times. Angot and Arago in the nineteenth century, and Aymard[31] more recently, invoke commonsense evidence such as the stability of vegetation limits and winegrowing areas in order to dismiss the possibility of any climatic fluctuation since antiquity. All they set against the headlong exaggerations of climate fiction[32] is a rigid fixity. But this type of criticism overshoots its mark, and rules out even the possibility of a scientific history of climate.

But such a history is feasible, on condition that it frees itself entirely (just as climatological geography must do[33]) of anthropocentric prejudice, and does not try to force reality into the straight-waistcoat of cycles. On condition also, and above all, that in establishing its basic series of data it limits itself to facts which are strictly climatic. A migration, a famine or list of famines, and still more a graph of agricultural prices are not and cannot be facts which are strictly climatic. Migration results from extremely complex human motives and compulsions. Famine derives from adverse agricultural conditions, in which the climatic element can never be deciphered a priori, whether it is a matter of hail, frost, rain, fog, pests, scorching heat, drought, or meteorological incidents, which are sometimes of very short duration and little climatic significance.

On the other hand valuable climatic documents can be derived from studying early meteorological observations and, before these, records and texts of all kinds giving nonquantitative accounts of the meteorological nature of certain years, days, weeks, months or seasons. Other useful documentary evidence is in the dates of harvests (rather than their volume), and in references, descriptions and pictorial evidence concerning glaciers. All this, provided it has been critically examined and duly translated into quantitative terms, can serve as source material to the historian of climate—on condition, of course, that he works via the history of the various different meteorological factors in themselves: temperature, rainfall, and then, where possible, wind and barometric pressure, sunshine and cloud. It is on these conditions only that fictionalized history of climate can become scientific history of climate, just as alchemy eventually turned into chemistry.

An objection might be that this kind of research, these documents and methods, do not, as such, directly relate to human history, but only to a sort of physical history, a history of natural conditions. Does not the historian who devotes himself to such work risk betraying the vocation assigned to him by Marc Bloch in a famous passage: "Behind the tangible features of landscape, behind what are apparently the most frigid of writings . . . it is human beings that the historian is trying to discern. If he does not succeed in that he will never be anything, at the best, but a learned hack. But the true historian is like the ogre in the story: wherever he smells human flesh he recognizes his prey."[34]

Nicely put. But in spite of my immense admiration for Marc Bloch his definition has always seemed to me too narrow, not adequate to the true scientific spirit. The time has gone by when Greek philosophers and physicists spoke of man as "the center of the universe" and "the measure of all things." Since the pre-Socratics and Ptolemy there have been many Copernican revolutions.

If one took Bloch's metaphor of the ogre and human flesh literally, it would mean that the professional historian would systematically neglect a whole category of serial or qualitative documentation, such as early meteorological observations, phenological and glaciological texts, comments on climatological events, and so on. A strictly human historiography *could* take such documents into consideration, but never to work out completely and for itself their intrinsic climatic content, only to check some usually minute point in human history or local or specialized knowledge (for example, some detail on the history of the thermometer at a certain period, or the evolution of vintage techniques in a certain vineyard).

In fact, total lack of interest in these matters is the attitude which has prevailed so far. Almost no historians have addressed themselves to the early series of climatological documents.

This has serious consequences. Even if these documents do not essentially relate to human history, according to Bloch the only history worthy of interest, the search for them and their subsequent analysis and application is a job strictly for the historian and for him alone. If the qualified workman does not present himself, the document remains unexploited and abandoned. At the worst—and the worst often happens—it is lost. How many are left now of the dozens of eighteenth-century manuscript meteorological registers listed by Angot in his great article in 1895?[35]

Naturally, the worst does not always happen. In many cases other qualified workers step into the breach left by the historians.

In England, for example, a country much smiled on by the gods in the matter of history of climate, meteorologists and geographers have lent a hand. They themselves have collected early texts on extremes of weather and built up excellent series from them.[36] The medievalist John Titow,[37] too, has discovered in the manorial archives of Winchester hundreds and hundreds of unpublished texts on the meteorology of the thirteenth century.

They have ordered this matter differently, however, in France. For a long time the absence of historians has been unbroken and unremedied. Is it by chance that historical phenology had made little progress since Angot? That since Mougin, Richter and Allix practically no early texts of any significance have been dug up concerning Alpine glaciers, and no meteorological series been built up for the eighteenth century in France since the work of Renou?[38] No, it is not by chance. Science has stood still over the fascinating question of climatic fluctuation because after the initial work of the pioneers and inspired amateurs, the specialists who might among others have taken up and advanced the research—the medieval and modern historians—have shirked it. They were only interested in human history, and to deal with natural phenomena as such seemed to them unworthy of their humanist vocation.

Unless, therefore, one wants to let a whole province of possible research lie fallow, one must, if not contradict, then at least modify and fill out what Bloch said. It is mutilating the historian to make him into no more than a specialist in humanity. The historian is the man of time and archives, the man to whom nothing which is documentary and chronological is alien. This being so, he can and still most of the time will be Bloch's charming anthropophagous ogre. But he may also in certain cases be interested in nature for its own sake, and make known by his own irreplaceable methods nature's own special Time and, in particular, the rhythm and recent fluctuations of climate.

In other words, just as there is a human geography and a physical geography, a *morphology* and a *geography*, so there can also be, for the so-called historical period which is covered by written documents, a human and a physical history—a *geohistory*, to use the word in more restricted sense than that given it by Fernand Braudel in distinguishing from *history* proper.[39] Both are parts of science; they are not of equal importance, but both are served by the same methods, and both concern the historian, with his traditional rules governing the rigorous use of documentation, and his new norms of quantitative elaboration.

Climatic history, thus defined as independent in its subject of research but related to human history by its methods, has to face various "frontier" problems.

First of all there are the frontiers with the other disciplines which up till now have occupied themselves, alone and on their own account, with the "recent" evolution of climate: meteorology, geography and morphology, geology, palynology, dendrochronology, archaeology, glaciology, radiocarbon dating.

The historian of climate must begin by learning all there is to be learned from these researches older than his and already built up into bodies of theory and results. He looks to them for the general framework of his own inquiry. Also for basic information on already well-marked-out climatic fluctuations and their climatological interpretation. These "requests for information" have been made for a long time now: thus it was only the glaciologists' and geomorphologists'[40] observations of recent moraines that made possible a correct interpretation of early texts on glaciers—texts which were brought to light by local historians, scholars and archivists.[41] Similarly it was as a result of comparison with the systematic observations of meteorologists that the phenological and wine harvest date series took on their full scientific meaning.[42]

But the historian is there to give as well as to receive. The very first results of his documentary researches enable him to export invaluable information across his frontiers, and give the other disciplines interested in the evolution of climate the thing they most need: accurate chronology, exact date. Thus the moraines of Chamonix and the Tyrol show glaciologists, by their pedological recentness and through carbon dating, that there was a considerable glacier advance between 1550 and 1760, or more exactly still between 1600 and 1710.[43] But texts from archives not only confirm this advance but also *date* it much more precisely than any carbon dating, which is only correct to within about a century. These texts show[44] on the one hand a permanent state of glacial advance in the seventeenth century, and on the other hand, against this background, individual glacial maxima in 1600–1601, 1643–1644, and 1679–1680.

Another significant fact is that when meteorologists, geologists, and biologists wanted to find out about the climate of the eleventh and sixteenth centuries, they invited a dozen rural and economic historians to their meeting. And they entrusted to them the task of establishing most of the series for the periods concerned, the historians having brought along with them the continuous, annual, quantitative, homogeneous lists of figures which the scientists had need of.[45]

So the relations between the history of climate and the related disciplines involve a fruitful mutual exchange, a constant flow of information in both directions. Hence the dual character of the present book, which consists partly of exposition, made necessary by the present compartmentalization of specialties, of results already obtained by related disciplines, and partly of a historian's history of climate.

Another frontier and another type of relationship still remain to be described: they are those between climatic and human history. Once the first of these is in possession of its own methods and initial results (from which, as has been said, all anthropocentrism must be banished), it can lead into the second. We then arrive at a second stage in our research, in which climate is no longer looked at for its own sake but "as it is for us," as the ecology of man. Climatic history would then become ecological history, asking such questions as whether the fluctuations of climate—or to put it more modestly, the brief fluctuations of meteorology—have reacted on the human habitat; on harvests and thus on economy; on epidemics and diseases, and thus on demography.

But, I repeat, this is only a second stage in the study of the history of climate, and once which is by no means a priori necessary. In this book I make no claim, and could not even attempt, to give a thorough account of all the successive stages. So I shall deal mainly with the problems of the first strategic stage, which itself was too long neglected: the initial, necessary stage which leads (for the period before which rigorous observations were made)[46] to the construction of a pure climatic history free of any anthropocentric preoccupation or presupposition. I shall only deal briefly and "laterally," without any pretense at completeness, with the problems of the second stage, which may perhaps lead to the establishment of an ecological history, a climatic history with a human face.

CHAPTER II

FORESTS AND WINE HARVESTS

The pre-meteorological period, before strict observations were kept, was also the last heyday of traditional societies (before 1800–1850). In these societies, mainly agricultural and dominated by the frequently difficult problem of subsistence, the relation between the history of climate and the history of man had, in the short term, an urgency it has now lost. The peasants of earlier times recognized this when they charged special saints to act as little rustic gods and protect farms and harvests against storms or against too much or too little rain. At Semur-en-Auxois (Côte-d'Or, France) there is a fine sixteenth-century stained-glass window in the church which depicts both St. Médard, the intercessor for rain, and Saint Barbara, protectress against thunder and lightning (and also patroness of miners). The window presents a scene worthy of the Marquis de Sade; one wonders what the congregation must have made of it. It shows the martyr with naked bosom, her luscious body whipped, torn with red-hot pincers, spitted on hooks, and finally burnt at the stake. After all this careful preparation or "cuisine,"[1] the body of the saint is ready to immunize human beings against thunderbolts and guide the miners' picks safely into the womb of the earth.

Thus the peasants, in their "pensée sauvage," provided themselves with shields and lightning conductors against the dangers of climate. But it did not occur to them to hand down a systematic, accurate and continuous record of the behavior of those elements which some years crushed them and some years gave them cause to rejoice. But this gap is irreparable, and it is no use regretting the fact that traditional societies have left behind no regular observations of temperatures or rainfall.

In the absence of such obervations, some nineteenth-century scholars[2] who prospected these early periods contented themselves with picking out here and there, without much attempt at method, such climatic events as had particularly struck people at the time: "terrible" droughts, "iron" frosts, "severe" winters, "deluges" of rain, floods, and so forth. Such a documentation could only be subjective, heterogeneous, sporadic, sparse, and episodic. One swallow doesn't make a summer, and a "series" of disastrous frosts several years apart does not necessarily indicate a real period of cold. Such episodic material can only become significant if it is tested, classified, and organized.[3]

So it was generally faith rather than the often somewhat unconvincing facts which sustained such primitive essays into meteorological history; the sort of robust faith which made Huntington explain the fall of the Roman Empire by a deviation in the path of cyclones and a drying up of the lands round the Mediterranean.[4] Underlying all such theories is the lazy but highly contestable postulate that climate exercises a determining influence on history.

In order to get out of the impasse of traditional methods, research had to find new paths, to turn to methods specifically suited to climatology—biological or historico-statistical methods which were positive and excluded all preconception, and set out in the first instance to establish series of meteorological data which were *annual, continuous, quantitative, and homogeneous.*

The first and most highly developed of the biological methods is "dendroclimatology" (see Appendix 16 for bibliography). The idea on which it is based is well known: a cross section of the trunk of a tree shows a series of concentric circles. Each ring represents a year's growth, so by counting up the number of rings one arrives at the age of the tree.

But while the total number of rings has this obvious chronological value, each separate ring also has a climatological significance because it reflects the favorable or unfavorable meteorological conditions affecting the tree's growth in that particular year. A good year produces a wide ring, a bad one only a narrow one, sometimes so thin as to be scarcely visible. So each tree ring is a sort of climatic note on the year which produced it, and by making a graph with years shown along one axis and thickness of the rings along the other, one gets the "growth curve" of the tree, and the fluctuations in this curve, rightly interpreted, show the meteorological fluctuations from one year to another.[5]

But the question arises, what does one mean by favorable or unfavorable meteorological conditions? And first of all, which is the more decisive factor, temperature or rainfall? Reason and experience give the same answer: it all depends on the place.

In semi-arid countries such as North Africa and the southwest United States, where there is a chronic lack of rainfall but warmth is rarely absent during the tree's periods of growth, a long series of rings which are mostly very thin indicates immediately a definite period of drought. Conversely, series of wide rings suggest periods which are comparatively wet. But these "indications," which reflect the specific correlations of dry, semi-desert countries, may be extremely complex and present different kinds of subtlety according to the type of tree used for research. Thus in the Mesa Verde mountains of southwest Colorado the Douglas fir, piñon pine, and Utah juniper each react differently to aridity stress. In the Douglas fir, meager growth and narrow rings usually correspond to very light rain from August to February, then to a combination of low temperatures and low precipitation from March to May, and finally to a mixture of drought and great heat during the June of the year the ring is formed. In the Utah juniper, if the ring falls below its average size for the Mesa Verde, it is because precipitation has been lower than usual between October and February, and because heat has been more intense than usual in October and November of the previous year and from March to May of the year in question. The piñon pine also reacts to excessive heat and drought with narrow rings, but the months in which these factors are operative are not the same as in the case of the Douglas fir and the Utah juniper.[6]

Another element which must be taken into account is the exact sites where the trees grow. In northern Arizona the trees most sensitive to climatic fluctuation are not those sheltered in the heart of the forest, but those which grow on its lowest and most arid borders, on the dangerous confines of the desert, where drought is more fierce.[7]

If we turn to northern latitudes, the factors affecting tree growth become quite different. In areas near the Arctic Circle—Scandinavia and Alaska, for example—the operative factor is no longer lack of rain but temperature, and one can say without fear of contradiction that a thin ring means a particularly cold year and a thick ring means one that was less cold.[8]

In moderate zones such as New England and non-Mediterranean western Europe, tree growth depends on both temperature and rainfall. Here the interpretation of growth curves is made more difficult by the interaction of various different factors. Harold Fritts, in an essay on the Ohio beech and the Illinois white oak, made a successful attempt to bring order into all this chaos.[9] He arrived at complex but clear correlations between temperature and rainfall in various months. But this is pioneer work which has still scarcely begun.

It is not by chance, then, that dendroclimatology has developed mainly in marginal climatic regions where growth curves can be read immediately and unequivocally. The areas most worked over by specialists are Scandinavia and Alaska on the one hand, and on the other the semi-arid southwestern United States (Colorado, California, and Arizona). The University of Arizona, first with A. E. Douglass and then with Edmond Schulman and Harold Fritts, has achieved notable results.[10]

In the early 1900s Douglass gave the decisive impetus to this new branch of research. He was an astronomer by training, and in 1901 he turned to the study of very old trees with the idea that the rhythm of their growth was somehow correlated to the fluctuations of sunspots. This hypothesis was useful in that it stimulated a new type of research, but was ultimately dropped because it was not borne out by the facts. Douglass' early efforts were greatly helped by the survival in the western United States of trees and groups of trees, conifers of all kinds and particularly sequoias, aged between five hundred and fifteen hundred years.

One of the earliest applications of Douglass' work was archaeological. Having used living trees to establish a rigorous chronology of outstandingly wet and outstandingly dry years from the fourteenth century on, he was able to match some characteristic sequences in the beams in Pueblo ruins in New Mexico. Thus, knowing the century the tree lived in from which the timber was made, and being able, from the last growth ring before the bark, to say precisely which year the tree was cut down, Douglass was able to date the ruins exactly.

It was on June 22, 1929, after a long process of experiment, that the dating techniques which constitute dendrochronology proper finally proved themselves. The almost military precision of specifying a particular day may provoke a smile. But the history of science has just as much right to its dates as the history that used to consist mostly of lists of battles. On June 22, 1929, then, Douglass was at Showlow, Arizona, where he had just finished some archaeological prospecting. He spent the afternoon in the shabby living room of the local inn, comparing with the aid of a magnifying glass the curves derived from very old living trees and the sections or "cores" which his colleagues had prepared from charred timbers taken from a very old local ruin. "We noticed with amusement," wrote Haury, one of the archaeologists on the team, "the ever enlarging smudge of charcoal on his nose as he repeatedly cross-checked the specimens against his paper records."

Finally Douglass raised his sooty face from his desk and said, "I think we have it. Ring patterns between 1240 and 1300 of the sequence I have derived from trees still living in the twentieth century correspond in all important respects to the patterns in the youngest part of the archaeological sequence [i. e., that derived from the fossil timbers]. This means that Pueblo Bonito [a site near Showlow previously excavated] was occupied in the eleventh and early twelfth centuries; the ruins of Betatakin and Mesa Verde are a little younger, mid-Thirteenth century." Thus at one stroke, thanks to the remains of a piece of wood consumed in an ancient fire, the missing link was found. The relative and indefinite chronology which archaeologists had derived from the tree rings in ancient timbers taken from ruined pueblos was now securely anchored in an absolute chronology supplied by the age-old forests of Arizona. And Douglass, with his phenomenal memory, was able to construct, tree rings in hand, the whole calendar of the prehistory of the southwest, coexistent in time with the European Middle Ages.[11]

These methods, now reinforced by the dating techniques based on the life of radioactive particles (radiocarbon dating), have made it possible to date exactly a considerable number of Indian pueblos.

But Douglass also saw the significance of all this in the study of the history of climate, and his work, continued today by his pupils, has led to the remarkable results shown in Figs. 1 and 2.

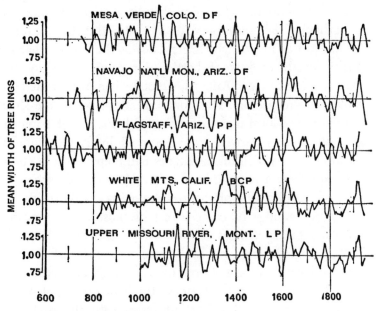

Fig. 1 *Dendrochronological Means*
        Running means of twenty years, based on five tree ring chronologies for the western United States: D.F.—Douglass fir; P.P.—Ponderosa pine; B.C.P.—Bristlecone pine; L.P.—Limber pine. (Source for diagrams and captions: Fritts, 1965 a, p. 877.)

Fig. 2
        (1) *Mean Tree Growth in the Southwestern United States (Twelfth–Twentieth Centuries)*: Annual figures showing departures from the multisecular mean. They relate to two localities in northern Arizona and another (Mesa Verde) in southwestern Colorado. Data derived from living trees is supplemented by series derived from excavated beams.
        (2) *Drought and Rain in the Western United States (Fifteenth–Twentieth Centuries)*: In the last seventy years, regional growth indices derived from trees sensitive to drought parallel the main and even the secondary fluctuations in the discharge of nearby rivers; they thus emerge as good rainfall indicators for earlier periods. (Sources: Curves 1, after Schulman, 1951, p. 1025; Curves 2, after Schulman, 1956, p. 472. N.B. In an earlier article I reproduced these curves, together with others, in a more complete form: Le Roy Ladurie, 1959.)

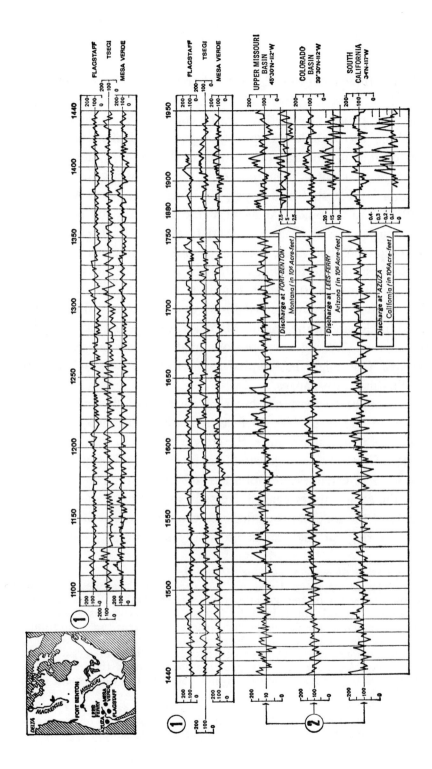

In the three curves in Fig. 2, the horizontal readings give times
—nearly a thousand years in all—and the vertical readings show the
relative thickness of growth rings. To construct these three curves
Schulman used data taken from three groups of conifers very sensitive to drought. The ages of the trees in these groups ranged from
hundreds to about a thousand years: two groups are in northern
Arizona, at Flagstaff and Tsegi; the third is at Mesa Verde in
southwestern Colorado. At Mesa Verde and Tsegi the trees used
are Douglas firs; at Flagstaff, a variety of pine. Each curve is of
great representative value, because it shows a mean constructed not
on the basis of one tree but on that of a group of trees scattered
through their respective areas. Episodes, such as disease, in the life
of each tree are compensated for so as to leave only the general
trend of climate in the region. Moreover, the likeness between the
three curves gives a general idea of climate over a whole geographical area. One might be tempted to say that the trees in question
are "natural rain gauges."

One might be tempted—but such an expression would be exaggerated and inapposite. In fact, as Harold Fritts has shown in
a series of remarkable studies, the growth of trees in the arid southwestern United States is governed not by one variable (rain) but by
two series of variables: precipitation and temperature. In a very hot
year the temperatures cause increased evaporation, reduce the humidity of the soil, and thus decrease the tree's intake of water; so
the tree ring for the year is necessarily narrower.

Fig. 3 *Maps Showing Variations of Climate*
These maps are based on the decennary departures from the
multisecular mean of tree growth in the western United States. Positive
departures (vertical lines) indicate (figures in italics) areas temporarily
cool and humid in the decade in question. Negative departures (horizontal
lines) indicate hot dry areas. H (high growth) and L (low growth) apply
to corresponding areas and refer correlatively to strong or weak growth
in the trees. Areas are marked when variations are more than ±0.6. The
numbers of the stations (ordinary figures) are omitted when there is no
data for them.
The decades 1576–1585 and 1931–1940 are very dry. The decade 1746–1755 is rather humid. The decade 1641–1650 offers a geographical contrast. (Source: Fritts, 1964 and 1965 b, pp. 433 ff.)

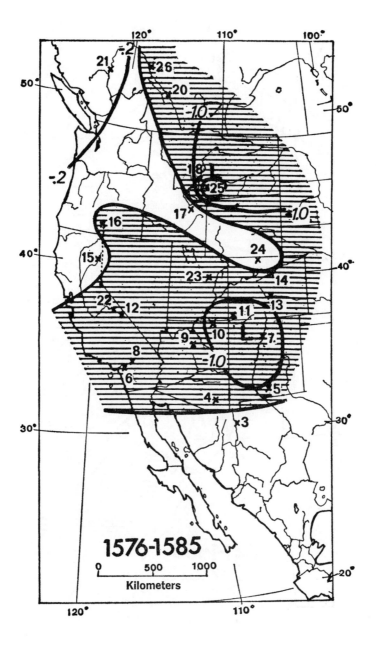

1576-1585

0 500 1000
Kilometers

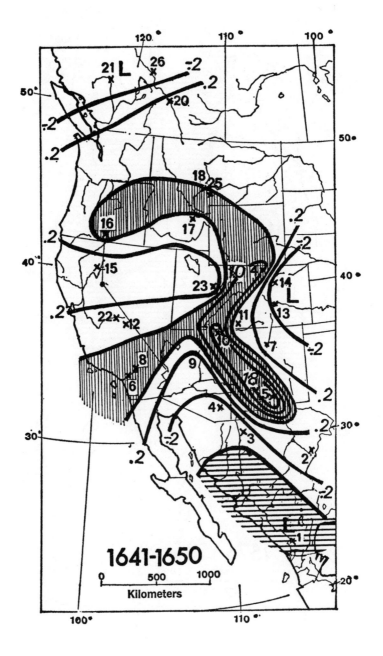

1641-1650

0   500   1000
Kilometers

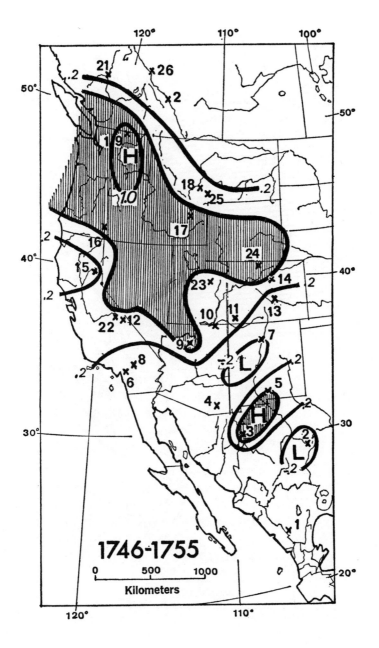

1746-1755

0   500   1000
Kilometers

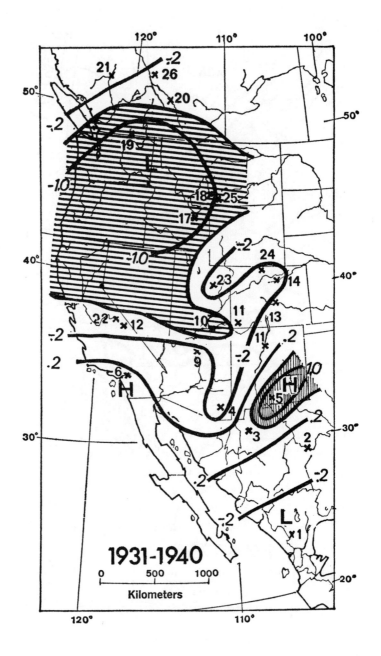

1931-1940

0   500   1000

Kilometers

To put it differently, growth of forests in the arid southwest is a function of the aridity index, which combines, with variable coefficients, the factor of dryness and the factor of heat-cum-evaporation. Fritts, working from tree data, has given a series of formulae and maps[12] which indicate variations in aridity over the whole of the far west for five centuries (Fig. 3).[13]

So much for the methods. But what conclusions do these authors draw from their patient researches? The first conclusion is the general stability of climate in the last thousand years, and in fact for the last two thousand at least: "Timbers cut seventeen hundred years ago have tree rings exactly like those of trees of the same species growing today on the same sites." Other authors come to the same conclusion on the basis of other data.[14]

But a second idea, much more interesting for the historian, concerns the existence of largish meteorological fluctuations, in this case of rainfall. Over periods that can be as long as twenty or thirty years, and sometimes even a hundred, the curve departs from the average to indicate prolonged waves of dry or wet. One of the most marked of these fluctuations occurs around the 1300s. "One gets a definite impression," writes E. Schulman, "from the study of average conifers, that in the southwest after the 1300s a very dry century was followed by a century of almost uninterrupted rainy years. This very wet interval may have been the longest in this region for the last two thousand years." The curves derived not from living trees but from the timbers of the Indian pueblos tend to confirm this opinion.

So it would seem that in the west of the United States the thirteenth century was dry and the fourteenth wet—at least if one goes by Schulman's results, though these are now quite old. In fact, the researches of scholars who have continued his work tend to focus attention on the most marked phase of dryness in the thirteenth century, that which stretches at Mesa Verde for example from 1271 to 1285. The length and aridity of this period of twelve very dry years were not to be equaled in this region by any similar episode in modern times, from 1673 to the present day.[15]

Another important dry period was that which affected the end of the sixteenth century. According to Schulman, the last twenty-five or thirty years of the sixteenth century in the south-western United States were characterized in general by a severe deficit in the growth of trees, precipitation, and runoff of rivers, a much more marked deficit than in the famous droughts of 1900 and 1934 (which ripened *The Grapes of Wrath*). Data from over-age trees tends to show, in fact, that it was the worst drought since the century-long one in the 1200s. Schulman notes important regional differences in the distribution of this drought: it was very severe in California, where it lasted from 1571 to 1597 and the rainfall deficit as reflected in slowing down of tree growth was twice as great as any other registered between 1450 and 1950; it was also very severe in Colorado, where it lasted exactly from 1573 to 1593; and it was much less pronounced in the north, for instance in Oregon, which shows only a slight and intermittent deficit in tree growth between 1565 and 1599.

Such a geographical difference is important and has a general bearing: it is quite wrong to misapply to damp, temperate regions conclusions valid for arid zones. What is true for Los Angeles is not necessarily so for Portland; in Europe, what applies to the Mediterranean does not necessarily apply to the countries bordering the North Sea, still less those of the Baltic.

At all events, the long and severe drought at the end of the six-teenth century had considerable consequences for the region where it was most felt—the southwest of what is now the United States. This great drought had devastating effects on the backward and sparse economy of the Indians, who lived grouped together in pueblos and raised crops in irrigated plots in Arizona, Colorado, and New Mexico. At all events it was precisely in the years 1560–1570 that the Chichunìque Indians chose to emigrate en masse to the silver mines established by the Spaniards in the north of Mexico. Were the two phenomena interrelated—the repulsion exercised by an excessively dry climate, and the attraction of a new mining industry?

Great discoveries may be expected from United States den-drochronology; this branch of research certainly has not got "its future behind it." The efforts of the new pioneers of the Tucson School are aimed at present in various directions.

In the first place these scholars want to recapitulate and catalogue the whole accumulation of basic tree ring material which has grown up during almost half a century in their laboratories in Arizona. Huge inventories on all the tree ring series, including those from all the archaeological sites, are in course of publication.

Secondly, these researchers are trying to widen the field of correlations derived from tree rings. The "grand old men," Douglass and Schulman, showed by chronological indications derived from trees the existence of comparatively simple phases characterized by the variation of one climatic element (mainly aridity). But present-day dendrologists like Fritts and Stockton want to go further: they are trying to correlate tree rings with the occurrence of complex types of weather. In order to do this they have taken the state of Arizona as their field, and divided weather into nine possible types which they call climatic classes. Each class corresponds to a combination of two basic variables (precipitation and temperature), these variables being divided into Above Normal, Near Normal, and Below Normal. The table gives the details of the nine classes.

Combinations

Precipitation	Temperature	Class
Below Normal	Above Normal	1
Below Normal	Near Normal	2
Below Normal	Below Normal	3
Near Normal	Above Normal	4
Near Normal	Near Normal	5
Near Normal	Below Normal	6
Above Normal	Above Normal	7
Above Normal	Near Normal	8
Above Normal	Below Normal	9

These nine classes can of course be applied to each season of the year, giving nine categories of spring, summer, autumn, and winter respectively.

Fritts and Stockton are also working on a typology of tree rings for Arizona. The average thickness of tree rings ranges from very wide through medium to very narrow in different years. Here too nine categories have been established, in descending order of thickness. The object of all this is to answer this question: given the category of tree ring characteristic of a certain year, which of the nine climatic classes is it likely to correspond to in each season of that year? A certain type of tree ring is normally correlated, as we have seen, with a certain type of meteorology, which either hampers or encourages growth during the year concerned. Of course this kind of research can only be applied in the first instance to a recent period of time, for which both tree rings and the necessary meteorological observations are available. But once this first stage is completed and probabilities thus established, it becomes possible to write history retrospectively, and, to use Marc Bloch's phrase, to light up the past by the present. With the aid of probable relations between tree rings and weather which are solidly based on the twentieth century, it is possible to work backwards into centuries for which no written meteorological observations exist, and determine the probable occurrence of certain types of weather each season during the seventeenth, eighteenth and nineteenth centuries. And so it becomes possible to construct a probable, retrospective meteorology. A cartography of climatic fluctuation is also under way.[16]

But the school of Tucson is not content with working out with ever-growing precision the chronology of climate during the period already long explored by the tree ring specialist—roughly the last thousand or two thousand years. They also endeavor, with the aid of trees of enormous longevity, to throw light on even earlier times. The most remarkable work from this point of view concerns living and fossil specimens of bristlecone pine

(*Pinus aristata Engelm*). These trees grow in a more or less compact group at an altitude of ten or eleven thousand feet on the border of California and Nevada. They grow extremely slowly, the growth period each year being very short—between about June 26 and August 8. It is precisely their economy of strength and life which enables these gaunt old men of the mountains to defy the centuries and the millennia.[17] Thanks to them a tree ring chronology of 7117 years,[18] going back to the fifth century before Christ, has been constructed. Its author, C. W. Ferguson, even hopes to extend it further into the past: he has recently discovered a fossil specimen of bristlecone pine eight thousand years old, which may enable him to add another thousand years to his collection.

It is not always easy to interpret climatologically the curves derived from bristlecone pine tree rings. Many of these Methuselahs of the forest are so robust—they need to be!—that they are comparatively insensitive to climatic stress and those years of adversity which normally produce extremely narrow tree rings. Still, with their usual ingenuity, the Tucson researchers have managed to isolate among the groups of bristlecone pines certain individual trees which, mainly because of their site, have shown themselves more sensitive than others to drought, and for thousands and thousands of years have acted as reliable indicators of climatic change. We may therefore hope that in the fairly near future the climate in the mountains of California and Nevada during several thousand years before our era will gradually yield up its secrets.[19] Even now the multimillenary specimens of *Pinus aristata Engelm*, dated, with complete precision, through their thousands of tree rings, make it possible to check the validity of radiocarbon datings, and even in some cases to criticize and rectify them. "Dual dates, derived from the radiocarbon analysis of tree-ring-dated bristlecone pine wood, show that, for material in the period 4000 to 3000 B.C., dates obtained by the conventional Carbon 14 are about eight hundred years more recent than dates established dendrochronologically. There is at present a spirited debate concerning the possible direction and causal nature of this relationship, a point that may be resolved when even earlier wood is found and dated, as we expect it will be.[20]

Old and very old trees are not the only "reservoirs" of climatic information in the Tucson area of Arizona. In this subdesert and, as far as rainfall is concerned, ultramarginal region, there are many other indicators of climatic change: the rooting and consolidation of self-sown saguaro or giant cactus[21] vary with the (fairly rare) shade and rainfall, and the average age of a population of Arizona saguaros can thus point to the early existence of a phase which was not too arid, and which could produce a period of repopulation among the cacti.[22]

In western California, on the vivid but scorching sands of Death Valley, other students of early climate set up as geomorphologists[23] and trace the agelong fluctuations in humidity, minute as it is in this region, through the spasmodic flow of local, intermittent springs. The faint trace of these aquatic "discharges" can be seen in the fossil strata of the salt marshes fed by these streams.

In 1968, when I visited Harold Fritts and Tucson's various palaeoclimatic laboratories, I was able to see for myself the work being done there in dendrology and geomorphology, much of which is likely to produce important results. The scope of this book does not allow me to go into detail, but from what I saw I am convinced that the work that has been done in the last fifty years in the palaeoclimatic laboratories of Tucson, in the peaceful university city overlooked by the Santa Catalina mountains, is, from the point of view of the science of climate, of history, and of the earth, one of the great intellectual adventures of the twentieth century.

This work on rainfall and precipitation in the arid zones of the United States is paralleled by similar work on temperatures carried out in the Arctic regions near Canada. In these very cold zones the thickness of the tree ring is proportional to the total heat received during the growth period. Giddings[24] has thus been able to reconstruct five centuries of climatic history in the Arctic (1450–1945) from the evidence of groups of trees near the delta of the Mackenzie river (north of the polar circle, near the frontier of Canada and Alaska). Here again the essential conclusion is the great stability of the curve: it oscillates around the same average for centuries, only rising slightly from 1850 on.[25]

Ten-yearly and even smaller fluctuations can be seen clearly on the same graph: one of the most striking is the series of hot summers between 1628 and 1650. No doubt similar fluctuations as those in Alaska were felt in Europe, though not at exactly the same times, and produced, especially in the northern countries and mountainous regions, beneficial or catastrophic effects according to their nature.

Though tree ring study is not so advanced in Europe as in the United States, it is beginning to give very important results. True, different countries are at very different stages: in France, for example, dendrochronological studies have generally only just begun, but researchers like Polge, Pons, and Lucie Leboutet have already published excellent and important works[26] on tree rings. But all this work is rather "preliminary" in nature and aims, quite legitimately, rather at laying solid methodological foundations than at actually exploiting the first factual and chronological results of the study of a given period. One such period is the seventeenth century, the climate of which is likely to be illuminated in the not too distant future through the study of old timbers.

From the point of view of dendrochronological research itself, it is only fair to say that France has not yet reached the level of Arizona. But the as yet unpublished work being done by the laboratory of medieval archaeology at Caen, which is carrying out specific research on tree rings, suggests it may not be too long before this gap is filled.

In England things are much more advanced. There, with all the old churches, piously preserved monuments, and flourishing medieval archaeology, one of the great objects of dendrochronology is the dating and interpretation of tree rings from timbers of the Middle Ages. Pioneer work has been showing the way since 1957.[27] Professor Fletcher, who made an important contribution to a symposium on climate held at Oxford in 1969[28] is at present concentrating his highly developed research techniques on fourteenth-century timbers. This period, with its series of wet years and terrible famines, is one of the most fascinating possible for a medievalist, whether he is a historian or an archaeologist.

Basic tree ring research in the Soviet Union is also archaeological: since 1951 over twenty-five hundred samples of fossil wood have been collected in the town of Novgorod alone. Most of these fragments, taken from the soil or subsoil of the city, come from the ancient wooden pavements characteristic of the streets there in the Middle Ages. B. A. Kolchin, who is directing the research, has been able to use this valuable material to establish a solid chronology of local tree rings from A.D. 890 to 1410. Although the climatological interpretation of this curve has only been begun, it seems that Kolchin's diagrams correspond to what local chronicles tell of exceptionally inclement weather, such as acute droughts or cold wet summers.[29]

But with the exception of Finland (where G. Siren's work should be noted, though it interests the historian less directly), it is in Germany that tree ring research has produced the most spectacular results from the point of view of long-term study of the history of climate.

The undisputed master of German dendroclimatology is Professor Bruno Huber, of the Forestry-Botany Institute in Munich. His own work and that of his school, which are considerable, have not always met with the reaction they deserve from the international public interested in scientific history. It is therefore worth while giving a brief account here of certain aspects of Huber's work which most concern the historian. In the pages that follow they will be set beside the voluminous data extracted from archives by other researchers. I shall concentrate for my purposes on the problems of climate in the fourteenth century in Europe.

B. H. Huber began his researches in 1938, and by 1956 had

succeeded in measuring 250,000 tree rings from trees in various parts of Germany.[30] After 1956 new on-the-spot research and samples enabled Huber to set up a *master chronology* which gives the average tendency for tree ring width each year in relation to the average over several centuries.[31] This master curve is based on the oak, which as the wood most frequently used for building is now the favorite quarry of European dendrochronologists. Huber's master chronology extends over 1160 years, from A.D. 800 to the present, and concerns southern Germany.

In western Europe, and especially in southern Germany, the oak can manifest rather perplexing reactions. When the tree being examined comes from the floor of a valley, i.e., from an area where moisture is almost always adequate—the width of the tree ring tends to be proportionate to the warmth received during the growing season. This is particularly marked in the case of *Quercus robur*, the pedunculate oak. Conversely, if the tree grew on a steep slope, as is often the case with *Quercus petraea*, the sessile oak, the tree rings tend to vary in width according to the fluctuations in rainfall. As the water does not stay very long on a very steep slope, lack of rain very easily makes itself felt, even in a temperate climate like that of southern Germany, and in these conditions the deficit tends to produce a narrow tree ring, and vice versa.

Given the diversity of these reactions, Huber's master chronology, stretching over more than a thousand years, is not always easy to interpret. A series of wide rings, for example, predominating over a decade or more, may signify either a number of hot summers or a succession of wet seasons, or even an inextricable mixture of the two.

But a continuous reading of the curve reveals to the historian of climate, familiar with the various outstanding years, many well-known points of reference.

A first and most characteristic example is the famous *Sägesignatur* or "saw-tooth signature." This description applies to a very curious phenomenon which occurred during the decade 1530–1540, or more exactly from 1530 to 1541.[32] During these twelve years one thin ring alternated with one wide one in a regular biennial zigzag. Such a unique occurrence could only have been brought about by chance, according to Huber and Courtois' calculation, once every five thousand years. The *Sägesignatur*, first detected and recorded in 1949, has been since encountered again on many old South German timbers, which by definition were growing *during* and cut down *after* 1530–1541. The *Sägesignatur* is thus a great aid to dating.

What is very interesting is that this typical saw-tooth formation is found reproduced exactly on the graphs showing harvest dates in the French vineyards nearest to southern Germany; i.e., those in the Jura, the Swiss Alps, and the Midi. The wide tree rings of 1531, 1533, 1535, and 1537 correspond to the years when the wine harvest was late in the Jura, a symptom of cool springs and summers, which were probably also damp. The narrow tree rings of 1532, 1534, 1536, 1538, and 1540 correspond to early wine harvests, which reflect hot and rather dry conditions in the springs and especially in the summers.[33] The connection between these three classes of phenomena—climate, tree rings in southern Germany, and dates of the wine harvest in eastern France—is easily explained. The alternation which produced a hot dry summer every other year for twelve years also produced narrow tree rings in southern Germany because of the dryness of the summer, and, simultaneously, early wine harvests in eastern France because of its heat.

The same applies to many other periods both before and after 1530–1541. Among the remarkably hot years, which produced narrow tree rings and early wine harvests, and which are visible both on Huber's curve and on that for the French wine harvests, I shall quote only those belonging to the periods with which I am most familiar: the years 1393, 1503, and 1504, 1517, 1525, 1567, 1571 and 1572, 1574, 1583, 1590, 1603, and 1634–1637.[34]

Conversely (and although one cannot make a completely general rule because of the complex climatic ecology of the oak, previously mentioned) one often meets with wide tree rings which in their time were steeped in water, and which correspond to wet years or series of years noted for floods in the chronicles of the Middle Ages or the seventeenth century. This is probably true of the year 1673 and the decade beginning in 1690. It is certainly true of the decade 1310, or more exactly the years 1312 to 1319. On the dozens of trees examined by Huber and his team, these eight years are characterized by very wide rings, in marked contrast to the preceding periods (before 1312) and those which followed (after 1319).

In this case the interpretation admits of scarcely any doubt. What we are dealing with is a huge wave of humidity, a series of rainy seasons which, during the greater part of the decade 1310 and over vast areas of western Europe, made the trees shoot up, swelled the rivers, destroyed and rotted harvests, and doomed the people to famine, which reached its peak after the disastrous year of 1315.

For proof one need only read Weikinn's list of inclement weather for the west as a whole during the Middle Ages.[35] This list confirms the above chronology (with a few differences in the last few years): from 1312 to 1317 each year bears mention of floods. The widespread floods of 1315 are only the peak in a tendency apparent both before and after that year. It is only in 1318 and 1319, according to the evidence collected by Weikinn, that the floods disappear temporarily from the chronicles. The most exact study of the crucial years 1310 to 1320, so exact that it stands comparison with Huber's curves, is that undertaken by the English historian John Titow. Using the manorial records of the see of Winchester and working out season by season the meteorological data indicated by the bishop's bookkeepers, he notes as very wet the eight years from 1313 to 1320. Apart from 1318, which was so dry that the episcopal plows snapped off in the hard ground, there is not one season out of the whole thirty-two in the eight years in question which was considered dry by the people living at the time. On the other hand, the winter and autumn of 1313, the autumn of 1314, the winter, summer and autumn of 1315, the winter, summer and autumn of 1316, at least one season in 1317, the autumn of 1318, the winter, summer and autumn of 1319, and the autumn of 1320 were probably (in three cases) and certainly (in all the other cases) wet or very wet.

Of course, one alone of the rainy episodes recorded in the trees of Bavaria and the episcopal archives of Winchester would not have been enough in itself to cause famine. It was their repetition that brought about the well-known catastrophe of the two terrible years 1315 and 1316, which destroyed the European grain harvests. The Winchester data throws light on the actual mechanism of the disaster, only generally indicated by the tree rings. In Winchester the year 1314, in spite of a rainy autumn, was quite a good one for wheat (autumn rain, coming after the harvest, does not affect it). It was in the winter of 1315 (December 1314–February 1315) that things began to go wrong: from then on, winter, spring and summer of that year were wet. The hay would not dry, plows got stuck in the mud, eels swam out of the brimming ponds and could not be caught; the spring sowing was a failure; the 1315 wheat yield was ridiculously low, at 2.5 grains for each one sown. There was not even sufficient seed for the following year, when the 1316 harvest was jeopardized, first by the poor autumn sowing, then by the rains that fell in the winter of 1315–1316 and the summer of 1316. The winter of 1315–1316 was so wet the bishop of Winchester was obliged to have his oxen reshoed more than once after they had lost their shoes in the mud. And the summer of 1316 was so damp there was not even enough good weather to shear the sheep. The wheat yield for 1316 was even lower than for 1315: the Winchester figure is 2.11 grains harvested for each one sown in 1316, as against 2.47 for 1315 and 4 for a normal year. It is easy to see why contemporary commentators talk of famine. In Winchester the price of wheat rocketed to double and triple its usual value, based on the period from 1270 to 1350. No similar rise is observable during the whole of the rest of the period from 1245 to 1350, which underlines the altogether exceptional nature of the decade 1310, and in particular the years 1315 and 1316. Deaths from starvation followed a similar rhythm: on the death of each of his villeins, men, women and sometimes even children, the bishop of Winchester collected a tax called the *heriot,* and the annual number of these gives an idea of the number of deaths among the peasant population. This number was never so high during the century from 1245 to 1347 as during the two terrible famine years of 1315 and 1316. These hecatombs were not equaled until the great plague of 1349, when at Winchester they were amply surpassed.

The terrified people organized processions to pray for deliverance from rain and the return of abundance: Guillaume de Nangis tells how in 1315 he saw pathetic bands of people thin as skeletons in the sees of Rouen and Chartres: "We saw a large number of both sexes, not only from nearby places but from places as much as five leagues away, barefooted, and many even, except the women, in a completely nude condition, with their priests coming together in procession at the church of the holy martyrs, and they devoutly carried bodies of the saints and other relics to be adored."

So the wet period of the 1310s (1312–1319 in Germany, 1313–1320 in southern England) culminated in 1315–1336 in panic and frustration, famine and widespread death.

Now it so happens that this wet period of eight years is by far the most marked in the period covering the end of the thirteenth and the whole of the fourteenth century. The most diverse figures, whether from tree rings or archives, confirm this. On Huber's diagram, the wet years of 1312–1319, denoted by strong growth and wide tree rings, tower over the whole period from 1298 to 1353.[36] For the rest of this fifty-five-year period the tree rings are generally rather narrow, indicating dry or normal seasons.

The Winchester archives, looked at in the long term, show the same thing. Titow, who has tabulated all the seasons for the hundred and thirty years from 1270 to 1400, considers the period from 1270 to 1312 as very dry: with characteristic exactness he gives for these forty-two years four which are wet, eleven which are "ordinary," and twenty-seven which are "dry." He gives the following period, however, from 1313 to 1320, as excessively wet. The fifteen years from 1321 to 1336 form a series of seasons in general dry or even very dry. The period 1337–1369 is "rather mixed"; 1370–1398 "rather dry"; and it is only in 1399–1403 that one comes once more on a group of years that are generally wet—but they are less intense, less numerous, and less productive of hardship than those of the decade 1310.

Such, in themselves and in comparison with before and after, were the eight wet years of 1313–1320, when all the west was up to its ankles in water. They are interesting in two ways: first because at their worst they produced one of the severest famines in the Middle Ages, and secondly because they show so well the advantage of a multiple approach, at once traditional and based on documents, and revolutionary and dendrochronological, like that of Huber.[87] The historian of European climate in the Middle Ages must have two mistresses, dendroclimatology the new one and palaeography the old.

The trees of America suggest presumptions and working hypotheses for long-term movements of climate in Europe. For shorter movements, on the other hand, we have already wrested certainty from the systematic study of very old trees from northern Europe in the last thousand years. This has brought to light not so much oscillations over hundreds of years as series of mainly hot or mainly cool years within, over, or between decades[38]; these series are analogous to though not synchronous with series already discovered in America. When there have been more of such studies they may throw new light on agricultural and thereby on economic history. But it will still be necessary to avoid oversimplifying the effects of weather on harvests.

A last point: the data derived from trees would be of even greater interest if they showed us not only meteorological series but also some law governing their periodicity. It would then be possible to foresee the recurrence of these series, and the history of climate, like weather forecasting, would then have a fully rational as well as an empirical content.

Early work by Douglass did, it is true, give rise to the assertion that dendroclimatology showed climate to be affected decisively by a sunspot cycle of 11.4 years. The curve representing tree growth was supposed to show periodic oscillations occurring about every eleven years and thus corresponding perfectly to the solar cycle. The temporary disappearance of this cycle from growth curves between 1645 and 1715 even led Douglass to suppose that the reign of Louis XIV was distinguished by a lack of sunspots. It must be admitted that here the creator of American dendrochronology was not relying on pure deduction: in 1922 E. Maunder, working on documents from early observatories, had pointed out a remarkable absence of sunspots between 1645 and 1715. Even if Maunder was entirely right, which is not certain, it would be a mistake to get too excited about the parallel proposed by Douglass between solar and tree data. As Marc Bloch said, ugly facts sometimes destroy beautiful theories. The scarcity of sunspots between 1645 and 1715 (or, at least, the weakened intensity of the eleven-year cycle of solar activity) is a fact which is plausible even if not proven. But the correlation of that fact with tree growth in Arizona is extremely problematical. Though at one time sunspots exercised an absolute fascination over the minds of researchers, it is by no means sure they had a comparable influence on tree rings. Douglass' recent disciples are much more prudent on this point than their master. Commenting skeptically on Douglass' suggestions in the '50s, Edward Schulman wrote that examples which had been noted of direct parallelism between the solar cycle and the growth of certain trees were often quoted, and, indeed, could not be entirely due to chance. But after this doubtful comment, he went on to say what he really thought: the cycles found in the growth of trees seemed to be characterized by variability in length, amplitude and form; to appear and disappear without exhibiting any general law; and to happen in almost any order. A satisfactory physical explanation of these characteristics had not yet been forthcoming.[39]

Harold Fritts, who since Schulman's death has been one of the chief researchers of the Laboratory of Tree-ring Research at the University of Tucson, Arizona, has abandoned the study of "helio-dendrochronological" correlations. He has devoted himself solely to the study of tree rings in and for themselves, and is the author of some excellent maps.

So we must no longer look to tree growth curves for information about some universal law governing the cyclic evolution of climate. Climatic curves, like price curves, are for the moment purely empirical, and it is impossible to deduce them from any kind of periodicity.[40] They have to be separately established for each continent[41] and for every large region.[42]

Until reliable "dendroclimatological" series have been established for Europe, which will require considerable time and expense, there is a simpler and quicker method of getting some idea of the climate in western Europe since the sixteenth century. This method has been known in France for about seventy-five years, and is based on the dates of ripening of plants and crops.

This is what is known as the "phenological" method.[43] The principle is very simple. The date at which fruit ripens is mainly a function of the temperatures to which the plant is exposed between the formation of the buds and the completion of fruiting. The warmer and sunnier this period is, the swifter and earlier the fruit (or crop, in the case of cultivated plants) reaches maturity. Conversely, if these months have been cold, cloudy and dull, ripeness and harvest are held back. There is a close correlation, which has been verified with great accuracy in many plants, between the total temperatures of the vegetative period and the dates of blossoming and fruiting. These dates are thus valuable climatic indicators.

It is true that for the historian the field of research here is at once limited: the ancien régime left little evidence about the dates of the lilac and the rose. There is only one date that appears regularly every year in many registers, pronouncements, and local police records, and that is the date of the wine harvest, when it was fixed by public proclamation. This date, fixed by experts nominated by the town or village community, is naturally related to the ripeness of the grape, and so is a good indicator of the average weather during the vegetative period, from March or April to September or October. "The grapes are ripe enough and in some places even withering," wrote, on September 25, 1674, the expert "judges of the ripeness of the grape" and the nine advisers appointed by the community of Montpellier. And they went on to fix the harvest "for tomorrow," and "next Thursday the 27 inst." as the date of the first *cuvée*. These decisions, it is stated, were "carried unanimously."[44] It was on September 12, 1718, that the experts of Lunel declared the grapes to be ripe; they fixed the date of the harvest as the nineteenth.[45] The grape harvest was particularly early that year in all the vineyards of Europe, from Languedoc to the Black Forest.

Naturally, economic and social factors as well as purely climatic ones affected the date fixed for the wine harvest. In Burgundy at the beginning of the nineteenth century the producers of high-quality wines, who were usually well off and able to take risks, sought after quality and preferred late harvests. Producers of cheap wines were not much concerned about quality and tried to gather the grapes as soon as possible.[46] The date at which the grape ripened also depended on the variety of the vine. But in spite of these "interferences," Garnier has shown that there is a sound relation, if not a perfect correlation, between the phenological curve for wine harvests in Argenteuil, Dijon, and Volnay in the nineteenth century and the average temperatures from April to September in the corresponding years, as recorded by the Paris Observatory (Fig. 4).

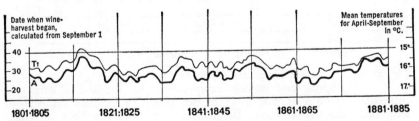

Fig. 4 *Wine Harvest Dates and Temperatures (Five-Year Running Means)*

N.B. To facilitate comparison with the phenological curve for the Île de France and Burgundy the Paris temperature curve has been inverted. (Source: Garnier, 1955.)

Figs. 4 and 5 authorize the principle: early wine harvests, warm year; late wine harvests, cold year, or rather cold vegetative period. So it is easy to see how important it is to know the dates of wine harvests for the periods where there are no sustained series of temperature observations. This applies in particular to the seventeenth century in Europe, a period when climatology is the more controversial because the facts are so little known.

Angot's great article[47] remains an essential source for the dates of wine harvests. He brought together the results of a national and European inquiry made by the French central meteorological

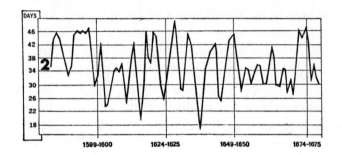

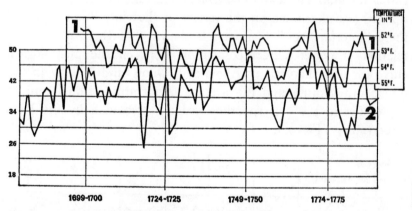

Fig. 5 *Wine Harvest Dates and Temperatures*
    (1) Spring and summer temperatures in England (March 1–September 1) on a running mean of two years. To facilitate comparison with the wine harvest curve, the temperature curve has been inverted.
    (2) Wine harvest dates (after Angot, 1883, and below, Fig. 6, curve for "general mean"). Dates calculated in days from September 1 (ordinate); running two-year means. (Source: Manley, 1953, pp. 242–558, and Angot, 1883.)

office about 1880. Many winegrowing centers provided Angot with copious information on the eighteenth and nineteenth centuries. For the seventeenth and end of the sixteenth century, Dijon, Salins, Kürnback in the Black Forest, and Lausanne, Lavaux, and Aubonne in Switzerland provided almost uninterrupted series. The series given by Dijon, Salins, and Lausanne go back to the beginning of the sixteenth century, though with some gaps. The Dijon series goes right back to the fourteenth century. We may sum up by saying that phenological data are copious for the nineteenth century, abundant for the eighteenth, adequate for the seventeenth, and fairly rare before that.

Thirty years ago Angot's series were augmented by the work of Duchaussoy on the wine harvest proclamations of the Paris area.[48] The Duchaussoy series starts in 1600. For the South of France, Hyacinthe Chobaut and I have each brought together material for a single southern series which also covers the period from the sixteenth to the nineteenth century.[49] (See Appendix 12.) Bennassar and Anes Alvarez have set up a series for the north of Spain, at Valladolid, the first for the sixteenth and seventeenth centuries, the second for the eighteenth.[50]

Lastly, in collaboration with Madame Baulant, who discovered some excellent early series of dates for Paris wine harvests, I have reconstructed a reliable curve for the sixteenth century (see Appendix).

At the end of his article Angot pointed out, without offering any explanation, certain fluctuations in the dates of wine harvests. I should like to call attention to these fluctuations.

The diagram in Fig. 6 which I have drawn up from the data published by Angot shows, vertically, the dates of the wine harvest, counted from September 1, in a number of places between the Alps, the Black Forest and the Massif Central. The years, from 1600 to 1800, are shown horizontally.

The first thing that strikes one is the very close, if not total, correspondence between the different curves every year. One example is the year 1675, when the wine harvest was very late all over Europe, from the Black Forest to the Languedoc. It was a

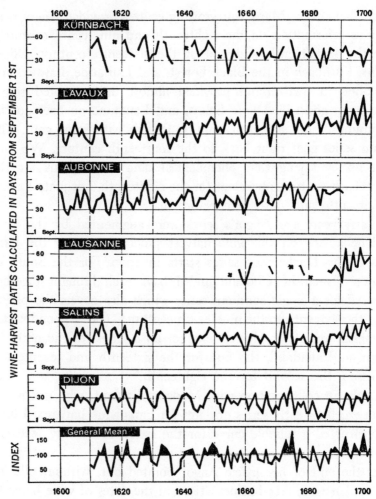

Fig. 6 *Wine Harvest Dates in the Seventeenth and Eighteenth Centuries*
      The local curves and the general mean have been constructed
from the figures given by Angot, 1883. The x-axis gives the years, and
the y-axis the dates of the wine harvest calculated in days from September 1.
Some missing dates have been interpolated, where this was possible, on the
basis of parallel curves relating to neighboring localities.
      The index 100, for the "general mean," corresponds to October
10; 50 indicates September 20, and 150 October 30.

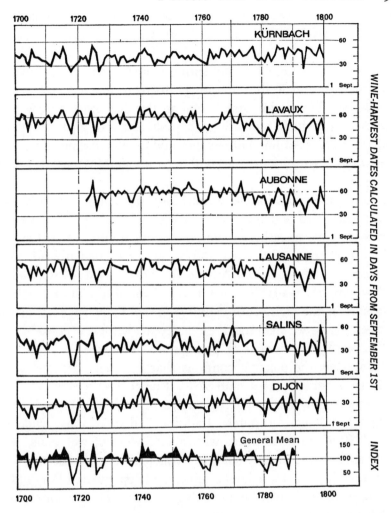

very cold summer, Madame de Sévigné wrote to her daughter, then in Provence.[51] "It is horribly cold, we have to light fires and so do you, which is certainly a very strange thing." This was on June 28, 1675. On July 3 the Marquise again refers to "an unusual cold"; on the twenty-fourth she says, "So you still have that north wind. My dear, how vexatious!" and she goes on to wonder "Whether the order of the sun and the seasons has not changed." Similarly the year 1725, when the whole summer was cloudy and wet, when wheat was dear and much of it rotted on the stalk, stands out for its very late wine harvests.

Conversely, 1718, when the spring [52] and summer [53] were very dry and hot, and wells and springs dried up and hay was scarce all over the Languedoc, stands out on the diagram for early wine harvests everywhere. Similarly with 1636, a year of great drought,[54] and 1645, a year of fine hot weather and excellent wine—a tremendous wine ("vin furieux") according to the curé Macheret in Burgundy.[55]

The parallel between the various localities appears to hold over short periods of between two or three and ten or fifteen years.[56] Series of years, or rather vegetative periods, that were particularly cold (1639–1643, 1646–1650, and so on) or particularly hot show up clearly on the diagram. Among the hot years were 1635–1639; 1680–1686; 1704–1710[57]; 1718–1719 (two of the hottest, driest summers in the eighteenth century); 1726–1728; 1757–1762; and 1778–1785. As we shall see, these last years witness overproduction and a slump in wines, and perhaps in wheat too.

Conversely the diagram confirms that between 1639 and 1643, and again between 1646 and 1650, France had a series of cold wet summers which might well have injured wheat production. Roupnel, relying on purely traditional documentary evidence, says: "After 1646 one meets with a succession of wet years, with chilly springs and stormy summers which everywhere destroy harvests already poor throughout a ruined and depopulated Burgundy." Having thus confirmed the phenological curve, he goes on to comment generally: "It will certainly be interesting one day to determine the character of these wet periods. Apparently, in the seventeenth century, the excess of atmospheric precipitation comes from the summer rains, i.e., the storms of June and July. Nowadays a wet year is good for forage crops and thus helps the farmer, though it is still a disaster for the vinegrower. But a dry year was more welcome in the Burgundy of the seventeenth century than it is with us, while a rainy year, with its storms and its hail, often destroyed the cereals which were then the staple crop and almost the only source of food." He does not hesitate to add:

"Six rainy years, from 1646 to 1652, brought about terrible hardship in the spring of 1652."[58]

Roupnel's comments are both categorical and impressionist.

Six bad harvests in a row is certainly something! No doubt Roupnel's intuitions found confirmation in various peasant proverbs which may be reduced to the assertion, on the one hand, that a year of rain is a year of pain, and on the other that a very dry year is one of good cheer.[59] But in order to check what Roupnel says properly we need to study tithes and establish the real curve for the harvests in Burgundy.

While it would be quite absurd to follow Roupnel entirely and "explain" the Fronde by the adverse meteorological conditions of the 1640s, one can agree with him that in a society which was in a state of latent crisis, "since 1630, 1635, 1637," agricultural difficulties arising out of climatic adversity might play a contributory role. The bad harvests were partly responsible for the "extraordinary cyclic peaks" of prices in 1647–1650. While they did not actually cause they did help to catalyze, as a secondary factor, the great "economic, social, and above all demographic upheaval" which found incomplete and ineffective expression in the rebellions of the Fronde.[60] The main catalyst, however, was nothing to do with climate: it was provided by the wars, the exploits of the reiters, and economic difficulties.[61]

Once having established these first data, it seems in order to set out the basis of a chronology of springs and summers in the sixteenth, seventeenth, and eighteenth centuries, based on the dates of the wine harvests.

I have worked out a series of wine harvest dates covering four centuries: in round figures, from 1480 to 1880 (see Appendix 12A). It concerns mainly the eastern half of France, and Switzerland. It makes use of data from several dozen vineyards (only ten or so in the sixteenth century; about seventy or eighty in the eighteenth and nineteenth centuries).

For the sixteenth century new series have been worked out in 1969 by Madame Baulant and myself, particularly for the Paris region. A comparison of the main curves (Paris-Chartres, Burgundy, Switzerland) gradually reveals the outlines of a periodization: the spring-summers[62] between 1510 and 1560, in round figures, were decidedly warm. They become first faintly, then definitely cooler between 1560 and 1600. These months of March to September in the last forty years of the sixteenth century, the cooling down of which is thus revealed by the wine harvest dates, are precisely the pilot months in which the annual melting or "ablation of glaciers takes place.

As these months, particularly the summer ones, grew cooler between 1560 and 1600, so the ablation decreased. The Alpine glaciers, for this reason among others, advanced considerably between 1595 and 1605.

With the seventeenth century comes a period of warm or average spring-summers between 1601 and 1616, without any major recurrence of cold. But from 1617 to 1650 there are outstanding cold episodes, spring-summers with low temperatures and late wine harvests. The culminating points are 1618–1621, 1625–1633, 1640–1643, and 1647–1650. Of course these thirty-three years do not form a solid climatic block. There are internal contrasts: in particular the 1630s had a heat wave between 1635 and 1639 which culminated in the scorching summers of 1637 and 1638. Nevertheless, if this period is compared with 1601–1616 or 1651–1686, there is no doubt that cold or cool episodes predominate. It is therefore not surprising that this new cool trend helped once more to produce spectacular glacial advances in the French, Italian, and Swiss Alps in 1643–1644 and up to 1653.

The following period, from 1651 to 1686, is somewhat different. Apart from a sizable cool episode in 1672–1675, the wine harvest dates are almost always early, or at least average throughout these thirty years. In the 1650s and 1660s the great spring-summer heats were accompanied by acute droughts. In the Midi of France, made desolate by drought for several years,[63] the inhabitants of Périgueux were reduced in August 1654 to visiting the relics of Saint Sabina and asking her to intercede for rain. At least one of the droughts of the 1660s had very far-reaching consequences. In 1666 the wine harvests were very early in the northern half of France. The weather was also warm and dry in southeastern England; the Thames was so low it spoiled the boatmen's trade and even threatened to ruin them.[64] In September 1666 the wood of which the Houses in London were built was so dry it needed only a spark to make it go up in flames. On September 2 Samuel Pepys wrote: "After so long a drought even the stones were ready to burst into flame." So according to Pepys the intense heat and drought of the spring and summer of 1666 were two reasons why the Great Fire of London took hold so easily and spread so fast.[65]

After a few years that were cool (1673) or very cool (1675), the 1680s brought a new period of hot dry summers. In England, where the first observations of temperature began to be made, as in France, where series of dates for wine harvests become abundant, the years 1676–1686 were swept by a great gust of heat and drought. In the Midi of France springs dried up, and in 1686 there was an invasion of grasshoppers from the south.[66,67] The 1690s, or more precisely 1687–1703, were completely different, with late wine harvests and very cold springs and summers. An examination of separate years illustrates the general tendency: almost everywhere the wine harvests of 1692 and 1698 were the latest known between 1675 and 1725, and neither in Salins nor in Dijon during this decade were there any wine harvests as early as those of 1684, 1686, or 1718.

Gordon Manley's excellent series of thermometric observations for England confirms this chronology: in England too the decade of 1690 was characterized by very cold springs and summers,[67] beating by almost one degree Centigrade all the records for cold during those two seasons for the "eighteenth century" (1680–1790). Contemporary chroniclers noted that after the cold spring of 1692 ("very cold and unseasonable weather; scarce a leaf on the trees on April 24th 1692"), there "commenced a series of extraordinary bad seasons: they have been traditionally referred to as the barren years at the close of the seventeenth century," wrote T. H. Baker, on purely documentary evidence and without benefit of thermometers or wine harvest dates.[68]

Comparison of French phenological series with English thermometric curves enables us to progress with a good deal of certainty during the eighteenth century. It has already been noted in connection with the 1690s that one of the earliest meteorological series, the one for central England published by Gordon Manley,[69] gives a temperature curve covering March 1–September 1 for the eighteenth century which is very similar to the curve for wine harvest dates (Fig. 5). The parallel does more than confirm the phenological series: it actually shows the history of thermic fluctuations during the springs and summers of northwestern Europe in the eighteenth century. The comparison adds much to the evidential value of the wine harvest dates, and enables us to use them with great confidence in the eighteenth century.

In France the three big groups of years with cold spring-summers between 1700 and 1800 are 1711–1717, 1739–1752, and 1765–1777. This more or less corresponds to what happened in England. For the third of these cold periods, 1765–1777, Von Rudloff's recent analyses based on German meteorological series are interesting: the springs in Germany were also cold during these years, especially between 1763 and 1772. The fact that the summers were very wet in the period 1763–1772, and very cool between 1767 and 1776, is accounted for chiefly by a situation of low atmospheric pressure. During these ten years the barometer fell and pressures were below normal in France, and especially in Holland and northern Norway, where the barometric deficit in summer reached 1.2 to 1.6 millibars. This absence of warmth and sun meant a very reduced ablation of the Alpine glaciers, which showed a marked advance, culminating in 1777–1778. Conversely, warm spring-summers and early wine harvests appear in eighteenth-century France from 1718 to 1737, from 1757 to 1763, and from 1778 to 1784. This last warm spell is found throughout Europe[70]: from Scandinavia and England to France and Switzerland the thermic averages for summer during the decade from 1773 to 1783 (the period of reference chosen by Von Rudloff) are higher than for any other decade since the beginning of accurate observations (i.e., since the beginning of the eighteenth century). In many cases they are the highest recorded up to the present day, exceeding the normal summer averages by over two degrees Fahrenheit.

Now we come to the nineteenth century, still seen through the wine harvest dates in France and Switzerland. The first decade, 1801–1811,[71] had hot summers; but from 1812 to 1817 there was a new series of cold spring-summers of which the like had not been seen in France since 1765–1777. About 1820 there was not un-naturally a thrust by the Alpine glaciers. The same cold springs, cool summers and late wine harvests are found again during the seven bad years from 1850 to 1856.

Conversely, the period from 1857 to 1875 is characterized by successions of warm and clear, or at least average, springs and summers.[72] The curve of the wine harvests in France corresponds closely to that for March-September temperatures, clearly trends, during these twenty years, toward an early date. Certain nonclimatic factors may have influenced the winegrowers, but there is no doubt that a chief factor in these early harvests was the warming up recorded for the three decades after 1856. The thermometric and barometric series which begin to multiply all over Europe during the time of Napoleon III and Bismarck give the measure of this trend. German wines were never so good, either, as between 1855 and 1865; particularly fine years were 1857, 1858, 1859, 1861, 1862, 1865, and 1866.[73] (See Appendix 12A.) One is struck by the coincidence between these good Rhenish vintages and the early wine harvests in France in exactly the same years. There had not been such a good time for wine since 1778–1784: the sun shone on Rhine, Saône, and Rhône alike. High pressure and fine weather dwelt over all central Europe. During the hottest and driest decade, 1862–1871, there was prolonged drought in northern Italy, Hungary, southwestern Germany, and the Rhine valley. The glaciers retreated more than a kilometer during the 1860s and 1870s, melted by the heat of successive summers, thus ending the expansion of over two centuries that had taken place during the Little Ice Age. The early wine harvests of the fifteen or twenty years after 1856 are one of the major fluctuations in the history of climate in western Europe.

One reason why it is interesting to continue the study of wine harvest dates right into the nineteenth century is that it then becomes possible to relate phenological data to the changing geography of the air, which finally determines the fluctuations of weather from one year to the other. The years 1822, 1834 and 1865, which had early wine harvests, suggesting warm spring-summers, correspond to anticyclonic situations. There were high pressures and anticyclones all over Europe in 1865; sunny weather and high pressure over northern Europe and Scandinavia in 1822; and in 1834, high pressure over Finland and Russia, and the föhn in Italy.

Conversely, late wine harvests, corresponding to cool spring-summers, often correspond to a cyclonic type of situation with predominant west winds, and the sort of cool, cloudy weather, with low pressure, which prevails over France, Ireland, and Scotland. This was true of 1816, which had a horribly cold summer, 1835, 1836, 1843, 1845, 1853, 1855, 1860, 1879, and so on. But it isn't quite so simple as this: some cold summers with late wine harvest belongs to a different barometric situation. In 1812 and 1813, for example, and partly also in 1816, it was high pressure centered over Scotland and driving north winds toward France which was responsible for a cold March-September and a corresponding delay in the wine harvest.

Such are some of the reflections arising out of a purely physical analysis of our series of wine harvest dates. It will be noted that the separate studies mainly concern annual tendencies, or tendencies within or between decades; in short, climatic changes extending over less than a century. One of the most remarkable of these intrasecular fluctuations is that of the spring-summer cooling down of the second half or last third of the sixteenth century, one of the consequences of which was the advance of the northern Alpine glaciers.

Is it possible to go beyond the intrasecular and discern longer and wider fluctuations by means of the harvest dates? If this did turn out to be possible, phenological data would supply indications of secular and intersecular trends, of really long-term movements of climate.

A superficial reading of certain phenological curves might indeed seem to reveal such long-term movements. Between 1640 and

1710, the wine harvests at Lavaux became later and later: in the first half of the seventeenth century they took place between September 20 and October 10; in the eighteenth century the grapes were not picked until between October 10 and 30.

This might be taken as a splendid example of a gradual cooling from one century to another, with the watershed during the reign of Louis XIV. But in that case how was it that the later harvest dates did not occur in Dijon until fifty years later? As for Salins and Kürnbach, the average date of the wine harvest there scarcely varied in the seventeenth century and the beginning of the eighteenth; if it changed at all, it tended to be earlier rather than later. Does that mean that the winegrowers of Franche-Comté and Germany worked under different climatic conditions from the winegrowers of Switzerland or Burgundy? Clearly not.

The close parallel between short- and average-term phenological episodes in vineyards far remote from one another shows that in the case of such agreement one single factor predominated in the decisions taken by village communities from Germany to southern France; and that factor was climate. Conversely, the obvious difference in long-term movements, even as between vineyards quite close to one another, cannot be due to climate. Late harvests at Lavaux or Dijon, for example, are due to human factors.

We know the reason. Sometimes in the seventeenth century, sometimes in the eighteenth, the winegrower, stimulated by a stronger and more discriminating demand, would aim at a higher quality wine than usual and one which would keep longer.[74] In Guyenne and Languedoc, he might burn his wine to make brandy. In both cases he would want to harvest later so as to get a grape richer in sugar and capable of giving more "degré" or "strength." The extreme case in the "pourriture noble" or "noble rotting" aimed at in the region of Bordeaux. This practice of delaying the harvest naturally tends to push the phenological curve upwards.

Delayed harvests, then, are a sign of a viticultural and not of a climatic revolution, and reveal an interesting fact of economic history: in order for the delay to take on meteorological significance and indicate a long-term climatic movement, it has to occur in all vineyards at once. But as we have seen, this is not the case here.

So in the seventeenth and eighteenth centuries the biological indicator does not reveal any long waves in meteorology over and above those short- and medium-term episodes which are perfectly clear and synchronous on all the diagrams.

It may still be that this long wave does exist. But it would have to be demonstrated by other, more sensitive and subtle means. At the present stage of inquiry the phenological indicator is too crude, too subject to influence by human factors, to provide evidence of movement on a secular scale.[75]

Whether or not wine harvest dates can furnish long-term descriptions, they do establish types of spring-summers: cool, with late harvests, or warm, with early ones. What are then the human and ecological consequences?

A first and typical example is the year 1816. That year the wine harvests were the latest ever known: in the Alps the grapes were not gathered until November; in northern and central France the harvest took place at the end of October or in November.[76] The thermic deficit in these regions was extremely marked from March to September. In Paris the temperature was 0.8 or 0.9 degrees below normal in March, April, and May 1816; it was relatively even lower during the summer of the same year: June was 3.5 degrees lower than normal, August 2.8 degrees, and September 1.6 degrees. At the same time, precipitation during the summer of 1816 in Paris was 211 millimeters as against normal figure of 150.[77] The cause of this singularity was an unusual distribution of pressure, revealed in the barometric observations then already relatively numerous.[78] In 1816 the subtropical high pressures which normally lap over on to Europe during the summer scarcely reached it at all. Low pressures therefore settled over central Europe, and this zone admitted the invasion of masses of cold Polar air which penetrated right to the south. Western Europe meanwhile came under the influence of the north winds directed toward it by the regions of high pressure over the North Atlantic. Once more, as in 1316 or 1675, all Europe spent the summer seeking refuge from the rain round the fire. It was under these conditions that Mary Shelley was inspired to write *Frankenstein,* and

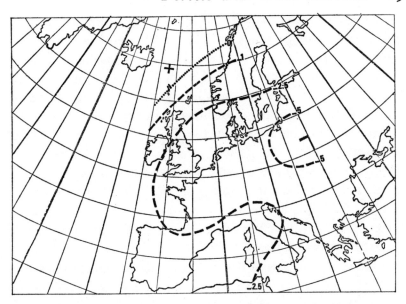

Fig. 6A Deviation of pressure during the cool summer of 1816 from the average for 1901–1950. (This cool summer badly damaged the harvest, causing grain scarcity in western Europe.) Source: Von Rudloff, 1967.

Polidori *The Vampire*. Both authors, together with Byron and Shelley, were in Switzerland, near Lake Geneva, that summer: "At first we spent our pleasant hours on the lake or wandering on its shores . . . But it proved a wet, ungenial summer, and incessant rain often confined us for days to the house . . . 'We will each write a ghost story,' said Lord Byron . . ."[79]

Mary Shelley was not yet nineteen when she invented Frankenstein, but although that summer may have been fertile in monsters it was not so fertile in cereals. A peasant of Brie in Seine-et-Marne, normally a rich wheat country, wrote: "The winter (1816) has been wet. The spring cold and late. The summer wet and late. We did not start cutting the wheat until August 20, which is very late. Because of the incessant rain we could not go on cutting two consecutive days. So wheat was both rare and dear—50 or 60 francs the sack. No wine. The vines were frozen where they stood. Such grapes as there were were gathered at All Saints', and the wine was undrinkable. As for the bread, it was damp and sticky, made as it was from wheat saturated with water; you couldn't eat it—it stuck to the knife as you cut it."[80]

But people had to live, or die, by this harvest for the next twelve months, from the summer of 1816 to that of 1817. "In the month of May (1817)," wrote another Brie peasant, "wheat had got so dear everyone thought to die of hunger. In June came revolutions in all the markets and towns, for bread was no longer to be had at the baker's."[81]

To sum up, this evidence, which is corroborated by many other witnesses, the cold weather and persistent rain of the spring and summer of 1816 hampered first the growth and then the ripening of the wheat, and finally made it sprout in the ear. Hence food shortages and social unrest.[82]

So the year 1816 acts as a model and throws retrospective light over three centuries of wine harvest dates, enabling us to see the human consequences of the climatic information they provide.

One certainty emerges: years or series of years which are, like 1816, cold and wet, with late wine harvests, are years in which the raising of cereals is full of danger. These years are not all disastrous for cereals by any means, but they do provide a considerable number of examples of famine or scarcity. And since, under the old regime, with its preponderantly rural population, wheat was a key product, it may be said that the whole fragile economy of the traditional societies of the temperate west (northern France, England, Germany) was vulnerable to these late wet years.

The phenomenon is already clear as early as the sixteenth century, when our wine harvest curve begins. There is a group of cold seasons with late wine harvests in 1527, 1528, and 1529, and the cold weather and heavy rain of these three years resulted in a series of catastrophic wheat harvests, accompanied by famine and riots all over France.[83] The great rebellion in Lyon in 1529, in which the protests of the hungry may perhaps have mingled with the first stirrings of the Protestants, was one of the high spots of this crisis. In England too, as a result of similar weather, the harvests were very poor in 1527 and still only mediocre in 1528 and 1529.[84] At Norwich, in 1527, the mayor's register for the year notes that "there was so great scarsenes of corne that aboute Christmas the comons of the cyttye were ready to rise upon the ryche men."

Another group of cool spring-summers and late or very late wine harvests occurs in seven successive years in the 1590s: 1591, 1592, 1593, 1594, 1595, 1596, and 1597. As has already been said, this is the coldest decade from this point of view since the beginning of the sixteenth century, and was prelude to one of the most violent glacial thrusts in the Alps in modern times. But the meteorological conditions were also reflected in the cereal crops. In northern France (leaving aside Paris, where the siege of the capital in 1590 by Henri IV caused a local disturbance in the cereal market), the years in which wheat was scarcest and dearest were 1594, 1595, 1596 and 1597.[85] This coincides closely with England in particular, and in general with what Professor Hoskins has called the great European famine of 1594–1597. There are records, he writes of "no fewer than four terrible years in a row. The rains fell incessantly all over Europe from Ireland to Silesia. The 1594 harvest was bad; 1595 was even worse; 1596 was a disaster; 1597 was bad too . . . The great famine extended all over Europe, lasting for some three years. In Hungary it was said that the Tartar women ate their own children. In Italy and Germany poor people ate whatever was edible, fungi, cats, dogs, and even snakes. In England there were food riots in many counties, the most serious perhaps being that in Oxfordshire."[86] In an attempt to reduce poverty the English authorities in 1597–1598 took temporary measures against enclosures.

In England as in France the second half of the 1640s was climatically somber: and on both sides of the Channel the cold rainy years which succeeded one another from 1647 on produced various adverse effects. W. G. Hoskins has skillfully disentangled the results of meteorology proper from those of the Civil War: "The 1640s," he writes, "opened with widespread agrarian disorders. But on the whole, despite the local ravages of the war years, it was a plentiful time down to the middle of the decade. The harvest of 1646 was, however, deficient, and the next three years were very bad, verging upon a dearth. The year 1648 was exceedingly wet: a bad harvest was accompanied by a widespread murrain among cattle. Food prices rocketed year after year, reaching a peak after 1650."[87]

There was another black decade at the end of the century. The years from 1690 to 1700 were extremely cold and often [wet, both during spring and summer and more generally during all four seasons. They give rise therefore not only, as one might predict, to late wine harvests, but also to several major food crises. One of these was among the worst famines in the whole seventeenth century: the failed harvest of 1693 caused an apocalyptic, medieval-type dearth which killed millions of people in France and the neighboring countries. No historian of the seventeenth century in France will say I exaggerate.

Of course there were certain nonclimatic factors underlying the famine. Since 1687–1689 all the kings of Europe had been at war with one another—and it was a real world war, cruel and devastating. Grain stocks for the armies were made at the expense of the meager supplies of the people. Moreover, taxes were increased, so that the peasants were too poor to buy seed when they did not produce enough of their own. In these conditions a poor harvest meant almost automatically that the next sowing, and therefore the next harvest also, would be inadequate.

So much for the nonclimatic factors. But the weather also played its part in making the 1690s so dreary. During the 1680s the growing season had been warm and dry and produced, in northern France and in England, a series of such superb harvests that the price of wheat had begun to decline.[88] But not for long. In 1687 began the dreadful cold seasons reflected in the late wine harvest dates of the last decade of the century. These cold seasons were particularly severe on the cereals of the high Alpine valleys, where even in a normal year they had some difficulty in surviving the rigors of the mountain climate. The Alps had cold, and often wet and dull, spring-summers in 1687, 1688, 1689, 1690, and 1692.[89] The effects on the local harvests are suggested by the comment of a Savoyard official writing in 1693: "Since the year 1690 most of the people of Tarentaise and Maurienne [valleys in the high Alps] have lived on bread made from ground nutshells mixed with barley or oat flour."[90]

Similarly, the *intendant* or administrator of Auvergne, writing of the three substandard harvests of 1690, 1691 and 1692: "This province has for three years been afflicted with great scarcity of corn . . . it and Limousin are the only two provinces in the kingdom to experience such great misfortune."[91] The very cold spring-summer of 1692 did nothing to mend matters. That year the wine harvests were horribly late—in Switzerland the grapes were still being gathered on the ninth and twelfth of November.[92] The dull cold summer months had a very adverse effect on crops such as chestnuts, buckwheat and grapes, on which poor people depended for their income or subsistence. In the autumn of 1692, Bouville, *intendant* at Limoges, wrote: "All the chestnuts are lost, together with most of the buckwheat . . ." He goes on to say that the weather is still so cold and wet "that the vines have suffered, and suffer still, so that there will be very little wine . . . ; and because of the shortage of buckwheat the people are like to go hungry by Lent."[93] The same gloomy reports come from the Île de France, Anjou, Normandy, Poitou, Béarn, and so on. A Brie peasant wrote in his diary: "The year 1692 has been a very late one. It was cold all summer. The grain harvest was very small."[94]

As October 1692 advanced, it turned out to be as cold as the preceding summer: six inches of snow fell near Paris on the twenty-seventh.[95] People grew more and more anxious, as the specter of famine loomed ever more clearly. On October 18, 1692, the *intendant* of Limousin wrote: "The frost has finished off the few chestnuts and the little buckwheat that remained. The vines look as if they had been swept by fire. The poor people are obliged to use their oats to make bread. This winter they will have to live on oats, barley, peas, and other vegetables . . ."[96] Finally, on October 24, comes a somber prophecy of a great famine in the following year; it was a famine that was to extend right over the next harvest too, from the summer of 1693 to the summer of 1694. "What is to be feared," writes the *intendant* of Auvergne, "is a great dearth of wheat not only next year, 1693, but also in 1694. Most of the land in Upper Auvergne has not even been sown. I do not mean they are left fallow or that there is not enough seed to sow; but *the bad weather has held back the harvest*

*and sowing so much that most of the men have not dared to sow,
and those who have dared are convinced there will be nothing to
reap . . .* The vines that were so promising have been quite destroyed by the recent frosts; it cannot be hoped the grapes will
ripen."[97]

The predicted famine duly arrived: the bad weather of 1692
was a sort of delayed action bomb. The year 1693, though somewhat less cold than 1692, suffered from a cool spring[98] which,
together in some places with blight,[99] was the finishing blow to
harvests that had already got off to a bad start in the autumn
sowing of 1692. So with the harvest of 1693 began one of the
worst famines western Europe had known since the early Middle
Ages. It turned France into a "big, desolate hospital without pro-
visions,"[100] a concentration camp the size of a kingdom, with
Louis XIV the commandant.[101]

In 1725 similar meteorological conditions (bad summer with
very late wine harvests,[102] cold and rain, and poor corn harvests)
again brought food crisis near in northern France, but this time
the worst was just avoided. (See Appendix 12A for late wine
harvest dates.)

But the eighteenth century offers much more cheerful examples
than the catastrophic ones we have just been studying, and from
this point of view one of the most interesting periods for the
historian of the vine and of climate is in the fifteen years that
precede the French Revolution. About 1780 one is drawn to make
a comparison between early wine harvest dates, the now abundant
thermometric observations, and the variation in the volume of
the wine harvest. There is plenty of data: there was by now a
considerable, active and reliable body of meteorological observers
both in France and abroad; wine harvest dates for the period,
correlated with temperatures, are known; and thanks to the labors
of Ernest Labrousse we know about the volume of the wine harvests. In northern France, and particularly in Champagne, where
the vine is at its northern limit, the wine yield depends partly on
the heliothermic index, i.e., on an adequate amount of sun and
warmth during the growing period (March–September).

Let us take up our position at about the center of our series, in the years 1780–1781, which saw the culmination of May–September heat, early wine harvests, and volume of wine produced.

The lesson of the curves is clear: in 1776, when the end of the spring and the summer were on the cool side, the pollen of the vines was washed away and the yield was only average.[103] In 1777 there was frost in April, a cool summer, a late and poor wine harvest: in Champagne and most other French vine-growing areas there was "neither harvest nor stock," and wine prices rocketed, reaching a peak in 1778. But from 1778 on there were four very warm years culminating in 1780–1781. The vines, rejuvenated, produced more and more and better and better. In Champagne the yield increased steadily from 1778 to reach its peak, simultaneously with that of temperature, in 1780–1781.[104] Vinegrowers on both sides of the Rhine, French and German, bear witness to the hot summers, early wine harvests, and superabundant wines of the years 1779–1781. In France[105] as in Germany[106] the wine harvest of 1781 was outstandingly early. But taken as a whole the summers of 1778, 1779, 1780, and 1781 were all increasingly hot and early, and taking as an example the quantity of wine produced in Champagne, one sees it pass from what was considered average in 1779 to abundant in 1780, and to exceptional and superabundant in 1781, "a year that surpasses an ordinary one by more than double."[107] It is only from 1782 on, when there was a cooler summer (in particular in the Paris region, the north and the east), that the spell was broken and the situation reversed.

It is quite clear in Champagne and Burgundy: as a result of the bad summer the wine of 1782 is "sour, weak, colorless, and absolutely bad." The grapes from which it was made had had great difficulty in ripening: after being attacked first by white frosts they had rotted on the branches just before the harvest.[108] The next year, 1783, which on the whole was not very cold, had fog and later cold spells which damaged the vines while they were in flower. The wine harvest in Champagne was 20 percent below normal. And so it went on. Lacking the stimulus of a sustained series of warm sunny years, wine yields declined; in 1784, in

Champagne, the volume was almost 60 percent below normal. Moreover, the winegrowers point out the poor quality of such wine as there was in 1783 and 1784.[109]

After an exception in 1785, which had a good yield, the volume of Champagne wine harvests plunges again after 1786. Here the empirical observations of the winegrowers coincide with the temperature figures: "Because of the long late cold spells and the summer of 1786, the grapes in Champagne have not been able to ripen and are beginning to rot on the vine." "The vineyards are a scene of desolation." In Burgundy too production fell in 1786.

The year 1787 was also cool from May to September and it too produced poor results. Stricken by frost and rain, the vineyards of Champagne gave under 40 percent of their normal yield.[110] While 1786, though less than half as good as 1785, was still an average year, 1787, down on the normal by two thirds or three fourths, was a very poor year, in quality as well as quantity. The grapes were over a month late and rotted before they could ripen. "The harvest will be late. No hope of starting before October 25 . . ."[111] So unfavorable meteorological conditions produced in Champagne in 1786 and 1787 wine which was late, scarce, and sour. Burgundy too was affected in 1787 by frosts, cold rain, a dull summer, and late harvests, and wine yield was very meager in comparison with the record year of 1785. After a slight improvement in 1788, which enjoyed a good May to September, the final catastrophe came in 1789, for which the thermometric curves show a very cold May to September. The vinegrowers' comments corroborate this: after the long and terrible frosts of the winter of 1788–1789, aggravated by late frosts in the spring and in the summer by cold and prolonged rain "the 1789 wine harvest in Champagne must be counted an utter failure." The situation was no better in Burgundy.[112]

To sum up. In spite of the complexity of the factors involved, it is clear that wine production in the north from 1778 to 1781 encountered conditions that were increasingly favorable to early harvests, quantity, and to a certain extent to quality too. All this culminated in the harvest of 1781, abundant, early, and delicious, and facilitated by a succession of mild winters, warm springs, and above all hot summers. The series of big harvests from 1778 to 1781 helped to bring wine prices down.[113] From being very high in 1777–1778, these fell steadily from 1778 to reach an extreme low after the huge harvest of 1781.

Then from 1782 on there is a change. Apart from exceptions like 1783 and 1788 the summers, or more exactly the periods of the year from May to September, become cool and adverse to the vine. As for the winters, which do not influence the phenological curve but are revealed by the thermometric one, the three severe winters of 1784, 1785, and 1789 stand out as without equivalent in the preceding period (1776–1783). None of these cold or cool episodes was in itself necessarily disastrous, but piling up as they did between 1782 and 1790 they prevented any very good harvests, with the exception of 1785. With that exception, every other year between 1782 and 1789 inclusive was either average, mediocre or poor in the vineyards of the north.[114] In the long run this deficiency in supply sent up prices again. They rose first slowly in 1783 and 1784, then went down again temporarily after the good harvest of 1785, then turned definitely upwards from 1786 and especially 1787 on.[115] There was thus a sort of dialectic between good harvests with low prices and bad harvests with high ones.

But here as elsewhere climate was of course not the only factor. It is certain that the increase in supply which brought wine prices down was also partly caused by excessive planting in the 1770s.[116] But as the extent of this excessive planting has not yet been measured, it is not yet possible to assess exactly the effect of this human factor on the overproduction of the 1780s.

So as far as the vine was concerned the hot early years around 1780 helped to accentuate the "intercycle" of lower prices. Was the same true of wheat? We have already seen that the wave of cool wet summers which culminated about 1770 and the wave of warm ones about 1780 coincided with bad and good cereal harvests respectively. This is shown by a comparison of northern wine harvest dates and the prices of wheat in Paris. The thermometric series for the Paris region and the figures collected by M. Morineau for the cereal harvests in Beauce confirm the connection.[117] After the foggy wet summers and spoiled harvests of 1766–1776, a series of good seasons culminated, for wheat as for wine, in two excellent harvests in 1780 and 1781. So for several years the sun of Beauce and Champagne bent the production curves, for wheat and wine, upwards, and the price curves downwards. The same trend can be seen at Cambrai on the curve for tithes on grain.[118] In short, the absence for several years of severe winters and catastrophic frosts, and a succession of brilliant summers, are the factors which cause a rise in supply and a consequent dive in prices.

It may be noted in passing that 1778–1781 or 1778–1782 are the years of the war in America, and it might be thought that the financing of the war, through the building up of stocks and the inflationist pressure of taxes, could, as during previous wars, have sent up prices. But this did not happen. There are many reasons for this, of course, but the high level of supply certainly helped to keep agricultural prices low. Louis XVI was lucky: not everyone who makes war can have good harvests and low prices, and be able to feed his soldiers, sailors, and subjects cheap.

But the warm dry spring-summers reflected in our series of early wine harvests do not always exercise the same sort of influence. On the whole they are favorable to grain, in northern France and northwestern Europe: here the dryness-warmth factor is the marginal parameter, on the presence or absence of which depends the volume of the harvest. But even in the northern areas of the Paris basin warmth and dryness can in certain cases be disastrous. A spell of dry heat at a critical moment during the growth period, when the grain is still soft and moist and not yet hardened, can wither all hope of harvest in a few days. This is what happened in 1788, which had a good summer, early wine harvests, and bad grain harvests. The wheat shriveled, thus paving the way to the food crisis, the "great fear," and the unrest of the hungry, when the time of the "soudure" or bridging of the gap between harvests came in 1789. So 1788 was decisive in its short-term effects. It is well described in the journal of a peasant vinegrower from near Meaux:[119]

"In the year 1788, there was no winter, the spring was not favorable to crops, it was cold in the spring, the rye was not good, the wheat was quite good but the too great heat shriveled the kernels so that the grain harvest was small, hardly a sheaf or a peck, so it was put off, but the wine harvest was very good and very good wines, gathered at the end of September,[120] the wine was worth 25 livres after the harvest and the wheat 24 livres after the harvest, on July 13 there was a cloud of hail which began the other side of Paris and crossed all France as far as Picardy, it did great damage, the hail weighed 8 livres, it cut down wheat and trees in its path, its course was two leagues wide by fifty long, some horses were killed."

It is worth setting this text point by point against the evidence of the best meteorological series for the Paris basin, those of Montdidier and Paris, and also against the wine harvest dates.

"There was no winter." At Montdidier, December 1787 and January and February 1788 were very mild, from one to three degrees above normal (it is well known that a too mild winter encourages weeds and pests and can prepare the way for a bad harvest in the following summer).

"It was cold in the spring." This probably refers to the short cool episode of March 1788: the temperatures that month were only a tenth of a degree above normal, and so gave the peasant of Vareddes the impression of being "cold." "The too great heat shriveled the kernels so that the grain harvest was small, hardly a sheaf or a peck, so it was put off." Here we have the withering process, which must have occurred in May, June, or perhaps the very beginning of July: according to the same text the grain harvest took place before July 13. If the text is checked against the Montdidier series one sees that these three months were in fact hot or very hot, with May 2.2 degrees and June and July 1.5 degrees above normal. So it is perfectly possible that a particularly intense wave of heat occurred against this background and withered the kernels. This summer heat was also the cause of the clearly marked early wine harvests of 1788.

Finally there comes the hailstorm of July 13 to provide the final disaster. The hailstorm is well known to historians. It has even hypnotized them in their attempts to explain the bad harvest of 1788, the historic consequences of which—scarcity, high prices, food riots—were so important in 1789. But this concentration on the hailstorm is perhaps excessive: it alone cannot account for the gravity of the cereals deficit of 1788. However intense such a storm may be it can only affect a small part of a district. Even the giant storm of July 13, 1788, which is supposed to have let fall 400,000 tons of hailstones (?), is only said to have damaged an area "two leagues wide by fifty long" (or, according to another source, 1039 villages) out of the ten thousand or so parishes at least in France north of the Loire. So the bad harvest of 1788 and the scarcity that followed in the spring of 1789 cannot be explained solely or even mainly by the hail. The fundamental cause of the grain shortage was the shriveling up of the wheat, before the hail, which withered the 1788 harvest *everywhere* in the last months before the early ripening of the wheat and the grapes.[121]

That being so, the hailstorm of July 13 was still the logical consequence of the implacable process which had led up during the previous months to the shriveling of the wheat. In the Paris region the most violent hailstorms always occur when an anticyclone from the Sahara coming via Spain has been stationary over the west. This is probably what happened in 1788, as in 1947 and at the beginning of May 1969. Then, during the period of oppressive heat reflected in the thermometric series, a gust of cool damp air from the Atlantic coming into contact with the hot air from the Sahara is enough to create a cold front and rapid rising of the air, causing the formation of enormous hailstones. The harvest of 1788 was doubly the victim of such an inflow of torrid air from the Sahara: first it was scorched by it, then much of what was left of the crop was battered down by the hail. As a consequence of this disaster, the next year, 1789, was to bring forth prices that send the graphs soaring,[122] the cornering of bread crops, and a plague of beggars and thieves. Thus did the hot summer of 1788 help to ripen the grapes of wrath, and bring about the "grande preur" of 1789.[123]

After this quick review of the problems involved in the climatic interpretation of wine harvest dates, it should also be noted incidentally that the vine produces climatic data of many other kinds. Not only the dates of the wine harvests but also the quality of the wines is an indication of the temperatures to which the vines were exposed. But whereas the wine harvest dates are influenced by the overall thermic conditions, cold, warm or average, which have prevailed from March to September, the quality of the wine depends essentially on the amount of heat during the summer months, i.e., from June to September.[124] If these four months are hot (and preferably dry), the wine is very likely to be good, and vice versa. Such at least are the results of the common experience of winegrowers, and also of a study which Angot made of the period 1811–1879, making a comparison between the thermic data and the quality of the wine.[125] This correlation lends great interest to the "wine chronicle" drawn up by K. Müller[126] for the German vineyards from A.D. 1000 to 1950. For each year which is documented Müller gives contemporary opinions on the quality of the wine. Thus in 1718, in the regions of Bodensee, the Upper Rhine, Kaiserstuhl, Ortenau, and the valley of the Neckar, the wines were variously described by connoisseurs and public as "abundant and good," or "abundant and very good," or "delicious," or "abundant and full-flavored," or "excellent," or "*extra gut.*" At the same time the winegrowers were enthusing over the exceptional warmth of the summer and autumn of 1718.[127] Conversely, in 1675, in these same vineyards of western Germany, the summer was cold and damp, there were even frosts, and the grapes, gathered in actual snow, gave an acid and vinegary wine. These German data for 1718 and 1675 are very relevant because we know from other evidence, particularly that of French wine harvest dates, that the summer was a very hot one in 1718 and a cold dull one in 1675. The enormous value of Müller's series lies in the fact that he gives data of this type for every year without exception from 1453 to 1950, relying, at least from the sixteenth century on, on evidence from several regions and various different accounts.

The series has been utilized at length by the climatologist Von Rudloff.[128] I shall confine myself here to one very important example: the period covering 1453 to 1622. This period is especially interesting because, as we have seen, its final phase, 1590–1610, was marked by a striking advance of the glaciers. This probably corresponds to a lack of ablation, and thus to a long period of cool summers[129]; one would expect these to coincide also with very bad years for German wines in their vineyards north of the Alps. The facts confirm this expectation. In the yearly and ten-yearly indexes I have constructed from the data given by Müller, from 1453 to 1552 good wines and bad wines alternate in years or groups of years. The sharp, acid wines only predominate in this pleasant century during a couple of black decades, 1453–1462 and 1483–1492 (the latter including the period 1485–1492, in which every year except 1486 was remarkable for the abominableness of its wines). But these two decades were exceptional. Generally, throughout the hundred years from 1453 to 1552, the summer sun shone normally and the German wine-growers had no cause for complaint.

But from the decade 1553–1562 on, or more exactly from 1554, and especially 1555, the trend changes. The summer skies over Germany are more often cloudy and cold, and endanger the delicate vines on the slopes of the Black Forest. Of course there are still isolated good years—1567, 1572, 1575, 1576 . . .—but not the dazzling series like 1493–1500, 1534–1541, and so on, which made the century which ended in 1554 such a successful one for wine. The ten-yearly indices for German wine are unrelentingly unfavorable during the five cool decades running from 1553–1562 to 1593–1602. (It will be recalled that the dates of the French wine harvest tell a similar story.) It is only after 1603, between 1603 and 1622 for example, that we find in Germany another sunny warm decade, good for wine. There is no need to underline the way this periodization corresponds with that of the Alpine glaciers nearby. The glaciers of Switzerland, Savoy, and the Tyrol all expand slowly during the four cool decades from 1554 to 1592. And then as the result of the extra cool decade from 1593 to 1602, they suddenly advance more quickly, crushing and burying the neighboring hamlets. And the ten-yearly index for bad quality in wine goes up at the end of the sixteenth century to "plus 26," beating all records for the period 1463–1622. (On this index and the way it is calculated, see Appendix 12A.)

So the viticultural and glacial indicators completely agree. This reassuring corroboration makes one hope that young researchers, whether in France, Hungary, or Switzerland, will work out the same sort of chronology on wine quality for their own countries as Müller has provided for Germany.

Finally, what with wine quality and harvest dates, it must be admitted that Bacchus is an ample provider of climatic information. We owe him a libation.

CHAPTER III

A MODEL: THE RECENT "AMELIORATION"

The previous chapter touched on the heart of the matter when it referred to secular movements of climate. What exactly is a secular history of climate? What scientific model can be used for the long fluctuations it tries to discover, and what length of time can legitimately be taken as a unit?

General data on the twentieth century warming up

It may be as well to start by establishing a vocabulary. Modern meteorologists distinguish three types of temporal fluctuation among their objects of study[1]: first, geological oscillations, which affect temperature and/or precipitation over thousands or millions of years; secondly, climatic and secular oscillations, affecting one or more centuries or a considerable part of a century (several decades).

This terminology may be supplemented by a few judicious borrowings from the language of economic historians, who in connection with prices, wages and so on talk of fluctuations which are intradecennary (less than a decade), decennary, interdecennary (more than a decade), and multidecennary. For longer stretches of time they use the terms intrasecular, secular, intersecular, multisecular, intramillenary, millenary, intermillenary, multimillenary, and so on. In other contexts they may refer to fluctuations which are short term (several years), medium term (one or two decades), or long term (a century, or at least several decades). In the case of these long-term secular fluctuations they often speak of secular tendencies or trends. All these expressions can pass directly into the vocabulary of the evolution of climate.

From taxonomic considerations we can now turn to the real secular problems, the preliminary problems of subject, content, and models.

For our secular subjects, contents and models we shall look in the first place to the rigorous disciplines of research bordering our own. For the last thirty years or so a large number of scientists —mainly meteorologists, but also glaciologists and others—have been diagnosing, recognizing, delimiting, and exploring the most recent climatic fluctuation: the tendency, in Europe and all over the world, toward warming up or amelioration. It is a sizable fluctuation which is now accurately known: the ultimate episode, the temporary conclusion, of the climatic developments of the last two millennia. It constitutes in itself a historical fact of the first importance. It forms part of a long-term fluctuation, of a secular and even, as we shall see, multisecular trend. It thus suggests the century (in certain cases the intercentury and the multicentury) as a chronological basis for our research. To put it more simply, it supplies a likely model—either homologous or symmetrically inverse—for earlier and much less well-known fluctuations, in the Middle Ages and after.

So it is worth going into it in some detail, and pointing out the main conclusions it implies for the historian of climate.

The recent climatic fluctuation (nineteenth-twentieth century) has been the object of many single studies. Various writers, often meteorologists, have conscientiously gone through the records of one or more local weather stations and noticed that during the last century the winters have got milder, and that other seasons, the summer, and the year as a whole, have got warmer, gaining sometimes a few tenths of a degree, sometimes 1 degree Centigrade or more.[2]

But the stage of separate studies is past. There is no science except through synthesis. In 1940 Arthur Wagner attempted an overall study bringing together all the series known and published throughout the world, and making it possible to map the "amelioration." Since then there have been at the same time new single studies, carrying on the thankless but necessary work of local knowledge, and also big synoptic syntheses that may be compared to those of general history. Among the syntheses may be mentioned, though the list is not a complete one, those of Ahlmann, Willett, Pédelaborde, Lysgaard, Mitchell, Veryard, and Lliboutry.[3]

I shall try briefly, within the limited scope available here, to give an up-to-date account of these syntheses, calling attention in passing to the most outstanding single studies.

Holland was the first country to produce very early meteorological series. From these Labrijn diagnosed a winter amelioration almost continuous since 1790; a less marked and more irregular summer amelioration; and a lessening of the thermic difference between summer and winter, suggesting a less continental tendency.[4]

English observations since the end of the nineteenth century are particularly trustworthy, as they were all made with the aid of a single type of apparatus, the Stevenson thermometric shelter. They cover twenty-nine stations, including ten in Scotland, and indicate a rise in winter temperatures from 1885 to 1940. This rise affects the figure for annual mean temperatures.

The series established by Manley carries these English curves back into the past. Beginning about 1690 and ending in the present, Gordon Manley's series is a monument of historical research.[5] He has patiently related early observations to one another and translated their measurements into uniform terms. What emerges is that in Britain the contemporary winter amelioration began around 1850, and was accompanied by a parallel but less marked and less continuous amelioration in spring and autumn. After 1930, the annual amelioration continues. In 1925–1954 it reaches 1.6 degrees Fahrenheit, compared with 1901–1925. But from 1930 the amelioration concerns mainly non-winter temperatures, and in particular summer ones.[6]

Iceland shows the same phenomenon in an even more marked form: there the winter amelioration was over 1.5 degrees Centigrade between 1870 and 1940. And the Iceland observations are

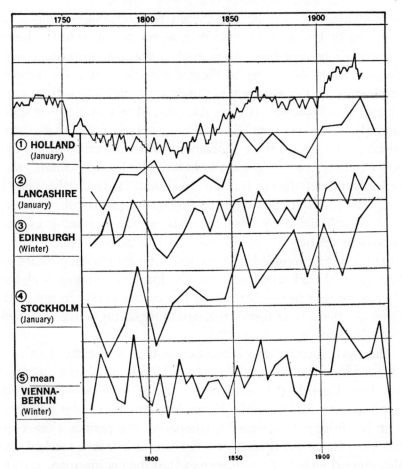

Fig. 7   *The Winter Warming Up*
Each vertical division is the equivalent of 1° C. Running means of ten, twenty, or thirty years have been used, by the authors referred to, to bring out the long-term movement. The diversity of the running means employed explains the variety of the curves in detail, though they agree with one another in terms of secular trend. (Sources: Graphs from Ahlmann, 1949, and Willett, 1950.)

particularly interesting because in the case of at least one major station there, Stikkisholmur, they are completely free from "city effect," the artificial warming up due to industrialization or heating in cities.[7]

The same trends appear in the Nordic countries, which have been carefully studied by Lysgaard, Hesselberg, Birkeland,[8] and others. Take for example the station at Copenhagen, which is dealt with by Lysgaard. The results here are presented in the form of running means of thirty years, each period thus covering about one generation. According to these thirty-year curves the annual temperature at Copenhagen is minimal around 1850—the "generation" of 1838–1867. It then rises very slowly and almost continuously up to our own time—the "generation" of 1930–1959, central date 1945. In three generations it has increased by 1.4 degrees C, rising from 7.1 degrees C to 8.5. The Danish winters grew markedly milder during this period. Even "today" (the period 1930–1959), in spite of a slight cooling during the last thirty years, they are still definitely less cold than at the beginning of the nineteenth century.

Lysgaard has rounded off his work by establishing statistics showing mild and severe winters, the former increasing and the latter decreasing in number during the secular period in question. These are the sort of figures that are most useful to the historian of earlier periods.

The most spectacular amelioration is that of the arctic or subarctic regions, the areas at the extreme limits of Nordic colonization, from Greenland to Spitsbergen. In the 1930s Scherhag[9] diagnosed a winter amelioration (November–March) of plus 5 degrees C at Jakobshavn (Greenland), comparing the periods 1883–1892 and 1923–1932. At Spitsbergen the winter increase reached the phenomenal height of 8 to 9 degrees C in the decade 1930, as compared with the "normal" for 1912–1926.[10]

In central Europe and northern France, temperatures, particularly in winter,[11] generally increased, though here the rise was more modest, with an average annual increase of under 1 degree C.

The Mediterranean area had only a faint amelioration most noticeable between 1920 and 1940.[12] In Andalusia the winters actually became colder during the period from 1881–1910 to 1911–1940.[13]

The tendency toward amelioration in the present century in the USSR, varying greatly in degree from the Ukraine to Siberia, was observed just before the last war,[14] and by Rubinstein in 1956.[15] Here it is mainly the winter months that are affected. The tendency is very pronounced in northern areas like the Barents Sea, the shores of the Arctic Ocean, and the estuaries of the Ob and Yenisey rivers.

Outside Europe, too, the amelioration, still concentrated on the winter season, is unequal but incontestable. It has been especially thoroughly studied in the United States: observations began at New Haven in 1820, at Philadelphia in 1825, and at St. Paul, St. Louis, and Washington in 1855, and these make it possible to draw up secular graphs.[16] In the United States there is a difference of 2 degrees C between the coldest decade (the one ending 1875–1876) and the warmest. Here again this is not to be explained in terms of city effect: at the Blue Hills station, far from any town, the winters have grown warmer by more than 3 degrees C since 1885.[17]

The trend toward amelioration is to be found in Canada too, but here it is less universal and more complex.[18]

In North Africa, and in what for simplicity's sake we may call the Middle East, from Athens to Cairo and from Nicosia to Jerusalem, the thermometric series are neither very old nor always very reliable. Such as they are, however, they show a rising trend of between plus 0.5 and plus 1 degree C from 1890–1900 until the decade 1940, and again in the 1950s.[19] The same tendency is established for India by the trustworthy series kept at the observatory of Colaba, near Bombay, since 1860.[20] The same trend is found again in Indonesia, according to various series of observations made since 1866 in town and country observatories and high-altitude stations.[21]

But generally speaking this recent climatic fluctuation is scarcely perceptible in large areas of the southern hemisphere. In the Antarctic, with its formidable glacial inertia, observations are rare and contradictory, and it is not certain that there has been any overall warming up in the last fifty years.[22]

The recent fluctuation seems to have affected mainly, among others, the countries bordering on the North Atlantic. It is an interesting fact for the historian that these particular coasts of Europe and of America should be zones of special climatic mobility. Nevertheless, the recent fluctuation is neither local, regional,

nor strictly Atlantic in the widest sense of the word, but relates
to the whole world.

Lysgaard, following Wagner, has attempted, at the end of what
is a standard work, to map this world context.[23] Callendar, Willett
and, following him, Mitchell have tried a somewhat different approach, and worked out the world means for the amelioration.
Their results agree in a remarkable manner, as can be seen from the
table below.

In addition to this table, Willett's work, as continued and
developed by Mitchell, also includes overall curves and different
curves for different latitudes, and maps. I shall briefly summarize
this very important work.

CHANGES IN ANNUAL MEAN TEMPERATURES FROM 1880–1920 to 1920–1950[24]

Zones	According to Callendar		According to Willett and Mitchell	
	Latitudes between:	Deviation in ° F	Latitudes between:	Deviation in ° F
World	60° N–50° S	+0.41	60° N–50° S	+0.37
Northern temperate zone	60° N–25° N	+0.70	60° N–25° N	+0.64
Tropical zone	25° N–25° S	+0.31	30° N–30° S	+0.35
			20° N–20° S	+0.39
Southern temperate zone	25° S–50° S	+0.25	20° S–50° S	+0.10
			30° S–50° S	+0.08

The Willett-Mitchell curves are constructed by "pentads," or
periods of five years, from 1840 to 1960, and use results from over
a hundred stations scattered over the various continents. But it is
only after 1880, and especially after 1900, that enough stations
are included for the curve to be representative for the world as a
whole. The curve has been weighted to avoid the statistical in-
convenience arising out of the unequal distribution of stations.
The final curves cover winter and annual temperatures, the temperate zones of both hemispheres, the tropics, and the world as a
whole.[25]

Mitchell's curves and maps lead to the conclusion that there is
indeed a world amelioration, particularly in the winter; it affects
principally the Arctic, and in the second place the cold and tem-

perate zones of the northern hemisphere, including the United States, Europe, and Siberia. Thirdly it affects the tropics, and finally, in a much less perceptible way, the temperate zones of the southern hemisphere. The only areas unaffected by the annual amelioration between 1900 and 1940 are northeastern Canada, a large part of South America, most of the southwest quarter of

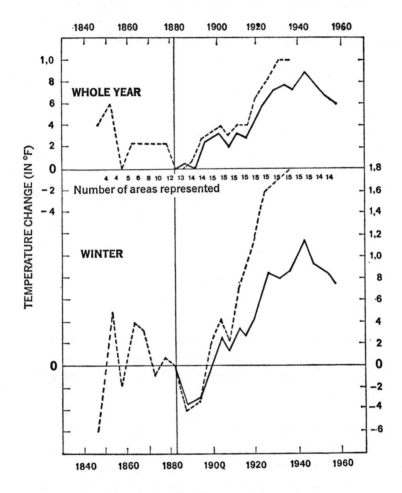

Fig. 8 *The Recent Worldwide Secular Warming Up*

- - - - - - : Uncorrected curve (mean for all stations)

―――――: Weighted curve allowing for unequal distribution of stations according to latitude.

(Source: Graph from Mitchell, 1963, p. 161.)

Africa, and certain regions of central Asia and Pakistan, and of the
Indian Ocean, southeastern Asia, and Australia.[26] It is important for a historian, who often finds himself by force of cir-
cumstance working on documents from Europe, to know that,
at least in the recent fluctuation, the European model conformed
to the world one; or more exactly the European fluctuation is
representative of a majority type of fluctuation that took place at
the same time throughout the world. So through the local climatic
history of Europe the researcher has a reasonable chance of coming
upon more or less universal truths.

Mitchell's other outstanding conclusion is that since 1940 the
amelioration has reached a ceiling, and even that a world period of
cooling has begun in both hemispheres and at many latitudes.
True, this cooling, quite marked in the Arctic, is a little less perceptible in western Europe.[27] It does not even appear on Lysgaard's *annual* temperature curves for Copenhagen, which cover the
period 1931–1960, perhaps because these are certainly kept higher
by the use of a running mean of thirty years.[28]

Where the cooling[29] does appear it is quite relative; there is no
return to the comparatively low temperatures of the middle and
end of the nineteenth century. It is simply that the temperature
curves in the various latitudes have ceased to progress, and fallen
by about $\frac{2}{10}$ of a degree F, or slightly over $\frac{1}{10}$ of a degree C.
They have fallen to their (rather high) level of 1920. It is still too
soon to say whether what we are witnessing is the beginning of a
cold fluctuation, the converse of the previous one, or merely a minor
interdecennary oscillation within a still rising or at least still high
secular trend.

Whatever may be the case, the recent cooling phase has already
had consequences for the relations between plant growth and temperature: "The average length of the growing season (daily mean
temperatures 6 degrees C or above) at Oxford in the 1930s and
40s was two or three weeks longer than in the last century; but
since 1950 it has fallen back one or two weeks."[30]

So much for temperature. What about precipitation? Taken
chronologically it is much less rewarding than temperature. The

curves show long, interregional and a fortiori world tendencies much less easily. Rain trends vary wildly, even as between places quite close together, according to the month and the season. Milan can become wetter while Rome gets drier, and vice versa.[31] Rainfall maxima and minima are much more irregularly scattered through the year in Scotland than in England or Wales.[32]

Such variability ought to make people very chary of generalizing.

To put it differently, temperature curves may well connect with general history. But rainfall curves belong in the first instance to local or regional history. This of course does not mean that they are not worth studying.

From the discordant tendencies of rainfall, Kraus has attempted to deduce certain general features.[33] It seems, at least in certain subtropical zones, that the beginning of the "heat wave" of the "twentieth century" (1890–1940) was accompanied quite often and quite suddenly (about 1890) by a decrease in rainfall, a scarcity of tropical cyclones, and an expansion of the arid zones.[34] The recent "cooling" (since 1940), on the other hand, seems to have been accompanied by heavier subtropical rainfall. In short, warming up of the temperate zone and drought in the tropics generally seem to go together, and vice versa. These parallels can clearly only be explained in terms of the general circulation of the atmosphere, a subject to which we shall return.

A monograph: The Von Rudloff model

Before coming to this final question, which will be treated at the end of this book, we must give a more detailed, more minutely subdivided, more regional and specific picture of the great recent fluctuation which culminates in the twentieth-century amelioration. Europe, specially western and central Europe, with its many series, forms an ideal hunting ground for the researcher, and it is here one comes face to face with the already considerable work of Hans von Rudloff.[35]

Hans von Rudloff's whole biography is that of a historian of climate by vocation and par excellence. He was born in 1922— "an extremely cold wet year," as he likes to point out. By the age of eleven he was keeping a sort of intimate journal recording the climatic events of the town where he lived, Fribourg-en-Brisgau. After training as a meteorologist he became a "langfrister," a specialist in the long term. His great book on the fluctuations and oscillations of European climate since the beginning of instrumental observations (from 1670 to 1965) makes it possible to place the recent amelioration in the context of the climate of the last few centuries.

Von Rudloff's purely meteorological history is completely original in its layout, the way it is written, and its dates, sometimes necessarily elastic and sometimes as precise as dates of battles. The story has a bad beginning, and not exactly a good ending. It ends with the recent cooling which began in Europe about fifteen years ago after the pleasant warm years from 1942 to 1953. The beginnings of Rudloff's series coincide with the icy end of the seventeenth century and the first regular instrumental series.

The great cold at the end of the seventeenth century was one of the important discoveries made by Gordon Manley in his archaeology of English temperatures, the conclusions of which are taken up by Von Rudloff. Around 1690 the springs, summers, autumns, and winters, measured by what were by then reliable thermometers, were nearly 1 degree C colder than they ever became again in the two and a half centuries of regular observations which followed, up till 1950–1960. As has already been said in connection with wine harvest dates, if there is a period in connection with which one may speak of the *pessimum* or Little Ice Age, it is certainly the decade 1689–1698, when the mean temperatures for every season were below normal.[36] Hence arose the exceptionally severe winters of 1693–1694 and 1694–1695, when the Bodensee had carts driven over it and Iceland was almost entirely surrounded by ice. Hence arose the great decennary increase in wheat prices: the grain failed to ripen, or rotted in the rain, or froze, and as a result there was the dreadful scarcity of 1693–1694 in western Europe and the famine of 1697 in the north, which caused the death of one third of the population of Finland.[37]

There was an appreciable amelioration in the first half of the eighteenth century. Springs, summers, and autumns warmed up noticeably; the depression tracks characteristic of westerly flow moved northwards.[38] Thanks to the work of Gustav Utterström,[39] the positive agricultural and demographic consequences of this warming up can be measured in Sweden. It culminated around the years 1730–1739, beautifully sunny and marked by a brief optimum. Only the winters seem to have resisted the thaw of the early eighteenth century. They remained very cold and snowy, their severity culminating in the terrible winter of 1709, which destroyed the crops and caused famine after the following harvest. Thousands in Europe died of hunger or as the result of epidemics.

Hundreds of texts have been found and published on the winter of 1709 and the famine which followed. In the Paris area[40] tens of thousands of people died, or emigrated, or failed to be born (through miscarriages, barrenness, or abstinence) as a result, direct or indirect, of the great cold and famine.

One recently discovered text is by the curé Lehoreau,[41] a priest of Angers who set down his impressions of the catastrophe.

"The cold," he writes, "began on January 6, 1709, and lasted in all its rigor until the twenty-fourth. The crops that had been sown were completely destroyed. So great was the disaster that eggs were twenty-five and thirty sous a dozen; most of the hens had died of cold, as had the beasts in the stables. When any poultry did survive the cold, their combs were seen to freeze and fall off. Many birds, ducks, partridges, woodcock, and blackbirds died and were found on the roads and on the thick ice and frequent snow. Oaks, ashes, and other valley trees split with the cold. Two thirds of the walnut trees died; a pint of nut oil cost seventeen sous. Two thirds of the vines died, among them the older ones. No grape harvest was gathered at all in Anjou, except a very little in a few small cantons . . . To tell the truth, in the whole of our country of Anjou there were not more than eighty pipes of wine made, and that undrinkable because the grapes were not ripe. I speak from experience, as I myself did not get enough wine from my vineyard to fill a nutshell!" Fortunately, he goes on, "God in his infinite mercy gave Normandy cider in abundance in 1709, and it was transported here and sold at forty livres the pipe, so that wine, though it was scarce, was not so dear as it might have been."

As for wheat, there was unrest in March 1709 in Anjou, Paris, and Rouen, with the poor gathering in crowds to stop the merchants from sending away what little wheat there was. But the riots did not do away with hunger, epidemics, and death. The people killed in France and elsewhere by the winter of 1709 were as many as might have died in a major war.

After 1709, although the winters still remained nearly 1 degree C colder than during the twentieth-century optimum, they did become slightly less severe for a time. This lull also reached its maximum in the 1730s.

After that a new cooling down began, perceptible in each of the four seasons in the 1740s. The year 1740 itself was uniformly cold in winter, spring, summer, and autumn, as we may illustrate by a brief detour in Belgium.

The Belgian winter of 1739–1740 was very severe and very long, with nine weeks of frost. The Meuse was frozen over, paralyzing river traffic, until March 15, 1740. The seasons that followed were not much better: "no spring, bad weather up till the middle of May, a short, cold, dull summer, a late wheat harvest spoiled by rain, wine harvests and fruit destroyed by the early frosts of October 8, 1740 . . ." This is the picture which emerges from the early Belgian chronicles, confirmed by thermometrical observations and recently summarized by Étienne Hélin. One might almost have thought one was back in the terrible years of 1315–1316, but that the more sophisticated economic and agricultural mechanisms of the eighteenth century made it possible to avoid complete famine and alleviate the food crisis.

To alleviate it, but not to prevent it. The price of wheat in Liège in 1740 went up again to the astronomical heights it had reached in 1709. The poor people of the town menaced the canons and others who had well-rounded bellies, threatening to ring their carillons to a tune they would not find at all to their taste. Prince Georges-Louis, the Governor, told the more prosperous citizens to "fire into the middle of them. That's the only way to disperse this riffraff, who want nothing but bread and loot . . ."[42]

So the year 1740 caused scarcity, though to a much lesser degree than in 1709. Specialists in demographic history[43] consider that the scarcity of 1740–1743 was the last food crisis worthy of the name to occur in agriculturally developed parts of France, such as the Paris basin. It struck a last blow at the population of the ancien régime. After this date, thanks to comparative progress of various kinds in agriculture, to improvements in roads and the growth of the wheat trade, there were no more real famines in the north of France causing massive numbers of deaths from sheer hunger.

At all events, the first forty years of the eighteenth century, up to 1739 inclusive, constitute a definite phase of warmth in comparison with the rigors of the end of the seventeenth century and with the 1740s. It was not enough to make the Alpine glaciers retreat—they were too well maintained by the persistent cold and snowy winters. But the moderate warming up at the beginning of the eighteenth century was sufficient to prevent the repetition of such human disasters as had accompanied cold and bad years in 1693 and 1709. Between the great scarcity of 1709 and the food crisis of 1740, a period of some thirty years, there were no major shortages of bread.

Despite a somewhat capricious chronology, the second half of the eighteenth century, as shown by Von Rudloff, presents no fundamental differences from the first half. The winters, for which there are now eight series, covering Scandinavia, France, Holland, Poland, and England, are still definitely inclined toward severity, much more so than our own contemporary period of mild winters from 1897 to 1939.

After 1750 the rest of the seasons, however, offer more variation. According to Von Rudloff's many diagrams[44] the springs from 1750 to 1800 were on the whole markedly cooler than those of the twentieth-century optimum. On the other hand the autumns were almost, and the summers quite, as warm as ours. True, there was a decade of cold summers between 1765 and 1775, which were often wet and dull as well. In the worst of these years some of the hungry peasants of Britanny could not wait for the harvest to ripen and cut their rye while it was still green. The resulting colics and intestinal infections ended in death for some of them. The food shortages caused by this series of bad summers were one of the causes of the failure of Turgot's wheat policy, and of the "flour war" of 1775, when the poor round Paris revolted.[45]

But round about 1780 there appears a magnificent summer optimum, with a series of hot summers never equaled since in terms of decennary mean temperatures. As we have already seen, from the early wine harvest dates and the excellent quality of the Rhine wines, this summer optimum caused a comparative abundance of wheat in France and almost all the west, and an absolute superabundance of wine. This in turn caused the collapse of wine prices, which has been so thoroughly gone into by Ernest Labrousse (see last chapter).

But these fine summers around 1780 were not enough to reverse the trend over the four seasons as a whole. The very cold winters, and generally cool or rather cool springs and summers of the second half of the eighteenth century, make the climate of Europe on the whole a few tenths of a degree Centigrade cooler than the means for the twentieth century. This coolness, added to factors up in the mountains which were more favorable to snow than they are today, allowed the Alpine glaciers to preserve those "sublime" dimensions which the first pre-Romantic engravers discovered around 1770–1780.

These somewhat negative thermal differences grow temporarily worse with the first half of the nineteenth century: all four seasons, and especially winter, become markedly cooler than during the period of reference chosen by Von Rudloff (1850–1950). A definite pessimum afflicts the decade 1810, particularly the winters, summers and autumns. Several bad scarcities are the result, especially in 1816–1817. Moreover this pessimum of the second decade of the century lays the foundation of the glacial thrust which affects all the Alps about 1820. It is in the period 1800–1850 that one encounters both the coolest summer (1816) and the coldest winter (1829–1830) in all the long series of European temperature observations.

After a last great decade of cool springs and summers, with food crises and glacial advances, ending about 1855, things begin to brighten up after 1856, and the great warming up is foreshadowed.

A first flash of the "optimum" lights up and lends a glow to the climatic landscapes of western Europe. The best years, for the summers, are the decade 1856–1865 (and also 1871–1880); for the autumns, the decade 1857–1866; for the winters, 1858–1867; and for the springs, later, 1862–1874. This group of warm dry seasons, with winters of little snow, culminates in 1858–1865, which were extremely productive agriculturally. And these hot decades, when for the first time for a long while both winter and summer grew warmer together, struck the first blow at the Alpine glaciers, which during the years 1860–1870 showed a retreat unparalleled in the two previous centuries. This time it really was the end of the Little Ice Age.

After this things took a slight turn for the worse, and the brilliant beginnings of the contemporary amelioration were followed by a return offensive of cold which was very appreciable in the decade 1880. Should we hold responsible, as Von Rudloff cautiously does, the huge explosion of the Krakatoa volcano in August 1883, throwing up eighteen cubic kilometers of dust that were to hang suspended in the earth's atmosphere for years, filtering and appreciably diminishing solar radiation? Climatologists are coming more and more to believe in the importance of volcanic eruptions as factors in the pollution of the atmosphere, and as a cause of negative and temporary fluctuations in the thermal elements of climate.[46]

However that may be, it is only at the end of the nineteenth century, from 1889 for the summers, from about 1890 for the autumns, and above all from 1897 for the winters, that the amelioration resumes its sway. Henceforward it moves more resolutely, though still with many hesitations and waverings, toward the twentieth-century optimum. (This chronology, as throughout this paragraph, applies mainly to western and central Europe.)

In the regions referred to, the contemporary optimum[47] of our youth is in a sense double-faced. As regards the springs, summers, and autumns—in short, the growing period—it culminates, after a more or less steady rise, in the splendid years (climatically speaking) from 1942 to 1954. Within this chronology each season has its own periodization. The springs, under the influence of increasingly marked zonal circulation, got steadily warmer from 1920 on. Spring temperatures reached their greatest height from 1942 to 1954, the culminating point occurring in 1947–1949. During the optimal years vegetation was stimulated by the spring warmth, which established a sort of "Pannonian"-type climate in central Europe and made the plants come into leaf earlier. Then, after 1954, up to 1965 and beyond, a cooling down of an average of about 0.5 degree C affected the European spring.

The same analysis, with some slight differences, applies also to the summers. These started to warm up a little later than the springs, i.e., in the decade 1930. Moreover the summer optimum moved slightly from north to south, sweeping over the continent from the Arctic to the Mediterranean. Summer heat culminated in Greenland in 1927–1937; in Lapland and the northern Baltic in 1932–1942; in northern and maritime Europe in 1937–1947; and in central Europe in 1942–1953. These ten years (and in particular 1942, 1943, 1944, 1946 and 1947), which were very good years for wine, were also the years of highest melting in the glaciers of the German Alps. The glaciers then lost annually from 10 to 15 meters in length, measured at their terminus, and as much as 60 centimeters in thickness. Some pessimists forecast the death of the glaciers, but they were belied by the cooling down which came at the end of the 1950s.

The summer optimum of the twentieth century culminates in 1947, when an anticyclonic ramification, a distant tentacle from the high pressures of the Azores, became stationary over Scandinavia, spreading warmth below. "The summer of 1947," writes Von Rudloff, "represents the peak, the extreme limit of present climate." Those who are too young to remember the summers of just after the war—1947, 1949, and in central Europe, 1950 and 1952—have not really known, climatically speaking, the "douceur de vivre."

After 1952, and especially from 1953 to 1962, the northern extension of Mediterranean and "Pannonian" climate ceased, and the warmer seasons of the year tended to become more cloudy and cool.[48] Temperatures grew more moderate. There were a few fine sunny summers from time to time, but they were isolated and did not come with the regularity which was a particular charm of the summer optimum between 1942 and 1953.

The autumns of the twentieth century in Europe follow much the same pattern as the springs and summers, according to Von Rudloff. The warming up of the autumns, which began at the end of the nineteenth century, resumed in force in northern and eastern Europe during the decade 1916–1925. This second wind came a little later in central and western Europe, beginning only in the decade 1920–1929, more precisely from 1926 on. From that date the autumns in the old countries east of the Atlantic went on and on enjoying high temperatures. Since thermometers first came to be used there had been no autumn warming up to compare in magnitude and duration, with that long golden autumn which began in 1926 and seemed as if it would never end. The autumn warmth greatly prolonged the growing period, retarded the fall of the leaves, and caused Europe to experience the "Indian" and "St. Martin's" summers which are usually more frequent in America. Chronologically, the autumnal optimum also moves from north to south and from east to west. It culminates in 1929–1938, from Russia to Poland and Lapland, where birch trees appeared in areas in which they had long been forgotten. In central Europe the autumn optimum occurs rather later, in 1942–1951, and coincides rather with the summer optimum. In southwestern Europe the culmination is even more out of phase with the other autumn optima. As for the ultimate cooling, which on the analogy of the other seasons ought to close these autumnal splendors, it occurs much later than for the summers. The autumns of the decade 1953–1962 were still quite warm, though the zones of strong frontal activity by the cyclones had already begun to move south, causing the beginning of a cooling down. But a definite cooling did not set in until the autumns of 1963 and 1964.

All in all, when the three seasons of glacial ablation are taken together (springs, summers and autumns), the optimum of 1942–1953, with its spring-summer temperatures 1 to 1.3 degrees C above the "normal" temperatures (for 1850–1950), was very hard on the Alpine glaciers. The year 1946–1947, with the winter of 1946–1947 almost without any snow in the mountains, and the spring, summer and autumn of 1947 all warm and dry, marks a particular paroxysm in the breakup of the glaciers.

The winters also warmed up, but following a periodization very different from that of the other three seasons. The milder winters characteristic of the long twentieth-century amelioration begin in 1897, a year in which both zonal circulation and the movement of oceanic air into western Europe began to increase considerably. The culmination of this zonal-maritime climate with mild winters occurs, in Europe, not in 1942–1953 as in the case of the other three seasons, but in 1910–1928 (or more exactly, about 1910–1920 for one group of about twelve stations; about 1920–1930 for six others; and about 1930–1940 for another ten or so observation posts or towns).

The relatively warm oceanic winters recorded during this early optimum are at the same time on the rainy side, because of the unusual flow of air from the west. During the quasi-optimal period of 1901–1930, the mean increase in January temperatures in Germany is about 1.4 degrees C over the "norm" for 1851–1930. The corresponding increase in precipitation is 17 millimeters. Yet, in spite of the increase in rain, there was very little winter snow in central Europe in 1925–1934, and this helped to restrict accumulation in the glaciers of the German Alps. So they were doomed to be eaten away at both ends: from above because of the lack of accumulation during the winter optimum of 1920–1930; from below because of the high ablation during the optimum of the other three seasons in 1942–1953.

But the very cold winter of 1939–1940 sounded the knell of the winter optimum, and marked a turning point toward a new long wave of winter cold. The phenomenon of *Abkühlung*, essentially limited to the three months of December, January, and February, characterizes the decade of 1940, and the hard winters of the war and after the war periods are in complete contrast with the simultaneous spring, summer, and autumn optima of 1942–1953.

Why have there been these new winter rigors in Europe in the last thirty years? Von Rudloff's answer is that the period of intense zonal circulation at the origin of the winter optimum ended about 1938–1939. (Ended, that is to say, as far as the winters were concerned, for the case was clearly different with the other three seasons.)

Von Rudloff's climatological analysis thus brings out the same phenomena that we shall find again at the end of this book when we come to H. Lamb's synthesis. "In the course of the winters since 1940," writes Von Rudloff,[49] "the depression tracks have tended to move farther south. As a result central Europe, in particular Germany, has come more often under the influence of masses of cold, Arctic, polar-maritime air, drawn in as a result of the creation of low pressures over the Mediterranean." High up, at a level of 500 millibars, or 4 or 5 kilometers up in the atmosphere, a north-south cold trough tends to be localized more to the west than it was during the winter optimum. Thus, from Norway to the Mediterranean down the western side of Europe, come, through this cold trough newly driven toward the west, waves of icy air which produce very cold winters. This is what happened, for example, during the "winter of the century" in 1962–1963. These phenomena represent to a certain extent (for the effects of the secular amelioration are far from being entirely wiped out) the end, in Europe, of the winter warming up.

The recent shrinking of the glaciers

The recent warm fluctuation, which in western and central Europe[50] culminates in the little optimum of 1942–1953, has interesting consequences. These consequences relate in the first place to those great climatic integrators and indicators, the glaciers. As will be seen, these are valuable aids to the long-term historian because they act as accumulators of climate. They also supply data in the form of texts, pictorial records, maps, and moraines.

In this connection it is worth recalling briefly the recent retreat of the glaciers between about 1870 and 1950. This phenomenon also provides a model for the understanding of previous glacial fluctuations, whether analogous or contrasting, in the Middle Ages, the seventeenth and eighteenth centuries. In this context to talk about glaciers is the same thing as talking about climate.

As far as our present subject is concerned, the six most important characteristics of the last fluctuation of the glaciers are:

1. The glacial recession of 1860–1960 is of *long duration*; it is a secular, or more probably, as we shall see, inter- or multisecular phenomenon.

2. It is on the whole *universal* affecting almost the whole world.

3. It is *climatically determined.*

4. It is *large-scale.*

5. It is *measurable and dated.*

6. It is *easily detected.*

"Long duration": in every instance when one can follow closely and chart adequately, over a century or more than a half century, the withdrawal of a glacial tongue, one notices that in spite of

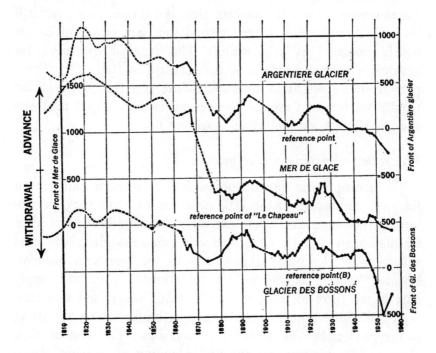

Fig. 9 *The Secular Withdrawal of the Great Chamonix Glaciers*
   The y-axis gives horizontal distances in meters; the x-axis gives years. (Source: Graph from Lliboutry, 1965, Fig. 18, 1, p. 720; data collected by Mougin and Bouverot.)

temporary reversals or halts, this withdrawal has a secular con-
tinuity and shows an authentic trend toward contraction through-
out the decades involved. This applies to the curves drawn up by
Mougin and Vallot, and continued by Bouverot, Lliboutry, and
Grove, for the glaciers of Mont Blanc; to Mercanton's curves and
maps on the Rhône glacier; Theakstone's curve on the glaciers of
Svartisen; Harrison's on the Nisqually glacier in the Rockies; and to
Heusser and Marcus on the Lemon Creek glacier in Alaska.[51]

On the whole *universal*. Of course all the maxima and minima
of one group, and a fortiori in those of different regions, are very

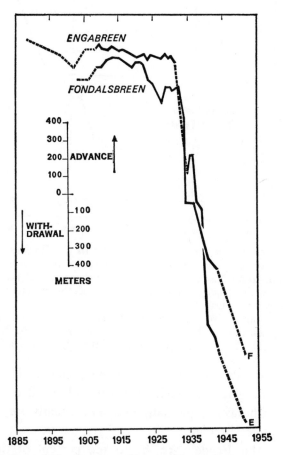

Fig. 10   *Glaciers of the Svartisen Group (Norway)*
This diagram, showing intensive withdrawal, is taken from
Theakstone, 1965, Fig. 3.

far from being simultaneous. An advance in one place coincides with a retreat in another, and so on. Moreover, the rates of withdrawal and advance vary greatly from glacier to glacier. Nevertheless, in spite of these diversities, there are regional tendencies which on the whole prevail (to the extent of 80 or 90 percent and more), in the Alps, Scandinavia, etc. And these tendencies in their turn tend to confirm one another, if not in the world as a whole, at least in most of the regions of the world.[52]

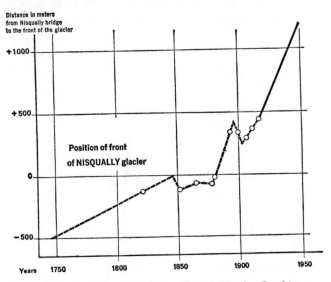

Fig. 11   *Withdrawal of the Nisqually Glacier in the Rockies*
The x-axis gives the years, the z-axis the distance in meters of the glacial front from Nisqually bridge. (The present site of the bridge was still covered by the glacier up until the end of the nineteenth century, hence the negative figures.)
This impressive withdrawal has now ceased, and since 1963 the Nisqually glacier has been readvancing (letter from A. E. Harrison, March 6, 1967). (Source: Diagram from Harrison, 1952–1956, p. 675.)

Thus in the Alps, in spite of passing incidents like those of 1920–1925, the glaciers, which have been observed continuously, have been withdrawing gradually and spectacularly for a century, up to a very recent date.

As late as the decade 1950, 80 to 100 percent of the glaciers were observed to be in regression in Switzerland, France, Italy, and Austria.

The same tendencies are to be seen in the French Pyrénées,

where the retreat observed at the end of the nineteenth century has been confirmed in the twentieth. One example among a dozen is that the glacier of the great amphitheater of Portillon almost entirely disappeared between 1912 and 1956.[53]

In Scandinavia, also since the end of the nineteenth century, there has been a simultaneous retreat of the glaciers of Jostedal, Svartisen, and Folgefonnen in Norway,[54] and that of Vatnajökull in Iceland; the latter has in some places receded a kilometer since 1904,[55] and between 1931 and 1947 J. Eythorsson was able to measure the retreat of thirty-one of its terminal tongues. The glaciers of the island of Jan Mayen in the Arctic Ocean have been retreating since 1880.[56]

In Greenland the recent retreat of certain glaciers in the northeast seems to indicate a lowering of the level of the continental ice sheet itself.[57] At Spitsbergen, steady retreats of up to 3 kilometers of the King's Glacier have been noted as far back as 1897 at least, 1906, and 1936, and as late as 1962.[58]

On Baffin Island in the north of Canada the glaciers have been withdrawing at the rate of 3 meters per year since 1860; many which stretched as far as the sea in the nineteenth century no longer do so.[59] The same tendency is seen in British Columbia[60]; and in Alaska many studies of moraines, accurately dated by the maximal age of the forests covering them, are evidence, in spite of minor fluctuations, of a marked withdrawal (several kilometers) over more than a hundred years.[61]

The ice sheets in the East (the Caucasus, Kurdistan, the Himalayas) have also shown the effects of this century. The Asau glacier in the Caucasus has receded 1.5 kilometers between 1849 (its advance in that year had destroyed a forest) and 1935. Bobek and Erinç have noted similar withdrawals in Anatolia and the mountains of the Middle East. In the southern Himalayas there have been retreats on a suitably phenomenal scale: certain glaciers have been receding by 40 meters per year for a hundred years.[62]

According to everyone who has written about them, the equatorial glaciers of Africa, in the mountains of Kenya, Kilimanjaro, and Ruwenzori, have been in a state of rapid recession ever since the beginning of the twentieth century. So much so that Spink, and then Whitow in 1963, have predicted that if the present rate continues they will disappear altogether in between forty and two hundred years. Equally marked is the recent deglaciation of the Andean ice sheets in Peru and Chile, after a maximum in about 1850. In New Zealand, between 1866 and 1919, withdrawals of over 2 kilometers have been observed.[63]

Apparently the only exceptions to this worldwide deglaciation trend are the Antarctic ice sheet, which no doubt has an enormous reserve of climatic inertia,[64] and the mountain glaciers nearest to it, such as the FitzRoy glaciers in southern Patagonia.[65]

Is this worldwide or almost worldwide glacial recession reaching its end? Or has it at least come to a stage of temporary readvance, like those noted in the Alps about 1920–1925? This is not impossible. Symptoms of a new advance have been noted in the last ten or twenty years (see Appendix 2): at Spitsbergen since 1957–1960 and since 1962 (the King's Glacier advanced 200 meters between 1962 and 1964, after having withdrawn 3 kilometers between 1897 and 1962); in the north of Norway (in the Jan Mayen glacier, since 1954); in the Rockies (since 1944, and particularly since 1952); and even in the Alps, in rare and scattered instances (the Argentière and Bossons glaciers, since 1952–1955). But the significance of all this must not be exaggerated. In 1955, during a marked thrust of this new glacial advance in the mountains of Switzerland, it was estimated, on the basis of accurate and up-to-date statistics, that 25 percent of all the Swiss glaciers had begun to advance again.[66] But the other 75 percent were still withdrawing! It cannot be considered a very successful "offensive" if three quarters of the troops are beating a retreat. Moreover, between 1957 and 1964 the percentage of glaciers that were advancing fell to 10–15 percent. In 1965, the last year covered by Hoinkes, whose results I am using here, the percentage goes up to 25 percent again, but this is still a long way from the advances made in the thrust of 1920, which itself was much smaller than the major advances of the Little Ice Age.

In 1969 I consulted on this subject R. Vivian, a specialist on the glaciers of the French Alps, and he agreed with me. In his recent journeys through Savoie and Haute-Savoie, he saw glacial fronts which were quiet and stable, very slightly advancing or very slightly retreating, but with no marked overall tendency.

This being so, are the small thrusts observed here and there recently a mere incident, or do they form a point of departure for a new secular fluctuation, the converse of the regression of the last century? It is not the historian's job to solve this problem. In any case, today the tongues of the glaciers are still well below their nineteenth-century maxima, and the great worldwide withdrawal continues to predominate.[67]

This is a phenomenon of major importance for the historian interested in the fluctuations of glaciers and through them in the fluctuations of climate. For the recent episode of the long withdrawal brings out an important fact. It is well known that the Alpine glaciers, and to a certain extent the Scandinavian ones, are almost the only ones whose history can be traced back through written documentation and illustration. But taken as a whole (and the last secular recession has proved this, as well as the great ice ages) these glaciers are in phase with those in the rest of the world, with the exception of the Antarctic. The parallel does not amount to complete synchronism, but it is a fairly general tendency.[68] And so the early movements of these Alpine and Scandinavian glaciers, their long-term advances or withdrawals, as known or deduced through texts and other methods, in the Middle Ages, or in the seventeenth century, are also very probably indicative of glacio-climatic phenomena outside the West European peninsula, which form part of an "intercontinental" glacial tendency.

A third essential element of our research is the fact that long-term glacial fluctuation is climatically determined. After Maurer, Ahlmann, and Wallen,[69] Hoinkes has again shown it with more force and accuracy than ever,[70] with the aid of all sorts of measurements, meteorological and glaciological, made on the spot[71]: of the two "budgetary" factors, accumulation of snow and ablation through melting, which determine deficiency or surplus, or in other words the retreat or advance of the glacier, it is the second, ablation, which has prevailed and caused the withdrawal of the glaciers in the last century. This withdrawal is only due in a minor way to a decrease in winter snowfall. The determining factor, caused by changes in the circulation of the atmosphere and symbolized by rising temperature curves, is the extension and intensification of the hot season, the season of ablation. On the one hand, there is a longer and more marked period of sunshine, which increases the amount of radiant heat absorbed by the glacier from the sun and the clear atmosphere. On the other hand, there are fewer incursions of cold air during the summer (these incursions determine summer snowfall and tend to increase the *albedo* [reflectivity] of the glacier, thus lessening the amount of melting).[72] These two major factors correspond, in the long term, to the well-known rise in temperatures in the summer, and thereby, through annual means, during the year as a whole; and it is their physical influence, combined, which modifies the thermic balance of the glaciers, changing their outlines and causing their termini to withdraw. There is a complex connection, a tissue of precise but now known intermediating factors, between the slow rise of mean temperatures and the gradual disintegration of the ice sheets.

Hoinkes' shrewd monographs are fully and entirely confirmed on the macroscopic scale. In the course of the last half century thermic and glacial fluctuation have generally coincided in the various parts of the world. They have been "in phase."[73] A simple adage, too simple to describe the facts fully, but useful enough to indicate the trend, has emerged as true: increasing mean temperatures give glaciers generally decreasing.[74] "Budgetary" studies have been made and graphs constructed which show (with a time lag or inertia of several years—three, six, or at the most ten—during which the original impulse is transmitted) the reaction of glacial termini to fluctuations in climatic conditions, and in particular to the rise in summer temperatures in the areas where ablation takes place.[75] Studies of this kind have been carried out by Haefeli and Zingg for the Alps, Callendar for the Norwegian glaciers, Chizov and Koryakin for Novaya Zemblya, Mathews for British Columbia, and Marcus and Heusser for Alaska.[76] The correlation these studies show leads to explanation and experimental verification such as that carried out by Hoinkes.[77] Conversely, explanation and verification finally confirm the correlation between temperature, ablation and glaciers; thus making this correlation available for use as such by the historian.

In the long term the correlation temperatures–glaciers is established both by such detailed studies as those referred to above, and by an overall view of phenomena A and B. The general retreat of glaciers (B) since the maxima of 1850 is attested by glaciologists all over the world, and corresponds to the universal rise in temperatures (A) since the minima of 1850. This rise in turn is witnessed to by dozens of local series, summarized in the work of Callendar, Willett, Mitchell and Von Rudloff. Thus all the detailed researches are brought together and invested with a historical significance: glaciers are the great long-term witnesses.

But they are not only integrators of climate. They actually magnify the information they give on contemporary meteorological conditions.[78] For the fluctuations of their termini, measurable and dated, are in the long term of very large dimensions; the glacial effect, if not disproportionate, is at least particularly spectacular in relation to the climatic cause. Take for example the recent fluctuation in climate, the "amelioration." As we have seen, it scarcely exceeds, and sometimes does not even reach, an increase of 1 degree C in terms of annual means. In the Alps, from 1900–1919 to 1920–1939, it was from 0.6 to 0.8 degrees F.[79] This is interesting, and significant, but when all is said and done it is small.

When, however, one takes the glaciers, this modest fluctuation (which continues, it is true, a previous warming up already begun) is translated into impressive substantial losses. The full extent of these can be guessed at from the evolution of recent reductions.[80]

Mougin, working from staff maps of 1853 and 1869 and on the surveys of 1885–1910, noted massive contractions from the middle to the end of the nineteenth century: the glaciers of the Isère basin in Tarentaise fell from 10,316 hectares in 1863 to

8664 in 1899–1910; those in the Arc basin, partly known in nine valleys, fell from 10,223 hectares to 8636. The 195 glaciers of Dauphiné and Provence (Galibier, Grandes-Rousses, Maurienne, Belledonne, Mont-de-Lans, Pelvoux, etc.) fell from 18,244 hectares to 15,921.[81]

In the twentieth century too, withdrawals of fronts, shrinkages of surface, and diminutions in thickness have reached very high proportions. In the Upper Haut-Arc and the Upper Haute-Isère seventy-seven glaciers, known from maps and aerial photographs, lost 36 percent of their surface between 1900 and 1956.[82] In the Romanche basin (Grandes-Rousses, Oisans, etc.) the losses reached 15 percent between 1925–1930 and 1952.[83] In Switzerland a comparison of the maps of 1860–1890 with those of 1927–1940 shows, over about sixty years, a decrease of glacier surface (reduced to the horizontal) equivalent to the whole of Lake Geneva: the glaciers have lost 469 square kilometers, or 25.3 percent, of their surface as it was in 1875 (1853 square kilometers); this is equal to 3.3 percent of the areas of Switzerland.[84]

In the Austrian Alps, eight typical glaciers studied by "photogrammetric" methods lost 17 percent of their surface between 1920 and 1950; at the same time annual mean temperatures in the same areas increased by 0.5 degree C.[85] The average diminution in thickness of these glaciers was 0.60 meters per year from 1856 to 1890, 0.30 from 1890 to 1920, and 0.60 from 1920 to 1950—in other words, a decrease of 0.49 meters per year for ninety-four years, or 50 meters in a century.

In Italy the 192 glaciers of the Val d'Aosta covered 236.91 square kilometers according to maps of 1884; 221.82 according to maps of 1929; and 190.54 according to those for 1952, which adds up to a reduction, in two generations, of a fifth.[86]

Of 239 glaciers in Lombardy studied since the beginning of the twentieth century, 66 disappeared between 1905 and 1953. The survivors have retreated considerably.[87]

Apart from the Alps, statistics of this kind are rare. Where they do exist they show equivalent or even higher regression percentages. The glaciers of British Columbia lost from 4 to 8 or up to 12 feet of thickness per year between 1911 and 1947. One covered

56,400,000 square feet in 1860, 48,800,000 in 1928, and only 34,300,000 in 1947—a loss of 47 percent in three generations.[88]

So the glacier exercises a magnifying effect, and though of course its "thermal scale" is not a precise one, it is one which in the long term is ultra-sensitive, strongly emphasizing secular climatic trends. For instance, long-term fluctuations in temperature (1 degree C in a century) are often very difficult to demonstrate, because of their smallness and possible changes of instruments, sites, and times and methods of observation. But glaciers translate them into a kind of peau de chagrin affecting 15 to 30 percent of the total surfaces; and by withdrawals of fronts which leave behind them great belts of moraine hundreds of meters or more wide. The multiplying effect leaps to the eye, even to that of a tourist who knows nothing about the long fluctuations of the thermometer.

So another element the historian can make use of is the fact that secular and large-scale fluctuations in glaciers are easily detectable.

In the century between 1860 and 1960, according to the very precise observations of Mougin, Vallot, and Bouverot,[89] and of the department of Eaux et Forêts, the front of the Mer de Glace has retreated, in horizontal terms, by 1100 meters, that of the Argentières glacier by 970 meters, and that of Les Bossons glacier by 600 meters in the sixty years from 1895 to 1955. Such movements of the termini produce very considerable changes and contrasts in the landscape. A critical comparison of maps, plans, prints, texts, and early and recent photographs brings out these contrasts. I shall just quote two examples, to which I shall often have occasion to return. The landscape of the lower valley of Chamonix has greatly changed since the time, during this century, when the retreat of the Mer de Glace made it cease to be visible from the "plain" of the Arve. Similarly, the appearance of the little village of Gletsch has become quite different since the Rhône glacier has withdrawn its imposing mass from its vicinity and become just a narrow tongue in the narrow upstream gorges, looked down on by the Hotel Belvedere.[90] (See Figs. 12A–12K,

Plates 9 to 12.) The pictures and engravings of the eighteenth
century and the photographs of 1850 make an eloquent contrast
when set beside the present-day reality of 1950 or 1960.

But the recent secular amelioration does not affect only glaciers.
It acts on the oceans, too. Since 1880 it has been warming them,
though in a very unequal manner.[91] It also increases their total
volume, since they eventually receive the water that results from
the melting of the glaciers.

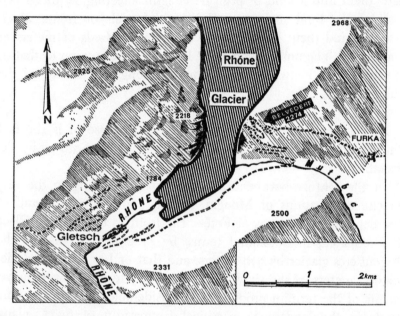

Fig. 12A This sketch of the Rhône glacier in its 1874–1882 position
shows the Rhonegletscher forming a *pecten* and barring the Muttbach.
Its front is about 800 meters from the village of Gletsch. (Source: The
map is drawn from the "synoptic map"—1/25,000—made at that time by
L. Held. It is very accurate and has been reproduced by Mercanton, 1916,
map 1.)

Figures 12A to 12K should all be compared with Plates 9 to
12. The figures and plates together give an account of the Rhône glacier
from 1705 to 1966. The account can be completed by reference to the
complete iconography in Mercanton, 1916; by consulting the Seiler collection at the Hotel du Glacier du Rhône; and from there, by going to see
the glacier itself.

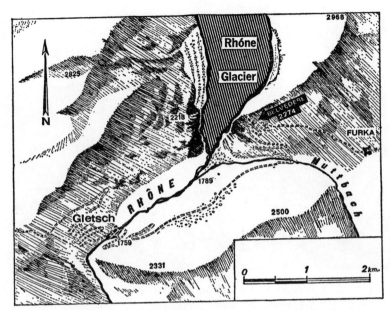

Fig. 12B This sketch of the Rhône glacier in its 1955 position is drawn from the topographical map of Switzerland (1/50,000; "Gothard," Section 50001).
On this map the Rhonegletscher, much withdrawn from the period 1874–1882, no longer bars the Muttbach. Its terminal point is 1850 meters from the village of Gletsch.

The warming up of the ocean is of great interest in itself, but, except perhaps for the difficult matter of fishing grounds, the historian of climate cannot make it one of his central studies: before the nineteenth century and the beginning of systematic observations by ships, there is little data about the temperatures of the various marine zones, except for certain very marginal areas like Iceland and Greenland, where variation in amounts of floating ice may provide early indications,[92] though these are sometimes difficult to transform into series.[93]

As for the "eustatic" or "glacio-eustatic" movement, by which glacial withdrawals result in rises in level of the sea, this seems at first glance much more exciting for the historical researcher. A multitude of texts, corroborated by pictorial records or archaeology, tell of subsiding coasts, once-inhabited shores swallowed up by the sea, or examples of spectacular emergence. The Breton coast, from

Fig. 12C *The Pecten of the Rhône Glacier Barring the Muttbach and Spreading over the Vale of Gletsch*
        The Muttbach is the first thalweg to the right of the picture. (Source: Drawing from nature by H. Besson, 1777, engraved by Niquet fils, published in Zurlauben, 1780, and reproduced here after Mercanton, 1916, Fig. 1.)

Fig. 12D *The Rhône Glacier Spreading in the Form of a Pecten over the Vale of Gletsch and Barring the Muttbach.*
        (Source: After a watercolor by Conrad Escher van der Linck, 1794, reproduced in Mercanton, 1916, ad fin.)

Fig. 12E *The Pecten of the Rhône Glacier Barring the Muttbach and Spreading over the Vale of Gletsch in 1848*
    In front of the glacier to the left are the concentric moraines which mark the glacial advances of the seventeenth century and of 1820. (Source: Watercolor by H. Hogard painted August 16, 1848, and reproduced here after Mercanton, 1916, Fig. 3.)

Mont-Saint-Michel to the Île de Ré, and the Languedoc coast from Aigues-Mortes to Port-Vendres, furnish many facts and traditions of this kind.[94]

Fig. 12F *The Giant Pecten of the Rhône Glacier Barring the Muttbach in 1849*
    (Source: Contemporary daguerreotype taken by Mercanton from the book dedicated, in 1893, "To the memory of Daniel Dollfuss-Ausset," and reproduced here after Mercanton, 1916, Fig. 4.)

It would be tempting, in short, to differentiate between tectonic and eustatic movements and put these historical phenomena, of-

Fig. 12G *The* Pecten *of the Rhône Glacier Barring the Muttbach and Spreading over the Vale of Gletsch in 1850*
(Source: An early engraving for which the author probably used a daguerreotype; the engraving has been recently reproduced by Photo-Edison O. Sussli-Jenny, Thalwill-Zürich.)

Fig. 12H *The Rhône Glacier in 1870*
The *pecten* barring the Muttbach is still there. The big building in the left foreground is the Hotel du Glacier du Rhône.

ten very well dated, in relation with the oscillations of sea level and with the melting (or expansion) of glaciers, or in other words with the fluctuations of climate.

But here too the historian must begin with the present, in this case the recent eustatic fluctuation, and see if it provides valid and sufficient models for researches into the past.

The recent glacio-eustatic, and thus climatically determined, rise in sea level over the last century is known through information given by marigraphs (self-registering tide gauges) scattered all over the world. From this information experts work out the increase at about 1 to 2 millimeters per year, or 3 millimeters at the maximum—in other words, 10 to 20 centimeters in a century of glacial melting.[95] In the case of accurate instruments like the marigraph, this rise, though small, is easily detectable. But for the historian, working necessarily on documents which are only approximate and date from before the systematic observation of the seas, the margin is much too narrow. While the secular oscillations of glacier termini are measured in kilometers, the eustatic ones, brought about by the same distant causes, have to be measured in centimeters. How then is the historian, dealing with apparent changes in sea level in the distant past and alterations in coasts and shores in historical times, to distinguish between the often considerable effects of tectonic movements, and those that can legitimately be attributed to eustatic fluctuations of glacio-climatic origin?[96]

To sum up, while the secular oscillations of glaciers are available, within certain limits, for use by the historian, the glacio-eustatic oscillations of the seas remain the province of the traditional sciences of nature. And until more is known about them, they cannot be made much use of by the historian applying his archival researches to the history of climate.

And what about life itself? Up till now I have only raised the problems and possible models offered to the historian by the recent climatic fluctuation in its purely physical aspects, such as the measurements of thermal variables, and effects on ice sheets and oceans. But the question of biological, and thus ecological

Fig. 121 In 1874 the *pecten* of the Rhône glacier has begun its great withdrawal; it is flatter than it was, but it is still there. The view is from the left side of the Rhône valley. (Source: an 1874 photograph reproduced here from Mercanton, 1916, Fig. 5.)

and human effects, cannot be sidetracked, for here too one may find possible models for the study of earlier occurrences of the same kind.

The recent secular amelioration has certainly determined some biological consequences in the arctic regions, which have been warmed by the new climatic conditions and in particular by the wave of hot summers. The limit of the forest has moved toward the north, the Far North.[97] In the tundra the population of insects and other invertebrates has increased, and areas once bare are covered with vegetation. "At least forty or fifty species of birds and mammals have immigrated into the lands of Eurasia north of the tundra forest . . . during the last half century."[98]

Bird migration has been especially well gone into in this connection.[99] Because of the warm springs and summers of recent decades the willow warbler (*Phylloscopus trachiloides viridanus*

Fig. 12J An 1899 photograph, taken from the Langisgrat slopes on the left bank. The Rhône glacier has made a spectacular retreat. A thick tongue still covers the rocky verrou, but for the first time in the whole pictorial record, so copious since 1705, it no longer bars the Muttbach, and no longer spreads in the form of a *pecten* into the vale of Gletsch. (Source: Reproduction taken from Mercanton, 1916, Fig. 6.)

has become quite widespread in Finland. The fieldfare (*Turdus pilaris*) has been nesting regularly in the south of Greenland since 1937.

Even more important from the point of view of human ecology are similar movements of fishes, though methodologically speaking it would hardly be permissible to use them alone as evidence of climatic fluctuation. But once this fluctuation is explored and recognized, as the recent one since 1850 has been, the northward movements of marine fauna may take on climatic significance: this applies to migrations of cod and plankton, which at the end of the nineteenth century moved from Newfoundland to the now less cold west coast of Greenland, where in certain years there are miraculous draughts of fishes.[100]

Fig. 12K   *The Rhône Glacier in the 1950s*
Note, in comparison with the previous figures, especially the 1870 photograph, the complete disappearance of the *pecten* and the withdrawal of the glacier to halfway up its rocky verrou. This process has become even more marked in 1966 (see Plate XII). (Source: Drawing from a photograph by Jules Geiger, Photohaus Films-Waldhaus.)

But man, even in the north, cannot live by fish alone, and the first interest of the historian of human societies is the effect of the recent climatic fluctuation on agriculture—on wheat, for example. Has the recent secular amelioration had any positive effects on the cultivation of grain which may provide a valid model for inquiry into earlier times?

In northern Europe this is quite probable. There wheat grows at its very northern limit, and its requirements as regards light and heat are far from being fully met every year, as they generally are in France, say, and particularly in the Mediterranean. So in Sweden and Finland temperature is the marginal parameter that largely dictates fluctuations in harvests. Wallen, and even more Hustich, have attempted to demonstrate this: setting aside the effects of improvements in agricultural technique, the gradual warming up of the growing season (spring-summer) in the twentieth century appears to have produced a corresponding rise in rye and wheat yields.[101] The parallel here between temperature and grain curves seems satisfactory and significant. During the 1930s, a decade of hot summers, the Nordic countries had superb harvests.

But Sweden and Finland play a very minor role in European cereal production. In England and non-Mediterranean France the limiting factor in wheat yield is excessive rain rather than lack of heat. So the thermal parameters which are mainly responsible for the recent secular fluctuation in climate have had little effect on the cereal stocks of the bigger countries of western Europe.

In England, for example, neither the recent mild winters nor the recent warm summers have had any marked influence on cereal yields. This is the conclusion arrived at by L. P. Smith, in a careful study which gives many statistics and graphs on the meteorological past of agriculture.[102]

On the basis of Smith's exemplary study and a few others, certain more general assertions may be made.[103] In the short or relatively short term (on an intradecennary, decennary, and in certain cases interdecennary scale), agricultural history is vulnerable to the caprices of meteorology which produce bad harvests and used to produce food crises. But in the long term the human consequences of climate seem to be slight, perhaps negligible, and certainly difficult to detect.

In subtropical countries the decrease in rainfall and the trend toward aridity which have sometimes appeared in the twentieth century may have combined with the contemporary warming up to harm local harvests. But works which have appeared so far on the subject have not demonstrated it. The great famines such as that in India in 1966 may well belong only to the short term, but on the other hand they may form part of the present climatic trend. The question has not been settled.[104]

On the matter of contemporary ecology, therefore, a quick review of the results obtained by climatologists reinforces the caution expressed in my first chapter, where it was laid down as a principle that contrary to the opinion or practice sometimes encountered, the historian must begin by collecting documentary data on natural phenomena and on the purely physical past of climate—temperature and rainfall parameters, phenological and glaciological observations, etc. The study of human consequences can only be a second phase, chronologically and methodologically quite distinct from the first and by no means indispensable to it.[105]

The facts about the recent climatic fluctuation, which is used as a model, bear out the need for caution. The fluctuation is now known thoroughly in its physical detail. Its chief parameters have been measured; it has been dated, judged, weighed, indexed, and exhaustively mapped. Yet in spite of the fact that this work has been going on for half a century, no clear picture emerges of its human incidence—as regards agriculture, economics, epidemiology, etc.—except in the case of certain very special activities, such as fishing, or in certain very outlying regions, such as Sweden, Finland, or even Greenland. If the transition from the physical to the human is difficult in the twentieth century, for which very full documentary series of all kinds exist, how much more difficult must it be for the historian of the twelfth or the seventeenth century, for which any alleged secular fluctuations, whether climatic or agricultural, meteorological or ecological, physical or human, are bound to be only approximately known.

In other words, it is perfectly legitimate and useful to look for historical data on climate in the periods before systematic observation began. Such research—especially on a secular scale, which is the most fascinating and least known of all—is in fact the main object of this book. But to attempt a synthesis, desirable as it may be, with economic or agricultural history, would for the moment be premature: the recent secular oscillation provides the historian with excellent meteorological, climatological, and glaciological models, but unfortunately it does not provide him with good and fully elaborated ecological ones.

Comparative history and prehistory:
The climatic "hypsithermal" phase

At this point the question arises: is the physical model of the twentieth century the only valid one, the only one the historian can project on the past two thousand years to see if he can find phenomena of similar scope, parallel or inverted? Or have there been, in historical times, as well as comparable climatic fluctuations, much greater ones, either in antiquity, the Middle Ages, or more modern periods? And in the first place, is there a *real* model for this hypothetical larger type of fluctuation, a model both convincing and valid?

It goes without saying that I eliminate, from the outset, the possibility of major climatic differences of up to or over 4 to 5 degrees C in comparison with the present, in the case of certain monthly means (July, for example), or even annual means. Roughly, 5 degrees C is the difference between present climates and those of the Würm or last ice age, 9000 years before our era.[106] In the two coldest episodes of the Würm, which occurred 55,000 and 15,000–20,000 years before the present era, the differences may have been even higher, as much as 10 degrees C less than contemporary mean temperatures.[107]

There is no chance of meeting with long-term differences of as much as 4 or 5 degrees C in the period covered by historical documentation. If there ever were such differences, they must just have occurred in exceptional and isolated years, as is still the case nowadays.

But there may in the historical period be long-term differences less than the 4–5 degrees C which separates the present from the Würm, but more than the 1 degree C characteristic of the twentieth-century secular fluctuation.

Prehistory offers a likely model in the climatic optimum or *Wärmezeit*, which palynologists and geologists call the "atlantic" or "hypsithermal" phase.

It is worth briefly summarizing the main characteristics of this phase, in the first place to bring out their value as examples and models, and secondly because the end of the optimum belongs to the historical period, the period of barbaric and classical antiquity. This optimum thus introduces climatic history proper, just as the recent secular fluctuation concludes it as far as we know it.

The existence of the *Wärmezeit* has been known for some time. Even before pollen analysis, two botanists, Blytt and Sernander, had made out, through the stratigraphy of swamps and lakes, a series of relatively warm periods which they divided into the *boreal*, the *atlantic*, and the *subboreal*; these are sandwiched between the post-Würmian deglaciation (*preboreal* period) and the rather cool climate of the recent era (*subatlantic*).[108] Von Post's studies on pollen have confirmed this first description: in southern Sweden and all northwestern Europe a warm botanical period with mixed forests of oak, hazel, alder, and lime preceded the forests of the historical period, in which fir, birch, beech, hornbeam, and pine predominate.[109]

Von Post suggested a threefold division of the postglacial era:
1. before the warm period;
2. warm or mediocratic period;
3. terminocratic period, in which the contemporary forests of trees thriving in cool temperatures were established.[110]

This division, and the contrast between phase 2 and phase 3 have often been corroborated since. But nevertheless palynologists working on these problems proceed with caution, for the factors arising out of climatic change proper become inextricably mixed in the last millennia B.C. with purely human variables due to the ever-expanding scope of neolithic agriculture.[111] Still, certain facts prove beyond dispute that climate, warmed up during the optimum and cooler again afterwards, exercised a particular influence. Thus in Sweden during the atlantic phase the hazel, together with other plants, spread as far as 64 degrees North, whereas in recent times it is rarely found farther north than 60 degrees.[112] This, according to Andersson, indicates summer temperatures 2.0 degrees C higher than the annual means. Other data is derived from fauna, fossil pollens, and geomorphology: for example, the spread of the pond tortoise (*Emys orbicularis*) and certain plants like *Hedera, Vis-*

cum, and *Ilex*; or the formation of travertines in the Karst areas of central Europe.[113] Among these facts those which concern animals and plants during the optimum period suggest winter temperatures for Denmark which are scarcely higher than those of the present day (plus 0.5 degree C), but summer temperatures more than 2 degrees C above those of the present. On this, Iversen's studies confirm Andersson's ideas. According to Godwin, who has measured the spread of heat-loving organisms like *Najas marina*, the heat maximum came in the atlantic phase between 5000 and 3000 B.C.[114]; more exactly, according to recent carbon datings, between 5475±350 years B.C. and 2975±224 years B.C.[115] Finally, a piece of decisive information comes from the Tyrolean Alps, where the juxtaposition of a peat bog and a glacier makes possible unusual and accurate cross references[116]: the fourth millennium B.C. (between 4000 and 3000) here emerges as the sunny millennium par excellence, with "optimal vegetation" and glaciers much reduced.[117] It will be recalled that it was in this fourth millennium, from 4200 to 4000 B.C. (radiocarbon dating), that the early agriculture of the Campignians and Danubians was established in the Magdeburg-Cologne-Liège area[118]: so the first cereals in this part of the world, which had come by various stages from the Middle East, were met on their arrival by extremely stimulating heliothermic conditions.

In the case of France the atlantic or "optimal" phase sees the predominance of the oak forest, mixed with lime and elm. Even the holm oak, a typically southern tree, spread widely as far as Normandy, from which it disappeared as if by magic in the subatlantic phase, with the advance of the beech. And did the first Norman cultivators, who have left traces right in the "atlantic" level of the peat bog of Gathemo, also benefit from the "favorable" conditions (for wheat) of the fourth millennium B.C.? It is quite possible, and after Elhaï's fine work[119] the question must at least be asked.

The millenary or intermillenary fluctuation of the neolithic optimum was to a certain extent, even more than its younger sister, the optimal secular fluctuation of 1850–1950, worldwide,[120] or at least intercontinental. It has been traced in the pollen diagrams for North America (Maine, Michigan, Newfoundland, Labrador, Alaska), and here there is no question of suggesting, as a hypercritical observer might in Europe, that the data may be affected by disturbances to forests caused by early cultivators. In Maine, for example, mixed oak woods preceded the present forests of beeches and pines, and there is an exact chronological parallel to this in Ireland, the corresponding climatic zone in Europe. H. E. Wright shows that the pollens of the lake deposits of Minnesota indicate an advance of oaks and pines and a retreat of other conifers from 6000 to 2000 B.C., which suggests a climate warmer (by 1 to 2 degrees C) and slightly drier than in the present millennium. In Alaska, numerous radiocarbon dates show that the glaciers were smaller between 7100 B.C. and 2200 B.C. than they are now. Palaeosoils and the forests then flourishing suggest that the climate in Alaska was at that time 1 degree C or more warmer than in the last three millennia.[121] A parallel warm phase is reported from central and northern Asia, with forests in what is now tundra (the evidence for this comes from the peat bogs), luxuriant steppes in Tibet, and so on.

It should be noted however that the warm phase encountered mainly in temperate latitudes does not necessarily correspond to a super-arid phase in desert and subtropical latitudes. On the contrary, the extremely complex researches of K. W. Butzer[122] on the Sahara tend to show that a slightly, and increasingly, humid trend prevailed in certain parts of that desert between 5500 and 2350 B.C., i.e., the time of the hypsithermal in Europe.

However that may be, palynological studies (see Appendix 3) have also diagnosed the optimum, or, to use R. F. Flint's terminology, the hypsithermal phase, in equatorial latitudes; they have established this through the stratified sediment of a lake 8500 feet above sea level near Bogotà, in Colombia.[123] It has not been established, as was at first thought, for New Zealand.[124]

But the greatest and most irrefutable evidence that the optimum was general comes from the oceans. Here the basic research is that of C. Urey, C. Emiliani,[125] and J. Wiseman, on the changing proportions of various oxygen isotopes contained in the massive deposits of carbonates ($CaCo_3$), especially in foraminifera shells.

These proportions vary with the temperature, and so provide a sort of "geological thermometer" giving the surface temperature of the sea at the time and place at which the shells lived. Cores taken from submarine deposits thus reveal strata of information on temperature stretching over millions of years.

It is not within the scope of this book to describe all that Emiliani[126] and Wiseman have contributed to the knowledge of the Ice and Quaternary ages. What interests us here is the fact that at the end of the curves they constructed on the basis of atlantic cores, after the intensely cold temperatures of the last ice age (Würm) there is a thermic maximum clearly culminating about 3000 B.C., and this maximum is rapidly followed by a swift climatic "deterioration." This deterioration lasts up till the present. The temperatures of the oceans, though more stable than continental temperatures, have apparently gone down by over 1 degree C since their maximum recorded five thousand years ago.[127] It should be noted, however, that this cooling down has its occasional "saw-tooth" movements, even in the sea: Wiseman's diagram for the Atlantic Ocean[128] shows a marked warm fluctuation in the twelfth century A.D. This corresponds to the "little optimum" described in a recent book (1966) by H. H. Lamb.

These two episodes, the great prehistoric optimum and the succeeding thermic recession or subatlantic "deterioration" of the Iron Age, are full of interest for historians, prehistorians, specialists in protohistory, geologists, geomorphologists, climatologists, and so on. The work of B. Frenzel and Émilienne Demougeot[129] may be mentioned here.

Frenzel, in a long article based on an exhaustive study of pollens in Europe, describes the subatlantic deterioration, which after the "subboreal" transition of the third and especially the second millennium B.C. brought to a close the propitious warm periods of the atlantic optimum (fifth and fourth millennia B.C.). According to Frenzel, the subatlantic combines two interrelated phenomena which are perceptible from 1500 B.C., and more especially from 900–700 B.C. on. On the one hand, and this is the important thing, there is a decrease in mean temperatures, winter and summer, in this last millennium B.C.; and on the other hand, again from about 1000 B.C. on, there is an increase in rainfall during the growing season. The changes involved, however, are slow, not of great magnitude, and liable always to be affected by shorter oscillations in the opposite direction.[130]

É. Demougeot starts from climatic data which are already solidly established (certain aspects of the atlantic and subatlantic phases may be disputed, but not the existence of the phases themselves). She then goes directly on to the second stage of this kind of research, the hypothetical human effects of climate. According to her brilliant but cautious article, the subatlantic deterioration of the first millennium B.C. may have helped to determine certain important facts of prehistory: the dying out, around 500–400 B.C., of the "thriving northern Bronze Age"; and the beginning of the first Germanic invasions of southeastern Europe by the Bastarnae and the Skires, from southern Sweden and the Baltic.

These significant conclusions have recently been supported by the evidence of the Tyrolean glaciers[131] the cold wave of about 450 B.C. is attested both by glacial advance and by pollens. But as the author herself agrees, her suggestions are only hypotheses. To make a comparison, who would think of connecting, as cause and effect, the cold oscillation in Europe of 1590–1650, well vouched for by the history of the glaciers, and the southern expeditions of Gustavus Adolphus?—though of course these were military excursions, not migrations of peoples.

Demougeot's conclusions concern prehistorians and protohistorians, but the historian of climate, whether medievalist or modernist, has quite a different attitude toward the contrasts between the optimal and subatlantic phases. What interests him is that between the *Wärmezeit* and the twentieth century, between 3000–4000 B.C. and A.D. 1960, an order of magnitude is established for such climatic fluctuations as are likely to be encountered.

In short, for three thousand years (since 1000 B.C.), since the real beginning of the subatlantic thermic recession, there have of course been many rises and falls in temperature, of varying magnitude and length; there have been many secular coolings down and warmings up (climatic fluctuation is not peculiar to the twentieth century). But none of these minor fluctuations has been such as to modify the pollen diagrams seriously or for any considerable length of time; none of them has driven back the beech forests, or brought back the hazel to central Sweden or the holm oak to Normandy.

To take only Europe, the main field of research in this connection, the thermic conditions[132] which prevailed there during the atlantic optimum (higher annual temperatures, because of summer temperatures regularly 2 degrees C higher than those of the present day) have never recurred in the course of the three millennia since, the twentieth century included, except so briefly (for example, during the "little optimum" of the eleventh and twelfth centuries) as to have no lasting effect on the composition of forests or make them return to what they were during the optimum.[133]

A number of summers, or even groups of summers, taken individually, have certainly reached the thermic level of 2 degrees C above the present norm which was practically constant during the great optimum: such, for example, were the scorching summers in France, in 1945, 1947, and 1959. But these high averages were never sustained, even during our own era of comparative amelioration, for as long as a whole decade. In England, for example, the hottest summer decade (1772–1781) since systematic observations began is only 2.5 degrees Fahrenheit above the "norm" for 1901–1930.[134] The relatively warm decade of 1940–1949 is only 1.3 degrees F above this same norm, whereas the temperature in England during the optimum would have been 3.5 degrees F above it:

WARMEST AND COOLEST SUMMER DECADES
(*Ten-year means over June–July–August in degrees F*)

	Lancashire	Holland (Utrecht)
Norm (1901–1930)	58	60.4
Hot decade (1772–1781)	60.5	62.7
Cool decade (1809–1818)	57.3	60.1
Climatic optimum	61.5	64
Comparatively warm decade (1940–1949)	59.3	61.8
Comparatively cool decade (1881–1890)	57.7	60.4

This emerges even more strongly if one takes periods longer than a decade. According to Gordon Manley, who has drawn up two and a half centuries of temperature observations for Britain, the difference in terms of annual means in recent times between coolest and warmest periods is 1.6 degrees C for periods of ten year, 0.7 degree C for periods of forty years, and 0.4 degree C for eighty-year or quasi-secular periods.[135] These differences define "the order of magnitude for recent variations in the English climate," and evidently preclude any spontaneous revival of optimal type vegetation, which would require positive and lasting thermic deviations of greater magnitude.

So, thanks to Manley, these are the modest variations[136] that the historian of climate may expect to find among the fluctuations of the last two or three millennia (which is identical with the subatlantic era, which is also the historical one and the only one for which there are written records and texts).

And so here, defined according to recent models, are the reasonable objects of research into the history of climate. No doubt they are not the only possible objects, but they are the most probable ones.

CHAPTER IV

THE PROBLEMS OF THE
"LITTLE ICE AGE"

We must now try to find these objects, and discover in the facts the proper study of our research. Our objective will still be a limited one: there is no question of going straight to the synoptic stage of world syntheses which experts in climatic fluctuation are now in a position to establish for the nineteenth and twentieth centuries.[1] I shall be more modest than this, and simply try to set up some reliable single series representative of the long-term movement, and concerning a definite and well-documented area.

But first one last preliminary question. How, before thermometers came into use, are mean secular oscillations of less than or equal to 1 degree C to be detected? In such circumstances the approach can only be indirect, and in the first instance we have to resort to those approximate but amplifying instruments, the glaciers. Glaciological models have shown there is no better introduction to our inquiry (a vast one, of which only a part can be represented in this book) than a detailed history of the European and especially the Alpine glaciers, carried out with constant reference to the texts. So this chapter will consist of a long commentary on the archives and early pictorial evidence on the Alpine glaciers. As well as being detailed, and constantly checking sources, this commentary will also be "interdisciplinary": the historical use of texts will be complemented by knowledge gained by glaciologists on the spot.

The first text to shed real light on the subject dates from 1546 (see Appendix 4). In that year the "cosmograph" Sebastian Münster gave an exact account of an Alpine glacier. Curiously enough, his description has been ignored up till now, even by such alert specialists as Kinzl, Richter, Mercanton, Lütschg, Mougin and Mayr. What does it tell us?

On August 4, 1546, Sebastian Münster was traveling from the Valais, the valley of the Rhône, and decided to go via the Furka pass to the St. Gothard, in order to "explore the difficulties." He was riding on horseback along the *right* bank of the Rhône, which he had to cross to get to the slopes leading up to the Furka. Crossing at the actual source of the river, he came upon the glacier, and here is his description, translated from the Latin: "On August 4, 1546, as I was riding toward Furka, I came to an immense mass of ice [*veniam ad immensam molem glaciei*]. As far as I could judge it was about two or three pike lengths thick, and as wide as the range of a strong bow. Its length stretched indefinitely upwards, so that you could not see its end. To anyone looking at it it was a terrifying spectacle, its horror enhanced by one or two blocks the size of a house which had detached themselves from the main mass. White water flowed out of it [*Procedebat et aqua canens*] so full of particles of ice that a horse could not ford it without danger." (On this point the German edition, published in 1567, is more precise: "A stream flowed out of it, a mixture of water and ice, which I could never have crossed on horseback without a bridge.") And Münster concludes: "This watercourse marks the beginning of the river Rhône."

His text is quite precise as well as charming. The glacier, the foot of which one could approach on horseback, was an imposing mass with a front from 10 to 15 meters high: the Swiss pike (*phalanga* and *spiess* in the Latin and German editions respectively) measured between 4.6 and 5 meters.

The frontal mass was at least 200 meters wide and perhaps more: the "range of a strong bow" could not be less in the sixteenth century in the land of William Tell. In front of the main mass there were several large séracs. The nearby bridge crossing the Rhône just below its source is probably the one which appears in the engravings of the eighteenth century.

Münster's description is of course very different from what the *Rhonegletscher* has looked like in the 1940s, 50s and 60s, since its recent withdrawal. I retraced Munster's journey in 1962, and again in 1966. The present glacial tongue is very thin, its height, and perhaps its width too, much less than those suggested by the sixteenth-century observer. It has withdrawn high up into the escarpments and the steep and slippery rocks of the mountainside. The stream flowing out of it runs first through a gorge, where it is sometimes invisible, and begins in a series of vertical waterfalls. No horseman, especially one approaching his sixties as Münster was, could approach the front and spur of the glacier as they are today and see water flowing out between séracs large as houses; still less could he come nearer, and either cross by bridge or ford the Rhône just as it issues from the glacier. All the riding manuals in the world could not teach anyone how to get a horse up that steep granite slope. The only means of access on horseback now would be on the left bank, along the path leading to the Belvedere Hotel, on the E.S.E. flank of the glacier. But the present path, which only dates from the end of the nineteenth century, only approaches the main glacier from a distance, on the left flank, and a long way upstream from the terminus. When he got to the present site of the Belvedere, Münster would have had to dismount and make a long and difficult approach, sometimes an actual climb, down the rocky cliffs to the terminus of the glacier and the source of the Rhône. So neither from the point of view of itinerary, description or topography does Münster's text correspond at all to the present state of the *Rhonegletscher*.

A first hypothesis, however, is that it may correspond to its state between 1874 and 1920, as it appears in many photographs and the detailed maps published by Mercanton.[2] (See Figure 12.) At that time the terminus of the glacier had already lost the majestic dimensions it had in the eighteenth century and other periods prior to 1850. Nevertheless, although it was then in slow and constant retreat, it was still much larger than it had become by the middle of the twentieth century. It still covered the rocky barrier (*verrou*) now left almost entirely exposed, and descended beyond it to the beginning of the vale of Gletsch (the first part of the valley of the Rhône), though it did not fill the vale with the *pecten* or immense "upside-down cockleshell" as it used to do before 1850.

The dimensions of the front or terminal rampart may well have corresponded, between 1890 and 1920, with those given by Münster (10 meters high by over 200 wide).[3] And this front together with the source and early course of the Rhône were much more accessible then than they had become by 1960: there was no rocky escarpment to forbid the approach of the traveler on horseback.

A second hypothesis[4] is that Münster's 1546 description may indicate an even more marked rise in the volume of the glacier than in 1890–1920: he would then simply be giving us a 1546 version of the "modern" state of the *Rhonegletscher* as it appears in pictorial evidence from the first known engravings in 1700, and after that without interruption to the daguerreotypes of 1848–1850 and even 1870. During that period the Rhône glacier presented itself in the form of a "huge mass" quite accessible on horseback,[5] just as Münster described it in 1546, with a terminal tongue over 200 meters wide.[6] The front often had huge séracs before it, and was quite close to a little bridge immediately downstream. For at least a century and a half (1700–1850) the *Rhonegletscher* existed in this impressive form, spread out over the plain in the shape of a *pecten*, as Mercanton was to call it. There are many pictures dating from this period and they all show the same thing; and it was surely like this too in the seventeenth century, during the maximum of the Alpine glaciers. So the probability is that Münster either saw the glacier under this "modern" aspect; or, to go back to the first and equally plausible hypothesis, he saw it in an intermediate form—bigger than nowadays (1930–1960), not so big as in the "modern" era of 1590–1850. This intermediate form was to occur again in the intermediate period of 1890–1920.

If one adopts the first hypothesis there is only one difference, though a big one, between the 1546 and the 1890–1920 pictures: in 1890–1920, the "intermediate form" comes *after* a phase of maximum advance (1818–1850) and *before* the new and even more marked retreat of 1930–1960. In 1546 the opposite was the case: then the "intermediate form" seems to have come before the long phase of maximal advance, well attested from 1570–1590 on. There is a sort of inverse symmetry: 1890–1920 is a phase of long withdrawal, while, judging by what followed, in 1546 the trend was about to become one of long advance.

Whichever hypothesis is correct, one thing is certain: in 1546 the *Rhonegletscher* was definitely much bigger than it was in 1930–1966. Neither in time (fifty years) nor in space was it very far from the maximal positions it would occupy at the very end of the sixteenth century. The data derived from the following decades, in the second half of the sixteenth century, do not contradict this assertion.

To take 1570 first. The excellent scholar and glaciologist V. Heim[7] recalled, following Reissacher, Simony and Posepny, the strange story of the medieval gold mines of Hohe Tauern, blocked by ice in the second half of the sixteenth century. In one of these mines, founded in the middle of the fifteenth century (the *Bartholomei-Erbstollen am Rauriser Goldberg*), the shafts were already covered with twenty meters of ice in 1570, and working had to be abandoned. By the eighteenth century (according to Heim's figures and sources, which could do with revising), the thickness in 1570 was 100 meters. In 1875 it was still 40 meters. In the 1880s, as a result of the retreat of the glaciers, some of the shafts were full of ice.

Still in 1570, Gudbrandur Throlasson's maps for the Glama and Hofsjökull glaciers in Iceland, just like later maps for 1840, show enormous glaciers longer by several kilometers than in the twentieth century.[8]

If we leave Iceland and the eastern Alps and look through the little-known sixteenth-century archives of the poor parish of Chamonix, we find understandable but not very revealing laments about the rigor of the climate and the poverty of the inhabitants: "It is a poor country of barren mountains never free of glaciers and frosts . . . half the year there is no sun . . . the corn is gathered in the snow . . . and is so moldy it has to be heated in the oven and the animals will not eat the bread made from it . . . the place is among mountains so cold and uninhabitable that there are no attorneys or lawyers to be had . . . there are a lot of poor people, all rustic and ignorant . . . so poor that in the said places of

Chamonix and Vallorsine there is no clock by which to see and know the passage of time . . . no stranger will come to live there, and ice and frost are common since the creation of the world."[9] The inhabitants also complain of the serious avalanches of 1559–1562, which destroyed "mountain buildings," probably chalets.[10] According to A. Poggi's recent article, avalanches would be a symptom of heavy snowfall and rather low temperatures.

A significant fact is that from 1530 to 1575 the inhabitants of Chamonix, while lamenting the *existence* of the glaciers which make their surroundings so cold, do not complain of material *ravages* caused by the glacier fronts;[11] so if, by a pure hypothesis, the advance of the glaciers is supposed to have already begun then, it still did not seem to involve much danger to houses or lands.

In 1575–1576 one of these unpublished texts gives an interesting topographical detail. In the course of a lawsuit over tithes (there was a sort of chronic series of strikes over paying them throughout this period), a farm laborer from Saint-Gervais who often came to Chamonix to buy cheese was asked by the local people to appear as a witness for them. He described Chamonix as a "place covered with glaciers . . . often the fields are entirely swept away and the wheat blown into the woods and on to the glaciers as the said witness has seen happen several years."[12]

The glaciers at this time, then, were so big as to be quite near the fields, which is not the case today. The Mer de Glace is now a long way off, and separated by a barrier of rocks from the cultivated or cultivable land.

It may be said that this text is rather vague. But four years later an inquiry made about tithes by Bernard Combet, archdeacon of Tarentaise, adds supporting and complementary information. It is worth quoting it in detail.

In 1580 Combet, coming from Servoz, entered the valley of Chamonix from the southwest. After calling at the priory (the present village of Chamonix), "we continued our visit," he wrote, "right up to the boundary of the glebe above les Tines; we saw the hamlets called les Frasserands, Montrioud (Montroc) and le Planet; and from there, retracing our footsteps, we had to go

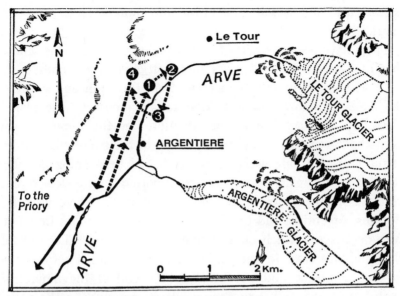

Fig. 13 *Route Followed by Bernard Combet* (1580)
Bernard Combet, visiting the upper valley of the Arvi above Chamonix, went first through his Frasserands (1), then Montrioud (Montroc), (2) then le Planet, and from there to Trélechamp (4). Then he turned southwest back to the priory. (Sketch from I.G.N. map of 1963—1/50,000.)

northwards, by a pass with the place vulgarly called Treléchamp . . . and from there we returned to the house of the said Aymon" (a representative of the community of Chamonix, and Combet's host).

After this description of his intinerary, Combet gives a more general account of the valley of Chamonix: "This valley," he writes, "is placed between the mountains; and to the right, as you approach from the south, these mountains are white with lofty glaciers, which even spread through rifts [*scissuras*] in the mountains themselves, and descend almost to the said plain [that of the Arve] in at least three places [*et descendunt fere usque ad dictam planitiem, tribus saltem in locis*]. One thing is clear: those rifts which people call moraines [*ruinas*][13] have sometimes caused unavoidable floods,[14] both in the regions through which the waters descend and in the middle of the valley, where they swell the stream said to rise in the Alpages du Tour [*qui ab Alpibus de Tour dictus est incipere*] and which then forms into quite a large river [the Arve]."

There have been two sets of comments on this text. One school, that of Charles Rabot,[15] gives it a restrictive interpretation. As we have seen, Bernard Combet saw from the plain and valley of the Arve only three glaciers "descending almost to the plain." Therefore, deduces Rabot, the situation was the same in 1580 as in about 1900–1910: at the beginning of the twentieth century, at the end of a long withdrawal which had been going on since 1860, only three glacial fronts of any importance were close to and visible from the valley or "plain" of Chamonix. These were Les Bossons, Argentière and Le Tour. As for the glacier Des Bois (the Mer de Glace), which once stretched as far as the village of Les Bois and was close to and visible from the plain, it retreated so far between 1860 and 1900 that by the end of the nineteenth century not only did it no longer "descend almost to the plain" but it could no longer even be seen from it. So, says Rabot, in 1580 at Chamonix it was still (*before* the advance of 1600) a period of low glaciation, and this situation was reproduced symmetrically in 1900–1910, *after* the advances of 1818 and 1850.

But various other experts, among them some of the foremost, like Mougin and Raoul Blanchard,[16] have not adopted Rabot's interpretation. According to Mougin and Blanchard the Le Tour glacier is too far from the "plain" proper and the main centers of the Chamonix valley for Combet to have included it in the three glaciers he saw. They say the three glaciers "descending almost to the plain" were Argentière, Les Bossons and the Mer de Glace. In 1580 the Mer de Glace was not withdrawn and invisible from the plain as in 1900–1960, behind the somber rocks of Les Mottets. In 1580, as during the whole period covered by early pictorial evidence (1730–1860), it spread out right as far as the hamlet of Les Bois, and was therefore clearly visible from the plain of the Arve.

The Mougin-Blanchard explanation seems to be the right one. For if, as I have done, one rereads the original manuscript of Combet's text,[17] one sees that both the paths he took and the expressions he uses indicate his complete ignorance of the "fourth glacier," the glacier of Le Tour.

To take his itinerary first: Combet did not go right up to the head of the Arve valley, to its last inhabited part, the hamlet of Le Tour, from which the glacier of that name was visible. He only went up as far as Montroc, then turned off toward Trélechamp, and then returned downstream. From these farthest points of his journey the *terminal* mass of the Le Tour glacier

Fig. 14 *Bossons, Mer de Glace, Argentière*
   "Descendunt fere usque ad planitiem tribus saltem in locis": "They (the glaciers) come down almost to the plain at least in three places."
   This engraving is the best possible illustration of Bernard Combet's account in 1580. The engraving is anonymous; under the frame are the words: *Verm. Gegenst., XLII; Mélanges, XLII; Miscell. Subj., XLII; Miscellanea, XLII.* Judging by the typography these words date from the first half of the nineteenth century, and the engraving cannot be much earlier. (Veyret, 1966, p. 233, considers the engraving to be by Linck at the end of the eighteenth century.)
   Reading from front to back one sees the glaciers Des Bossons, bristling with séracs; then the Mer de Glace, overtopping the Côte du Piget; then the massive Argentière glacier. Today the terminus of the Mer de Glace is invisible from this point of view. All one can see is the glacier Des Bossons and the Argentière glacier "coming down almost to the plain." (Source: From an engraving reproduced by permission of M. Snell, dealer, of Chamonix.)

and the beginnings of the Arve river were practically invisible to him, as I have several times verified on the spot.

Combet did not see the little district of Le Tour, and knew practically nothing about it. He just mentions vaguely, by hearsay, that the Arve rises in the Alpages de Tour (*qui ab Alpibus de Tour dictus est incipere*).

Mougin and Blanchard, though they were unaware of certain details I learned from the manuscript and went simply by their own intuition, are therefore probably right. Combet's "three glaciers" included not that of Le Tour, but the Mer de Glace, near the plain and visible from the hamlets of the Arve, as it always was during periods of great Alpine glaciation (the seventeenth and eighteenth centuries and the first half of the nineteenth). So the glacial advance had definitely begun by 1570–1580.[18]

Four years later, in 1584, comes the first account of a "cave glacière," a phenomenon which has interested speleologists.[21] The glacière in question is the Froidière of Chaux (Jura), which Bénigne Poissenot visited in 1584, publishing his description of it in 1586 in his *Nouvelles histoires tragiques*.[19]

It was in Besançon, on June 24, 1584, that Poissenot, drinking wine chilled with ice, was told of the Froidière of Chaux, the natural refrigerator which had supplied it. "Burning with desire to see this place filled with ice in the height of the summer," Poissenot arranged to be taken there on July 2. He was led through the trees, "by a twisting path," to the huge opening of the cave, a place so "terrifying" that he "was reminded of what is said to be St. Patrick's hole in Hibernia." Frightened but intrepid, he drew his sword and went right down into the depths of the cave, "which we found to be as long and wide as a big room, all paved with ice, and with crystal-clear water, colder than that of mount Arcadia Nonacris, running in a number of little streams, and forming small clear fountains in which I washed and drank greedily." He adds: "Whenever I looked upwards my whole body shuddered with fear and my hair stood up on my head, seeing all the upper part of the cave covered with great blocks of ice, the least one of which falling on me would have been enough to dash out my brains and tear me in pieces: so much so that I was like the criminal whose punishment in Hell was to have a big stone continually threatening to fall on him."

In short: a floor entirely paved with ice; many little streams because of the melting (it was the month of July); enormous stalactites of ice all over the roof. Such was an accurate description of the Froidière of Chaux in the Jura in July 1584.

Had it long worn this aspect in the sixteenth century? It is quite possible: the gentry of Besançon had long been in the habit of having ice brought on horseback at night from the Froidière to keep in their cellars to cool their wine. Every year, in honor of an old custom, the people of Chaux "were supposed to come and offer to the great church of Saint-Jean de Besançon a quantity of ice . . . which they brought to the town by night on horseback, so that it should not melt in the heat of the day."

On July 22, 1686, just over a century after Poissenot's visit, the abbé Boisot told in a letter to the *Journal des Savants*, a learned review, how "a few days before" he too had been to see the "famous glacière" at Chaux. His visit also took place during a fine summer, "of excessive heat." Moreover, in 1685–1686 the winter in the Jura had for once "not been severe," and all the "private ice houses" (stone-covered pits where well-to-do people preserved ice and snow from winter) "had been short of ice." But the famous Froidière of Chaux was still all right, and a line of mule carts formed up outside to take lumps of ice not only to the nearby towns, but even to the army camp on the Saône. Even on a very hot day during that summer of 1686 the cave produced more ice, by freezing the water that flowed into it from the local stream, than its customers from all over the province could carry away in a week. Boisot, like Poissenot, noted the big stalactites of ice on the roof of the cave: "Great pieces of ice hang down from the ceiling producing an effect of great beauty." So the Froidière piled up ice on its ceiling as well as its floor.

Still, two hundred years later, at the end of the nineteenth century, it is noted as normal and usual for the ice in the Froidière to be 1.2 meters thick, with tall pillars of ice rising above from the floor. And this in spite of the floods which every so often

melted the ice, and the fact that the cave was now being in-
tensively exploited industrially (192 tons of ice are said to have
been taken from it in 1901).[20]

Now that is all changed. After the floods of 1910 the ice never
formed again as it did before: the conditions which had maintained and renewed the little underground glacière of the Jura no
longer obtained, or else they had withdrawn, as in the case of the
big open-air glaciers of the nearby Alps. In other words, the
twentieth-century amelioration prevented the ice of the Froidière
from re-forming after the floods of 1910, though in the Little Ice
Age it had been replaced quite easily, in spite of intensive commercial exploitation. When I visited the Froidière in the summer
of 1963, all that was left of the impressive spectacle described by
Poissenot and Boisot were a few scattered sheets of ice on the
floor of the cave, in places about 20 cm thick. Very little melt
water to form streams. And not a single ice stalactite hanging
from the roof; no tourist was in the sort of danger Poissenot had
feared. The disappearance of the "stalactites," so high that it used
to be impossible to get at them with ice-picks, reinforces the idea
of a subtle change in the natural conditions prevailing there.

So, like its big brothers at the sources of the Rhône and of the
Arve, the Froidière as it was in 1580 bears witness to a much
greater glacial development then than in our present era of glacial
withdrawal.

In the last decade of the sixteenth century the problems begin
to emerge more clearly, and data suddenly becomes abundant,
accurate and significant.

The first symptom occurs in 1588, when the Grindelwald glacier "breaks through its terminal moraine": "strekt der Gletscher
d'Nasä i Bodä und drückt ä Hübel mit ämä Ghalt weg."[21]

In 1589 the Allalin glacier descended so low it blocked the
valley of the Saas. A lake formed at the barrier, the Mattmarksee,
but the barrier broke on September 8 and the waters of the lake
burst through and flooded the land below. The bed of the Saas-
Visp, a stream flowing out of the glacier, was so damaged that

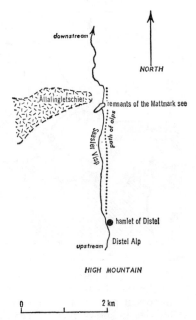

Fig. 14A The Allalin glacier (Allalingletschier) in its present withdrawal and the mountain pasture of Distel Alp. The figure shows how, during periods of the Little Ice Age, the Allalin glacier could bar the valley of the Visp, causing the Mattmarksee to grow larger and serving as a boundary between the mountain pasture of Distel Alp and the meadows farther downstream. (From the Swiss topographical map of 1960.)

the local people were obliged, at great expense, to construct a new one.[22]

On June 25, 1595, the Giétroz glacier in the Pennine Alps crashed into the bed of the Dranse, from which it is now a long way away. The same thing happened again during the glacial advance of 1817–1818.[23] The 1595 disaster, which has been well documented by O. Lütschg, temporarily submerged the town of Martigny and caused seventy deaths, according to a contemporary account. In 1926 there was still a house in Bagnes with a beam inscribed, "Maurice Olliet had this house built in 1595, the year Bagnes was flooded by the Giétroz glacier.[24] The hamlet of Ander Eggen was completely wiped out by the deluge of ice in 1595, and became a sort of waste land known as Gletschera (glacier).

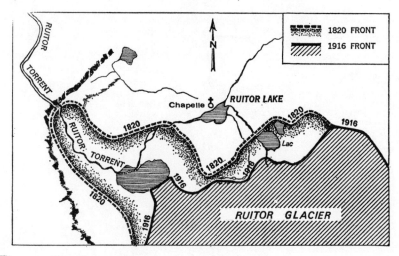

Fig. 15   *The Ruitor Glacier*
    This (simplified) sketch is based on the very detailed "Schema topografico della regione rutorina" to be found in Sacco, 1917, together with an account of the historical phases of the retreat of the glacier from 1820 to 1916. The sketch clearly shows that in 1820 the glacier covered the present site of the torrent flowing out of Lake Ruitor, and that the glacier thus barred the waters of the lake.

From 1594 to 1598, interesting things began to happen on the "Italian" side of the Alps. The Ruitor glacier, in the Val de Thuile basin, reached an extreme point of advance, almost a kilometer beyond the present front. It did not finally withdraw from this position, until our own day at least, before the 1860s on. In its extreme position the glacier barred the lake of the same name, which is now unbarred and almost dried up, but then amassed huge quantities of water. In summer a channel formed under the ice, and floodwaters of up to three or four million cubic meters would pour through and inundate the surrounding valleys.

The first time this happened and was clearly attested was in 1594,[25] though it recurred frequently in the seventeenth and eighteenth centuries. (The recent glacial withdrawal, culminating in the twentieth century, has meant that no such disasters have occurred in the last hundred years.) There was another ruinous and violent overflow at Ruitor in 1595; again in 1596; on July 15, 1597; and another, less violent one in August 1598. The people of Thuile and Aosta, and the authorities, took alarm. Experts like Simon Tubinger and Jacomo Soldati were called in, and they proposed, for ten thousand ducatoons, to lead the overflow of the lake away through a tunnel in the rock, or to block the dangerous channel under the ice with stone and wood! Their plans were examined, and tenders for the carrying out of the work were invited between October 1596 and January 1597. Needless to say there were no offers.

In 1599–1600 the peak was reached throughout the Alps; it was the great historical maximum of the glaciers. Richter diagnosed it as early as 1891, using just the Grindelwald and Otztal documents. His discovery was confirmed and generalized when the early archives of the Chamonix valley were first brought to light twenty years later.

A text from the Chamber of Accounts of Savoie shows how the effect of the glaciers was felt at Chamonix as early as 1600: "At the time of the tallage reform (i.e., 1600: this tax was reformed in Savoie by an edict of May 1, 1600) the glaciers of the Arve and other rivers ruined and spoiled one hundred and ninety-five journaux[26] of land in divers parts of the said parish (Chamonix) and in particular ninety journaux and twelve houses were ruined in the village of Chastelard of which only a twelfth part was left. The village of Les Bois was left uninhabited because of the glaciers. In the village of La Roziere and Argentier seven houses were covered by the said glaciers, whose ravages continue and progress from one day to the next . . . two other houses were ruined in the village of La Bonneville . . . because of all this ruin the tithe was greatly reduced."[27]

This text is significant in several ways. Topographically, because the places affected are now over a kilometer from the glacial fronts. Chronologically, because of its specification of dates. And finally, because of its detail: the glacier's effects are not limited to the floods and the fall of the séracs; either it or its moraines cover seven houses "and progress from one day to the next" over them. Nothing is known of when these seven houses were built, but as glaciologists who have commented on these texts have said, they must have been put up in the belief, founded on experience of some length, that they would be safe from such incursions.

In other Chamonix texts, published and unpublished, which I have discovered among the papers of the community and the priory, there is further information. These texts, like the previous ones, are interesting from three points of view: they show the specific action of the glaciers; they suggest a topography for their termini; and they give a chronology for the glacial thrusts.

First, the action of the glaciers: in the 1610 inquiry[28] the witnesses make a definite distinction between the damage done by the glaciers and the much more minor destruction wrought at the same time by streams, the river Arve, and avalanches or *lavanches*.

Thus Nicolas Moret, a laborer of Servoz questioned as an impartial witness in April 1610 by François Bertier, officer of the Chambre des Comptes, replied, when asked about the "falls from the glaciers," that "the streams have spoiled seventeen journaux of land since the tallage reform (1600) . . . the Arve has spoiled eight . . . and the glaciers two hundred and four and a half."[29] Gervais Vian, a laborer aged fifty-six from Saint-Gervais, confirms these figures, commenting reasonably enough that "it was the glaciers which caused the worst and greatest damage." It should

Fig. 16　*The Mer de Glace and its History*
　　　　━━━━: Terminal limits of the Mer de Glace in 1958.
　　　　– – – –: Maximal moraine for the seventeenth century (probably
　　　　　　　　1641-1644), according to Lliboutry, 1965, and Mougin, 1912.
　　　　⊕　　　Approximate site of Le Châtelard, according to Rabot, 1920.
A comparison of all written sources suggests that Le Châtelard must have been somewhere between Les Bois and Les Tines. In my own opinion (judging from the texts and from what I was told by M. Couttet), Le Châtelard was a few hundred meters more to the southwest than Rabot placed it.
　　　　⊙　　　Approximate site of Bonanay (according to Rabot, 1920).
　　　　B　　　Present site of the "wood of Bonanay" on the survey map of
　　　　　　　　1945.
　　　　▨▨▨　Côte du Piget.
　　　　(Source: The map is drawn from Lliboutry, 1965, II, Figure 18-3. After Lliboutry and other sources I have included on it the various information as noted above.)

I. LE CHATELARD at the end of the fourteenth century

Reference to the village of Le Châtelard, which was to be destroyed at the beginning of the seventeenth century by the advance of the glacier Des Bois. Reference also to Les Bois and Les Praz. All in the earliest accounts of the priory of Chamonix (1380–1390): *de exitu decime de castellare* . . . (2 bichets of rye) . . . *decime de memoribus* (5 bichets) . . . *decime de pratis* . . . (5 bichets).

(Source: ADHS, G 226–265, first accounts. Photo: Archives of Hérault.)

II. SAINT-THEODULE DU CHATELARD

The chapel of Saint-Théodule, "called of *Le Châtelard* or of Les Tines"
(see below, Appendix 7), still exists, a distant souvenir of the old Châtelard
destroyed by the glacier Des Bois. It is said that the villagers built the little
church against an epidemic of plague; perhaps it was also to exorcise the then
menacing advance of the glacier Des Bois. It was built around 1640–1650*
below the already heavily damaged site of Le Châtelard, from which nevertheless it took its name. The chapel soon became identified with the new
hamlet of Les Tines, which was built nearby as replacement of and successor
to Le Châtelard; hence the significant toponymy in the texts of 1693 and
1702: "the chapel Saint-Théodule of Le Châtelard known as of Les Thines."
Plate II shows the naïve and damaged picture over the altar of the chapel,
probably painted around 1640–1700: it depicts Saint Théodule, bishop of
Sion, a basket of Easter eggs at his feet, making the devil take a paschal bell
to Rome.

(Photo: Madeline Le Roy Ladurie, 1966.)
*According to a notice in the chapel.

A. *Das Eyß oder Gletschen so vom Boden auff wachset, und alles presch flißt mit ungestühm und vilen Krachen* B. *der fluß Lütschinen so under dem Eyß harfür quillet*
C. *Wohnungen mit welchen man dem Gletscher bald weichen müssen* D. *Hochgeburgen mit Ewigem Schnee bedeckt*

III. GRINDELWALD in 1640

About 1640 the lower Grindelwald glacier, swollen and bristling with séracs, stretched far beyond its present positions and occupied entirely the gorge and lateral valley enclosing it, right down to where the valley opened into the Grindelwald valley proper. (Engraving taken from Merian, 1642.)

In the foreground is the White Lutschine, which flows out of the glacier. To the left is the Mettenberg; in the middle, the Fischerhorn; to the right, the Hörnli, a small extension of the Eiger.

(Photo: Bibliothèque Nationale.)

Grundriss der Eisthäler und Gletscher im Grindelwald.

IV. MAP of the GRINDELWALD GLACIERS in 1686

Note, right upstream in the lower glacier (the one on the right, debouching not far from the village), the rock of the "hot leaves" or *Heisse Platte*. The rock, which is shaded dark, emerges from among the séracs at the confluence of the two streams of ice which together form the lower glacier.

(Drawing by Herbord, 1686, reproduced in Gruner, 1760 and 1770.)

Aussicht der Eisgebirge und Gletscher im Grindelwald
im Cant: Bern.

V. GRINDELWALD in 1720

About 1719–1720, the two great Grindelwald glaciers came down to the main valley.

(Drawing by F. Meyer, about 1720; engraving from Gruner, 1760 and 1770.)
(Photo: Bibliothèque Nationale.)

VI. GRINDELWALD in 1748

In 1748 the lower Grindelwald glacier came down much lower than it does now. It came up quite close to the Marmorbruch or marble quarry (F) on the right of the valley, whereas today its terminus is scarcely even visible from that point. It could be seen, downstream, overflowing the wooded ridges on its right bank (to the left of the engraving), while in 1966 the visible part of the downstream glacier adjoins a right bank completely bare of trees.

(Source: Altmann, 1751.)

VII. GRINDELWALD about 1775-1780

Lower glacier still in advanced positions.

(Engraving by Besson.)
(Photo: M.L.R.L., from the copy of the engraving at the Marmorbruch inn at Grindelwald.)

VIII. GRINDELWALD in 1966 (View from the village)

The lower glacier has retreated, and is no longer visible in the gorge, as it was in the seventeenth, eighteenth, and nineteenth centuries.

(Photo: M.I.R.L., 1966.)

IX. THE RHONEGLETSCHER in 1705

Downstream from where the séracs fall (1), the glacier forms a *pecten* (3, 4, and 5) in the vale of Gletsch; the Muttbach flows up to it and runs beside it on the left (10, to the right of the picture). The source of the Rhône is "bicorn," or pronged.

(Engraving from Scheuchzer, 1723, Vol. II.)
(Photo: Bibliothèque Nationale.)

X. THE RHONEGLETSCHER about 1720

About 1720 the Rhône glacier, in the form of a *pecten*, spread over the vale of Gletsch. (Drawing by F. Meyer, about 1720; engraving reproduced in Gruner, 1760 and 1770.) Like Scheuchzer in the previous engraving, Meyer foreshortens in order to show not only the Rhonegletscher but also the lateral glacier, known as the *Muttgletscher* (6).

(Photo: Bibliothèque Nationale.)

XI. THE RHONEGLETSCHER in 1778

In 1778 the Rhône glacier, in the form of a *pecten*, barred the valley of the Muttbach and filled the vale of Gletsch. (Engraving taken from Bourrit, 1785, who did the original drawing from life.)

(Photo: Bibliothèque Nationale.)

XII. THE RHONEGLETSCHER in 1966

Photograph taken in summer 1966 by Madeleine Le Roy Ladurie. The sky was unfortunately clouded at the time (in the course of various visits we have never seen the Rhonegletscher under a clear sky). The front of the glacier has shrunk back to the top of an almost entirely bare rocky verrou. The *pecten* in the valley is only a memory. The Muttbach, which appears as a thin thread of water to the right of the picture, emerges from the lateral valley to flow directly into the Rhône, unhindered by the ice which stood in its way in previous centuries.

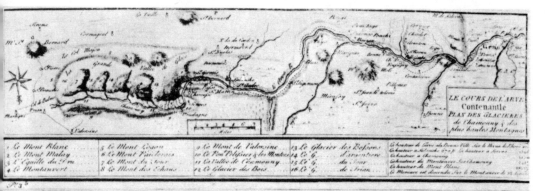

XIII. First pictorial evidence on CHAMONIX (1742)

These sketches, done on the spot by Martel in 1742, are taken from the Cabinet des Estampes of the Bibliothèque Nationale, dossier "Etats sardes et Savoie," cotés V b 1. (V b 1 and V b 2 have many convincing pictures of the "Little Ice Age" in the Alps.) Although clumsy, this sketch is conclusive as evidence: the glacier Des Bois (12) is visible "from the church of Chamouny ...in the year 1742." It already shows the classic contrast with the present situation.

(Photo: Ed. Flammarion.)

XIV. THE CHAMONIX VALLEY about 1770

This is, as far as I know, the third engraving in order of date which represents the glaciers of Chamonix (including the strange drawing, perhaps a mirror image, in Gruner, 1760; see Plate XIII b). The Mer de Glace spreads out widely, as a piedmont glacier, over a landscape from which it has now disappeared. In the background are the Argentière and Le Tour glaciers, also very developed.

(Source: Bourrit, 1773, p. 80.)

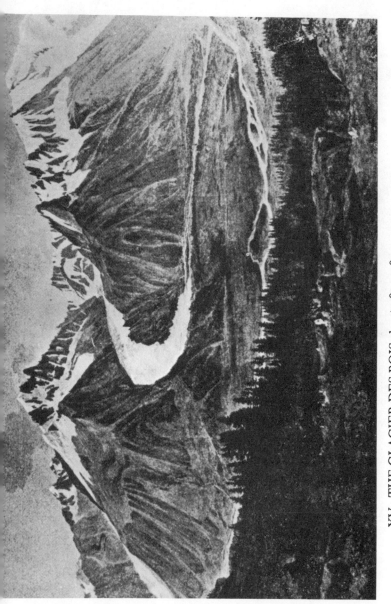

XV. THE GLACIER DES BOIS about 1800–1820

Painting by Linck still preserved in the town hall of Chamonix and reproduced here from Mougin, 1912, Plate VII. In the foreground are the Arve and the Arveyron. The Mer de Glace, "curving like an arm" (Hugo), descends "almost to the plain." In the photograph Mougin took of the same scene in 1911 (Mougin, 1912, Plate VIII), the Mer de Glace has entirely disappeared, leaving behind only moraines where its terminal tongue was in the early nineteenth century.

Mougin, 1912, p. 165, after a good critical study, dates the picture 1800 to 1823. I could not get permission to photograph the original.

XVI. The same landscape in 1966

   The Mer de Glace, in secular débâcle, has retreated toward the Montenvers. The old glacial curve (see caption to Plate XV) has disappeared and been replaced by the torrents and cascades of the Arveyron.

(Photo taken from La Flégère by M.L.R.L.)

XVII. THE MER DE GLACE from the outskirts of Chamonix, about 1840

This anonymous engraving (probably by Anton Winterlin, 1805–1894) belongs to the Romantic era. (It is reproduced by permission of M. Snell, of Chamonix; the photograph is by M.L.R.L.) In the foreground is the priory church (the present village of Chamonix). The glacier Des Bois, which has now completely disappeared from the scene, spreads out widely toward the village of Les Bois and overtops, on the left, the Côte du Piget. Lower down, below the glacier, can be seen the valleys of the Arve and the Arveyron. Right in the background, between the trees, a little steeple is discernible.

In this engraving the Mer de Glace has the characteristic dimensions of the Little Ice Age (see, for these same dimensions, all the reproductions in Mougin, 1912, for the period 1770–1880).

(Photo: M.L.R.L., 1960.)

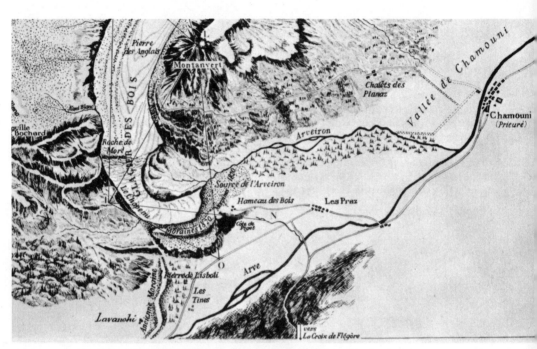

XVIII: First scientific map of the MER DE GLACE (1842)
From Forbes, 1943.

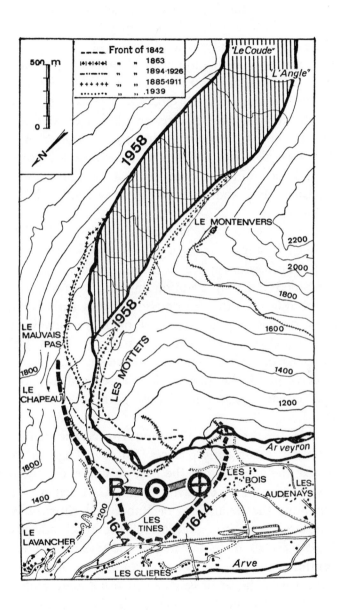

"Le Coude"

L'Angle

1958

LE MONTENVERS

2200

2000

1800

1600

1400

1200

LE
MAUVAIS
PAS

1958

LES MOTTETS

1800

LE
CHAPEAU

1600

Ar veyron

1400

B ▨ ⊙ ⊕   LES
                       BOIS   LES.
                                   AUDENAYS

1200

1644

LE
LAVANCHER

LES
TINES

1644

Arve

LES GLIERES

be noted that this kind of damage was never mentioned by witnesses in the various inquiries carried out at Chamonix during the sixteenth century.[30] Next, the topography of the termini. Nicolas de Crans, deputy commissioner of the Chambre des Comptes, carrying out an inquiry at Chamonix in 1610, on the complaints of the inhabitants, wrote, probably referring to the damage done by the glacier in 1600–1601: "We recognized the ruin brought about by the glaciers and the river Arve in various places in the said Chamonix, including the glacier called Des Bois (the Mer de Glace), which is terrible and frightening to look on and has ruined a good part of the land and all the village of Chastellard and quite carried away another little village called Bonnenuict."[31]

Where exactly were Le Châtelard and Bonnenuict (*Bonanay*, as it is called locally), the two villages wiped out by the Mer de Glace? The names survive in certain pieces of land. Charles Rabot recorded the oral tradition in 1920,[32] and I continued his work in 1960. All this information checks with factual data in modern surveys and in the "old map" or survey of 1730, which shows a chapel "du Châtelard" near Les Tines. Le Châtelard, the ruins of which were still visible in 1920, stood about midway between the hamlets of Les Tines and Les Bois. The hamlet of Bonanay was immediately north of Le Piget slope. (Fig. 16).

The placing of these two *Wüstungen* of Savoie confirm what is suggested by certain early pictures and engravings, in particular the "colored line engraving of the Mer de Glace, drawn from the life by M. Jalabert and engraved by G. G. Geissler" (1777),[33] which confirms the deductions of Rabot and Kinzl, and before them Saussure, as to the position of the most advanced moraines.[34] At the peak of its maxima of the seventeenth and eighteenth centuries, the Mer de Glace, spreading out like a piedmont glacier, swept over the small heights which normally protected the villages of Les Bois and Les Tines; it also jumped over the Côte du Piget and overhung the slopes leading down to the Arve and Les Tines. This was the extreme position reached by the glacier Des Bois between 1600 and 1610. After its last leap over the northeast edge of Le Piget, the glacier could, to use Nicolas de Crans's expression, "sweep away" the most exposed site, Bonanay, of which today no trace remains, except a place name, attached to "a piece of forest bordered on the south by an old moraine."[35] It also hung over and ravaged Le Châtelard, "ruining" it with the séracs which broke off and the floods that poured from its terminal mass. An excellent text bears witness to how the Mer de Glace hung for a long time over the villages of Les Tines and Le Châtelard. In 1616, ten or fifteen years after the first catastrophe at Le Châtelard, Nicolas de Crans, an objective witness, went through the ruins, where a few poor people still lived, threatened by the glacier that towered over them nearby: "Went to the village of Chastellard where there are still about six houses, all uninhabited save two, in which live some wretched women and children, though the houses belong to others. Above and adjoining the village, there is a great and horrible glacier of great and incalculable volume which can promise nothing but the destruction of the houses and lands which still remain."[36]

The threat was enough. At some undetermined date between 1643(2) and 1700 the ruins of Le Châtelard were evacuated, and nothing more was heard of them.

So in 1616 the glacier Des Bois "adjoined" the ruins of Le Châtelard, a kilometer above its present terminus (1960). Similarly, at the beginning of the seventeenth century, the Argentière glacier reached the village of Argentière and the hamlet of La Rosière, now separated from it by woods and a distance of at least eleven hundred meters. The text of 1605, referring to the destruction of 1600–1605, had spoken of how "in the village of La Rozière and Argentière seven houses were covered by the said glaciers, whose ravages continue and progress from one day to the next." In 1610, Bertier talks of the glacier "called of Largentière or of La Rosière."[37] The term glacier of La Rosière, no longer in use today, only makes sense in the context if the glacier of Argentière was then very close to the hamlet of La Rosière.

This proximity turned out to be catastrophic on June 22, 1610. "Since the inquiry conducted by M. Bertier in April 1610, it happened that on the twenty-second of June 1610, by the overflowing of the glacier of La Rosière, eight houses and forty-five journaux of land were completely destroyed."[38]

The 1610 catastrophe was clearly distinguished by witnesses from that which occurred about 1600. In 1616 Nicolas Granjean said that "the glacier of La Rosière entirely crushed eight houses and five barns," and expressly differentiates "the damage done fifteen or twenty years ago" (about 1596–1601) and that done "six or seven years ago" (1609–1610).[39]

Three years later, in 1613 or 1614, the "overflowing" of the Arve, swollen by the melt water from the glaciers, finally finished off the hamlet of La Bonneville,[40] which had already suffered so much in 1600–1605 and 1610. The Arve in fact changed course and swept the wooden houses of the hamlet away. Oral tradition, which of course must be treated with caution, seems to confirm the texts. According to M. Coutet and the guide L. Ravanel in 1960–1961, La Bonneville was situated near the Arve, between Les Tines and Argentière, or, more exactly, between the present hamlets of Le Grassonnet and Les Chosalets (Fig. 25).

In 1616, alerted by the complaints of the villagers, Nicolas de Crans came to make his inquiry, and his text provides details of the first importance about the glacier of Argentière. It had spread out over the plain to a quarter of a league beyond its twentieth-century positions; and the south side of its frontal mass adjoined the hamlet of La Rosière. All the Crans texts of 1616 agree on this: "La Rosière above which is a great and grim glacier which brings down great stones which have covered and spoiled much land . . ."[41]

"La Rosière which is still in great danger . . ."[42]

"The great glacier of La Rosière every now and then goes bounding and thrashing or descending; for the last five or six years . . . it has been impossible to get any crops from the places it has covered . . . the witness (Michel Faure, of La Rosière) has lost a house and a barn entirely submerged, it being impossible for him and for others who have suffered the same loss to oppose or prevent the fury of the said glacier."

And finally the decisive text which clearly gives the topography of the south front of the Argentière glacier in 1600–1616:[43] "Behind the village of Les Rousier, by the impetuosity of *a great and horrible glacier which is above and just adjoining the few houses that remain,* there have been destroyed forty-three journaux (of land) with nothing but stones and little woods of small value, and also eight houses, seven barns, and five little granges have been entirely ruined and destroyed." So in 1616, years after the disasters of 1600 and 1610, the Argentière glacier still "adjoined" the houses of the unfortunate village it had partly destroyed.

In 1616 it had, but in a more accentuated and advanced form, the forked shape which is still just visible in an engraving of 1830.[44] One mass descends toward the village of Argentière, the other toward the hamlet of La Rosière. Between flows the subglacial stream, the Arveyron or Arbéron. Here is the explanation of the expression used by all the 1616 witnesses, which so much intrigued Kinzl: "The river Arveyron descending from the top of the mountain between two great glaciers."[45]

Another merit of the 1616 text is that it gives the exact configuration of the village of La Rosière just as it is today. An old moraine, a scree of "great stones,"[46] borders the village on the north; above the moraine grow fir trees several hundred years old; it marks an old southern limit of the Argentière glacier. A stony path from the northwest disappears among the boulders of the moraine (Fig. 17), which obviously flowed across and blocked it. Local tradition, as reported to me by Louis Ravanel, says it is the old path from Les Chosalets to La Rosière. On its southern edge and east corner, just where it meets the moraine, the path is bordered by clearly visible bits of foundation wall half-buried in the scree—perhaps the ruins of the little granges destroyed by the glacier. (This suggestion of oral tradition checks with the 1616 texts and with the topography.)

This waste land, now calmly reabsorbed by the forest, and these ruins together probably represent the stigmata of the catastrophes of 1600–1610. And thus we see confirmed the suggestions put forward by Crans: the Argentière glacier, at its maximum dimensions in the modern era (1596–1616), really did not only approach the village of Argentière, but also reached and bordered La Rosière farther to the south.

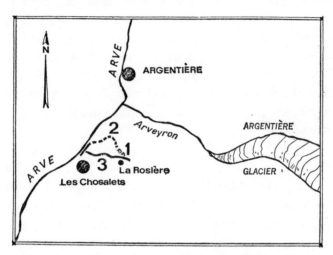

Fig. 17   This sketch (1/32,000) of the Argentière–La Rosière area reproduces in reduced form the 1949 topographical map (1/20,000). I have added indications 1 and 2 to the map; I noted them on the spot in the company and at the suggestion of M. Louis Rosière.

3: Present path from Les Chosalets to La Rosière
2: Old path from Les Chosalets to La Rosière
1: End of this old path and ruins of barns, all buried in the early terminal moraine (1600–1620) of the Argentière glacier.

With the Mer de Glace as far advanced as the Côte du Piget, and the Argentière glacier adjoining La Rosière throughout the first decades of the seventeenth century, both are a good kilometer in advance of their present fronts. This sizable and prolonged advance explains the sixteen years of disasters and catastrophic floods[47] which entirely destroyed three villages dating from the Middle Ages (Le Châtelard, Bonnenuit, and La Bonneville) and severely damaged a fourth (La Rosière).

But in all this, someone will say, we have been neglecting the glacier Des Bossons. Was it too very advanced in 1600–1610? Probably. In the course of the 1616 inquiry, Nicolas Granjean said that "about forty years ago he used to frequent the said parish and valley of Chamonix in which he saw and remarked that the impetuosity of the torrents, glaciers and rivers in that valley of Chamonix had from time to time caused great and incalculable damage . . . he noticed as worst that damage hereafter declared to have occurred six or seven years ago and in the first place in the village of Le Foulier (Le Fouilly, commune of Les Houches) as far as the priory; the river Arve and the many torrents and storms issuing from the glaciers had caused great damage."

There is no doubt possible here: the only glacier near the plain between Les Houches and Le Prieuré (the present town of Chamonix) is the glacier Des Bossons.[48] In 1610 it is represented as the most dangerous and destructive of all, and it is very likely that then, as again in 1640–1645, it was in a state of marked advance.

The Chamonix archives are also interesting in that they define some important points in the chronology of the glaciers. It is true no text gives the exact date of the first catastrophe at Chamonix, the one which occurred about 1600. But the inquiry conducted in 1605 by the Chambre des Comptes (quoted above, p. 173) sets a *terminus a quo* of 1600 and a *terminus ad quem* of 1605. In addition, the evidence of Nicolas Granjean (see above, p. 176) places the episode between 1596 and 1601. By crosschecking one thus arrives at a first maximum some time in 1600 or 1601, which coincides exactly, as we shall see, with the contemporary maxima at Grindelwald and Le Vernagt.

That at least was the chronology I felt justified in putting forward in the first edition of this book in 1967. But since then M. Roger Couvert, an excellent scholar of Chamonix, has discovered a new text among the private archives of a local family. The gist of it is that on April 6, 1600 a notary called Blanc, who came from Aosta, on what is now the Italian side of Mont Blanc, was visited by one Jacques Cochet from the village of Les Bois, near Chamonix. Cochet was thus a near neighbor of the Mer de Glace. In the name of his fellow inhabitants at Les Bois, he asked the notary two questions:

1. Was it true that the glaciers on the Italian side, in particular that of La Brenva, had recently retreated, to the great relief of the people living nearby?

2. Was it true that the parishioners of Courmayeur had recently sent a delegation to Rome to ask the Pope to pray God to make the glaciers withdraw once and for all?

The notary's answer to both questions—corroborated by several witnesses—was in the negative. The people of Courmayeur had preferred to apply to God without going through an intermediary. And unfortunately the Brenva group of glaciers had not retreated and were as threatening as ever.

This text is of the first importance. It shows once and for all that on both the French and the Italian sides of Mont Blanc the glaciers had extended so far by the beginning of 1600 that the people were in a panic. So the chronology deduced from the Crans dossier and the papers of the Chambre des Comptes is completely confirmed by this newly exhumed document.[49]

The Chamonix archives also suggest the possibility of a more long-term chronology. La Rosière, Bonnenuit (Bonanay) and Le Châtelard were hamlets dating back a very long time, and their destruction therefore implies a glacial maximum which exceeded any that had occurred for some centuries.

More exactly, two presuppositions emerge:

1. When the hamlets were founded the glacial fronts must have been comparatively far off and have seemed to present no danger.

2. In the interval, the glaciers probably approached quite close to the three villages on one or two occasions. But these approaches must have been much less dramatic than the one which took place between 1600 and 1610 and ended in the destruction of the three hamlets.

All this is only theory, of course, but it is a theory which can lead to a firm chronology. For there are plenty of documents, mostly unpublished, which describe the life and history of these villages before the fatal incursion of the glaciers (see Appendix 7).

First Le Châtelard. The name, from low Latin, suggests a

fortified place guarding the pass of Les Tines. There is a possible reference to it in 1289; it is certain that from 1384 to 1640 the accounts of the priory of Chamonix, which are in a state of fair preservation, mention it among three neighboring hamlets (the other two being Les Bois and Les Praz) paying tithes in rye or cash. The tithe paid by Le Châtelard is usually higher than that paid by the other two places. In 1458, 1467, and 1521 there are texts giving the names of some of the tenants of Le Châtelard (Jean Audran, for example, and the four Gaudins), and confirming its position under the slope of Les Bois (i.e., below the Rocher des Côtes, or the Côte du Piget). In 1523 there is mention of cultivated fields below the village, and possibly a mill. In 1528, 1530, 1540, 1544, 1554, 1561 and so on, there are many land transactions between the local families, the Gaudins, Lechiazes, Simons, Cachats, etc. In 1559 the people of Chamonix refused to pay the canons of Sallanches the first fruits of a measure of rye, so the canons drew up a list of the heads of families responsible. Le Châtelard, Le Rosière, and La Bonneville, three places which half a century later were to be destroyed, in 1551 had twenty-one, twenty, and thirteen families respectively—as many as, or more than, villages like Argentière (twenty families) and Les Bois (sixteen), which survived.

In 1562 another refusal to pay tithes called forth a description of Le Châtelard. Jean Faure, one of the inhabitants, had lent his house to the canons to store wheat. His neighbors, furious at such an act of treason, had broken into the house, though they had not set fire to it for fear of setting the whole village alight. The village is referred to as being built "en lattes." It can therefore be deduced that Le Châtelard was a collection of wooden chalets with stone foundations, probably built very close together. Another interesting fact is that among the ten or so witnesses who were questioned about the affair, all householders at Le Châtelard, several were about eighty years old.

In 1564, 1565, and 1570 a number of peasants acquired houses or land at Le Châtelard either by purchase or exchange. They do not seem to have been at all put off by the idea of any danger from the glaciers. These transactions go on until the end of the century, until 1600 itself, when there is a record of houses being exchanged among the inhabitants of the hamlet. But in 1602, nothing: the catastrophe which occurred some time between 1600 and 1605, probably in 1601, intervened. In 1616, as we have seen, a few poor people still lived there in a few dilapidated houses, but the owners had left. In 1622 I came across the purchase of a "storehouse at Le Châtelard" by an inhabitant of Le Lavancher, which had been spared by the glaciers. But there is no mention of houses in the village.

After that there is nothing: Le Châtelard is just a place name, a tithe subdivision lying between Les Tines, the Arve, Le Lavancher, and the moraine or "ravine" of the glacier Des Bois. In 1693 and 1702, the chapel of Saint-Théodule du Châtelard, built about 1640–1650, is considered part of the village of Les Tines—which is where the former inhabitants of Le Châtelard probably settled. It is as if the advance of the glacier Des Bois pushed the hamlet of Le Châtelard on to the more sheltered Les Tines.[50]

The fate of Bonanay (Appendix 7) is also typical: it enjoyed a long period of security, followed by a sudden irruption.

It had always been a tiny out-of-the-way place with less than a dozen houses. In 1458 only a few fields are mentioned there, *terra Bonenoctis* (*Bonanay* was then thought to be the vernacular equivalent of *Bonnenuit*; the name, which suggests a sheltered place out of the wind, does not seem to be very old). This reference is repeated in 1467. In 1523 there is an allusion to farm buildings (*curtilis* or *curtinis*). In 1556–1557, one Hugo (Hugo *de Bonanex*, who also owned property at Le Châtelard) took up residence there. By 1562 there were one or more houses: Jean Faurefuye, being questioned in connection with some tithe disturbances, said, "he did not know who had committed these outrages, because at the time he was in another house at Bonnenuyt." In 1571 a farm (*courtil*) and its dependencies was divided up there. In 1591 there still seemed to be no fear of any catastrophes, since for twenty-seven florins Georges Gaudin bought a piece of land that stood "below Bonnanech."

After the disaster in which the Mer de Glace swept over Le Piget and the site of Bonanay, nothing was left but the place name. Texts of 1679 simply say it is near "a moraine, ruin or ravine."

La Rosière does not seem to have been mentioned before the early land clearance in the area. There is a possible allusion to it in the Chamonix texts in 1315, and it certainly appears in the tithe accounts from 1390 on: at that time the amount of grain handed over by La Rosière was greater than that of the neighboring hamlet of Argentière. In the fifteenth century one of the inhabitants of La Rosière was among the eleven syndics or representatives of the community of Chamonix. And the life of the hamlet continued throughout the sixteenth century.

The tithe accounts complete our information on these three hamlets (see Appendix 7).

In the sixteenth and at the beginning of the seventeenth centuries the Chamonix tithe accounts reflect two diverging kinds of development. In those hamlets which are not much or not at all exposed to the glaciers, the tithes, paid in florins, tend to rise because of the contemporary inflation and perhaps also because of an increase in real production. But in the case of hamlets directly affected by the incursions of the glaciers (Les Bois, Le Châtelard, La Rosière, Argentière, Le Tour), the nominal value of the tithes remains generally stationary in the sixteenth century, especially in the second half, and in the first thirty years of the seventeenth. This nominal stagnation during a long period in which prices are rising suggests a decrease in real production. At La Rosière, which was so much damaged by the catastrophes of 1600–1610,[51] the tithe revenue fell from fifty florins in 1577–1600 to thirty-two in 1622. At La Bonneville, the ancient hamlet which existed in the fifteenth century and was destroyed in 1614 by the flooding of the Arve, the tithe dropped from fifty florins in 1570–1602 to twenty-six florins in 1625. At Les Bois and Le Châtelard the tithe remained stable, but the disappearance of the latter is reflected in the fact that the tithe, referred to from 1520 to 1590 as the tithe of Les Bois and Le Châtelard (*Nemorum et Castellarii*), is referred to only as "the tithe of Les Bois" from 1600 until the tithe records cease in 1625.

In contrast with the stationary or failing tithes of the threatened hamlets are the rising ones of Vallorcine, a parish completely sheltered from the glaciers: ninety-two florins in 1520, a hundred and sixty in 1578–1590, three hundred in 1600, and four hundred in 1622–1625.

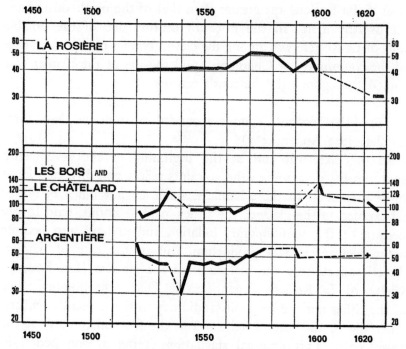

Fig. 18 *Chamonix Tithe Returns (in Florins)*
In the glebes immediately below the glaciers and so affected by the advances of the latter between 1580 and 1620. The scale is semi-logarithmic; the x-axis shows the years, the y-axis the amount in florins. Note the stagnation (and in the case of La Rosière the decline) in the nominal value of these tithes. Such a trend occurring during a long phase of strong inflation is a sign of a decrease in real crop yields.

This first attempt at a chronology, though still far from complete, enables us to put forward a tentative outline which will be developed more fully later.[52]

Phase A—During a period of relative glacial retreat, similar to or more marked than that of the twentieth century, settlers come and clear land and build houses in what appear to them to be safe places, in spite of the big glaciers only about a kilometer away.

Among these sites is the group formed by Le Châtelard, Les Bois, and Bonanay below the Mer de Glace, and La Rosière below the glacier of Argentière. This phase A probably occurred in the "Middle Ages," a vague but useful expression which will be defined more exactly later.

Phase B—This is a period of relative advance, in which the

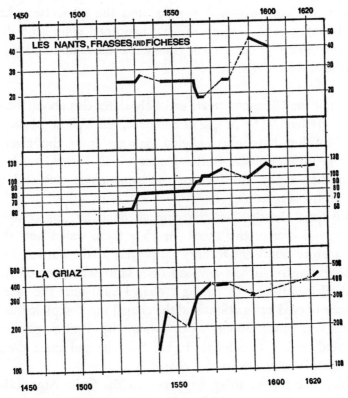

Fig. 19   *Chamonix Tithe Returns (in Florins)*
In the glebes not immediately below the glaciers, and so not directly affected by the advances of the latter between 1580 and 1620. The scale is semilogarithmic. Note the healthiness of these curves compared with those in Fig. 18.

glaciers probably reached positions beyond those of the twentieth century. It is in this period that the "approach" of the glaciers begins. The exact date when the glaciers began to advance in this way is still not exactly determined, but in the case of the Alpine glaciers there is good evidence for placing it between 1546 and 1590.

Phase C—A maximal phase which began between 1600 and 1616. During these fifteen years the villages built several centuries earlier, in phase A, were partly or completely destroyed by the glaciers, which after their final advance occupied positions, like the Côte du Piget and the outskirts of La Rosière, completely unforeseen by the builders of phase A. Phase C in this history of the glaciers of Chamonix—and perhaps also the other two phases—may be taken as an example of what happened more generally, for the maximum that occurred at Chamonix was reproduced throughout the whole of the Alps.

One of the Chamonix texts expressly states this about the other Savoie glaciers: "The people of Samoëns, Sis, Vallon, Montpitton, and Morillon have also suffered damage from the glaciers."[53]

Again in Savoie, the Ruitor glacier which barred, much lower down than today, the lake of the same name was still giving trouble in 1603–1606.

We have already seen the damage done by this complex of glacier and lake in the 1590s. In 1604 there was a new danger, and a procession went to offer up prayers at the barrier. In 1606 a young huntsman had a vision of a saint who ordered him to ward off catastrophe by having a chapel built near the lake. The villagers obeyed instructions and the chapel was completed in 1607. Every year until about 1870, when the Ruitor glacier had retreated so far as no longer to bar the lake and cause danger, a procession was made to the chapel.[54]

The same sort of thing happened in the Swiss Alps. Like the Argentière glacier, the Mont-Durand glacier covered an old road,[55] and an official decree of January 28, 1599, orders a new one to be built.

At Grindelwald the "Cronegg" or local chronicle notes:[56] "In

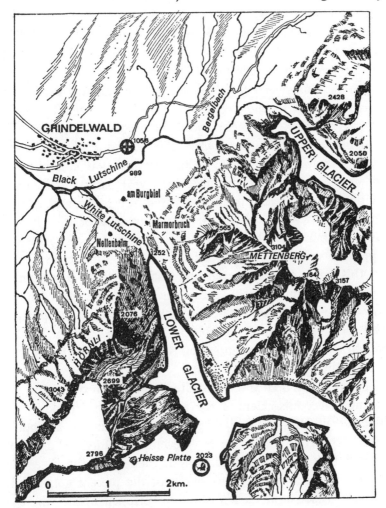

Fig. 20 *The Grindelwald Glaciers*

The contemporary outlines of the terminal tongues are marked with a thick black line. I have also marked as follows:

⊕   Grindelwald church.

X   Site of fossil larch and cembro pine trunks in right lateral moraine of lower glacier.

O   Rock of "hot leaves," or *Heisse Platte.*

(Source: From the Swiss topographical map of 1956—1/35,000.)

the year 1600 the upper glacier advanced near to the lower Bärgel bridge over the Bärgel stream[57]; and two houses and five haylofts had to be evacuated because the glacier covered the places where they stood. The lower glacier went as far as the Burgbül . . . , and the Lischna river left its normal course and was swollen by the glacier . . . The whole community tried to struggle against it, but it was no good, the *kälter* had to be evacuated, together with four houses and many other *kälter*. The flood spread over all the land, eroding and spoiling it." This description is confirmed by a fine drawing which Mérian made forty years later (see Plate 3) showing the lower glacier much larger than it is today, and, nearby, several "houses which had to be moved because of the glacier." In 1603, certain inhabitants of Grindelwald whose possessions had disappeared under the glacier were exempted from taxes.[58]

At this point occurs the old and charming *pons asinorum* of the chapel of Sainte-Pétronille. Like that of the Savoyard hamlets of Le Châtelard and Bonanay, the history of the chapel provides useful topographical and chronological references. W. A. B. Coolidge, an excellent scholar and Alpinist, wrote a little-known pamphlet on the subject.[59]

Coolidge examined and "tabulated" all the texts, maps and eyewitness accounts. Here is the gist of what he found.

The cult of Saint Petronilla, believed to cure fever, was popular in the Alps from at least the eleventh century. The chapel dedicated to her at Grindelwald was at the foot of the Eiger (see map, Fig. 20) on the left bank outlet of the present gorge under the lower glacier, near Nellenbalm (or Petronellenbalm, the cave of Petronilla). There was a bell in the little chapel which was transferred somewhere else and long survived the building it was originally made for. In 1780, Besson claims to have read the date 1440 on it; later authors made it out to be 1044! On June 30, 1520, the council of Berne ordered the "people of Grindelwald" to nominate a "brother," probably a hermit, at Saint-Pétronille. So the chapel still existed in 1520. It was probably destroyed not long after 1534, when the "town and republic" of Berne ordered its citizens to destroy all minor religious buildings in the name of the Reformation.

So far all this has no connection with glaciology. But now comes Hans Rebmann, poet and prior of Thun, near Grindelwald, born 1566, died 1605. He wrote a poem which refers to the glacial thrust at the end of the sixteenth century in very similar terms to those of the "Cronegg." "On the side of the mountain (the Eiger), at Sainte-Pétronille, there was once a chapel, a place of pilgrimage. But a great glacier now hangs there (the lower glacier of Grindelwald), and has entirely covered the site. The houses had to be moved away. As if by miracle, the glacier drove before it earth, trees and houses." This account was later embellished by local tradition, as recorded by Mérian in 1642, Scheuchzer in 1723, and Gruner in 1760. Where Rebmann merely says the glacier occupied the site of the chapel, tradition has it that the glacier actually destroyed it. Gruner even says that on a clear day one was supposed to be able to see the chapel door through the ice! (See Appendix 7.)

This story enables us, as in the case of Chamonix, to distinguish two phases for the lower glacier at Grindelwald: a phase of lesser glacial development which the existence of the chapel places between 1440 and 1520, and a phase of glacial extension, about 1600, when the site (probably already vacated) of the old chapel was occupied by the frontal lobe of the glacier.

After 1600 the Swiss glaciers, like those of Chamonix, kept up their drive. According to the local chronicler, Trümpi, the Glarner glaciers were "very developed" in 1608–1610. In 1620 the Grindelwald glaciers were still very close to their terminal moraines.[60]

There was a similar and concomitant thrust in the eastern Alps (the Tyrol, etc.). The first symptoms are reported in 1595 in connection with the Vermont glacier in the Silvretta mountains. A text of that year says: "This glacier or *Ferner* (*Gletscher oder Ferner*—a Tyrolean dialect word for glacier) becomes longer year by year, and rougher, colder and more biting. Moreover it splits and throws out séracs, and its crevasses can be seen growing wider and wider."[61]

Five years later, in 1600, comes the catastrophe of the Vernagt

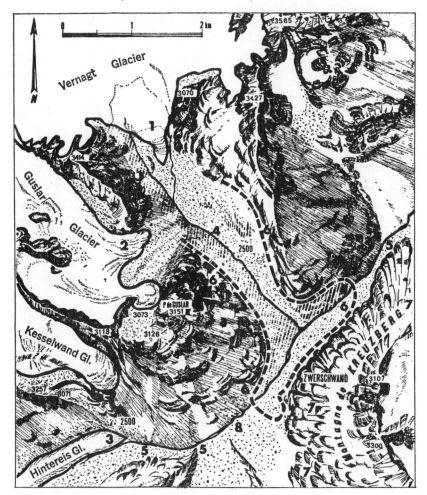

Fig. 21   *Vernagt Glaciers (Vernagtferner) and Rofen Valley*
        1. Vernagt glacier
        2. Guslar glacier, the terminal tongue of which mingled with
that of the Vernagt glacier in periods of advance (1580–1860)
        3. Hintereis glacier
        4. Vernagt torrent
        5. Torrent flowing out of the Hintereis glacier, which lower down
becomes the Rofen
        6. Maximal position of the Vernagt glacier in periods of extreme
advance, as in 1600, etc. (dotted line)
        7. Kreuzberg mountain and Zwerschwand
        8. Site of barrage lake, determined by the advance of the glacier
when it reaches position 6
        (Source: This sketch is based on the map—1/50,000—Otztal
Alpen, Pitztal, Kaumertal, p. 43 [Kompass Verlag and Kartog. Institut
H. Fleischmann, Innsbruk].)

glacier in the Tyrol. Richter, an excellent geographer and scrupulous archivist, wrote a very good study of it, using the chronicle of Kuen, an early map, the papers of a correspondent of the Fuggers, and the numerous contemporary administrative reports.[62]

The Vernagt glacier,[63] the movements of which are of great magnitude, is today several kilometers distant and totally invisible[64] from the Zwerschwand, where the stream flowing out of it joins the narrow valley of the Rofen (Rofenthal). But during periods of maximum advance the glacier reaches as far as the opposite slopes of the Zwerschwand and bars the Rofenthal, thus causing a lake to form (*Eissee*) which tends to break through or overflow onto the land below. This happened frequently, as we shall see, in the seventeenth and eighteenth centuries; the last occasion (before our own time) was in 1848. The first recorded instance was in 1600.

From 1599 on the advance of the glacier was such as to cause disquiet. In that year and in 1600 the glacier reached the Rofen valley, where it formed a barrier, and behind the barrier a lake. On July 20, 1600, as the result of heat and the pressure of the water, the dike of ice burst. Masses of water poured through into the Rofenthal and, beyond, into the Otzthal. Bridges, fields and paths were destroyed; the damage amounted to twenty thousand florins. But the break in the barrier was only temporary. The glacier began to grow again, till in 1601 it was even bigger than the year before. According to a text of 1611, the lake which formed again in 1601 was 625 fathoms (*Klafter*) long, 175 wide, and 60 deep.[65] In 1601, here as in all the northern Tyrol, everyone feared the worst again. But, fortunately, between July and September, the water seeped away through a sort of arch "half a pike-length high" which had formed under the barrier.

Was the Vernagt catastrophe of 1600 the first of its kind for a considerable period? It is very possible. The Fugger text of July 9, 1601, says the glacier has formed a barrier lake where there used to be "fine meadows."[66]

A judge's report from the Tyrolean valley of Schnals in 1601 says "the people have much more difficulty than before in getting in the harvest, for every year more land falls into disuse; and because of the growth of the glacier many fields and meadows are abandoned and spoiled."[67] But in 1602 this glacier began to retreat.

Thus, about 1600, the glacial thrust is universal in the Alps, and such a widespread phenomenon raises specific questions of cause and effect which call for a short digression. The basic question is, to what kind of meteorological conditions, what "climatic complex," was the 1600 glacial advance a reaction? And in the first place, are there any relevant and accurate models, directly applicable to *small* glacial thrusts in the Alps in the twentieth century, which can be projected into the past in such a way as to account for the *big* glacial thrust at the end of the sixteenth century?

The clearest answer to this question has been given by Hoinkes, in a recent article.[68] This Austrian glaciologist has tried to throw light on the meteorological factors which determined the small glacial advances in the Alps in 1890–1895 and 1912–1920. These two recent thrusts were much smaller and briefer than the offensive of 1600, but, like it, they were reproduced all over Switzerland and Savoie. And they can be explained by means of meteorological series.

According to Hoinkes, the first thrust, in 1890–1895, resulted from the predominating influence of a group of cyclonic summers between 1886 and 1890; these summers were damp and cloudy, with relatively low temperatures unfavorable to *ablation*. But the corresponding winters (those of the 1890s) did not have particularly heavy snow. So the *accumulation* factor was not particularly pronounced, and the outcome was a fairly modest expansion after 1890, relatively limited in both space and time.

In comparison, the glacial advance of 1912–1920 seems much more marked. For several years contemporaries even had the impression—though they turned out to be wrong—that the trend toward glacial retreat which had been so apparent since 1860 was about to give place, in 1920, to a new secular advance.

The intensity, brief but undeniable, of the 1912–1920 advance was the result of two factors combined. On the one hand, very snowy winters were abnormally frequent between 1900 and 1920; there was therefore heavy accumulation. On the other hand, the summers around 1905–1910 were cyclonic, cold, and unfavorable to ablation.

In short, the little advance of 1890 is the result of cool summers, and that of 1912–1920 is the result of cool summers plus snowy winters. This gives us something to go by as to the probable causes of the thrust of 1600.

Hoinkes' studies also provide chronological information which may be applied to other periods. The combination of meteorological factors favorable to glacial advance (cool summers plus snowy winters) culminated around 1907–1908, then again around 1917–1918. The consequences, seen in the advance of glacial fronts, were seen in 70 percent of Swiss glaciers from 1912 on (i.e., after a time lag of about five years); and again in 1919–1920 (i.e., after a time lag of two years). By analogy, and on the basis of known facts, it is possible to reconstruct the probable causes of the thrust of 1600. Then too the glacial advance was due both to a series of snowy winters (1580–1600) and to a group of cool summers (1590–1600). But there are essential differences between the periods in question: in 1595–1600 the Alpine glaciers attacked from original positions which, from 1580, because of a cooler secular climate, were much more advanced than those at the beginning of the twentieth century. This explains why the thrust of 1600 was so spectacular and so disastrous.[69]

It is probable that after their exploits of 1600 and 1610 the Alpine glaciers, apart from such minor fluctuations as that noted at Schnals in 1602, remained in fairly advanced positions.

There was another catastrophe at Chamonix some time between 1628 and 1630, which according to a contemporary estimate (probably exaggerated) destroyed "a third of the cultivable land." Three slightly later texts (see Appendix 8) refer to "those falls of snow and glaciers" of 1628–1630, thus dating the episode, but unfortunately giving no other details. The text which is most precise from the chronological point of view dates from 1640, and merely alludes to a great flooding of the Arve, caused by glaciers "twelve years ago," i.e., in 1628.[70] As so often happens, the dates coincide: in 1630 the lake at Ruitor burst through the glacial barrier again—evidence that the glacier there was advancing once more, and much larger than it was in the Twentieth century.[71]

The same thing was going on with the Allalin glacier,[72] the permanent descent of which into the valley of the Saas was always setting up barriers and causing frequent overflows from the Mattmarksee: floods occurred in 1620, 1626, 1629, 1630, and on August 21, 1633. This last is one of the best documented and also one of the most destructive. According to an eyewitness, Gaspard Bérody, a local notary who died in 1646, it destroyed eighteen houses, a bridge, and about six thousand trees, torn up "by the root."[73] The peasants of the Visp valley explained the Allalin series of disasters as the result of a succession of seven cold damp years, around 1627–1633, when the wheat did not ripen in the summer. French and Swiss wine harvest dates[74] confirm this in more detail. During this period the grapes were generally picked rather late: my phenological curve shows no very early year between 1617 and 1629; 1623 and 1624 were fairly early, and the other eleven years were either average, or more often late or very late and grouped in series (especially 1617–1622 and 1625–1629). The years 1632 and 1633 were also late. Repeated late wine harvests mean groups of cool spring-summers, and therefore diminished ablation and bigger glaciers.

In 1626 the bursting of the Mattmark was catastrophic. The valley of the Saas was ravaged, and according to contemporary estimates, probably exaggerated, "half the land in the valley" was washed away and "half the inhabitants" obliged to emigrate to Italy, France, Switzerland and Alsace.

The catastrophe of the Zermatt glacier of Bies or Weisshorn in 1636 is also characteristic: it was repeated in 1726, 1736 (or 1730 or 1737), 1786, 1819, 1848, and 1865. As Richter has shown in a chronological and geographical study, these episodes correspond to a stage of advanced development of the glacier. When the tongue reaches a steep wall of rock which occurs at an altitude of about 2200 or 2400 meters, it ejects huge falls of séracs which not only accumulate at the foot of the slope but also hurtle down into the inhabited valley of Randa, below at a height of 1350 meters.[75] In 1636 the people there thought the whole glacier was coming down on top of them and there were about forty people killed; in 1819 it only seemed as if a large lump of the glacier had broken off.

To this textual evidence on the period 1600–1640 may be added other evidence of different origin. In the summer of 1961, I, a historian, and my friend Jean Corbel, a glaciologist, set out to find and date fossil trees from moraines on the French side of Mont Blanc, as others had already done in the case of Alaska and Switzerland. We worked systematically through all the glaciers descending from Mont Blanc, and Corbel at last found, on the edge of the Tacconaz glacier, tree-trunk remains of *pinus cembro*, stripped of bark and firmly fixed in a high lateral moraine. The trees were probably contemporaneous with the formation of the moraine, since they were all twisted by the pressure of the rocks between which they were caught.[76] (See Appendix 8.)

The radiocarbon dating carried out in 1962 by Professor H. Œchsger in Berne gives the following times for the death of the two samples:

Sample 1 (B349): 330±80 years before the present era. In other words, about 1630.

Sample 2 (B350): 280±80. In other words, about 1680.

Sample 1 is the more firmly embedded in and twisted by the moraine. The tree therefore probably died at about the same time as the formation of that bank of the moraine, which is the highest part of the whole moraine and seems to correspond to a definite maximum of the Tacconaz glacier. This maximum must have occurred around 1630 or 1680; certainly some time between 1550 and 1760.

At all events, it was from their already advanced positions of the 1630s that the Alpine glaciers began, between 1640 and 1643, a new assault which, like that at the beginning of the seventeenth century, submerged the land that was most exposed.

Fig. 22   *Tacconaz Glacier (Terminal Tongue)*
Position in the right lateral moraine of samples 1 and 2 of fossil wood found by J. Corbel. (Source: I.G.N. map of 1949–1952 (1/10,000), Aiguille du Midi.)

As early as March 1640 a "notice to his Royal Highness" from Chamonix recalls the catastrophes of 1628–1630: "a third of the land lost about ten years ago . . . through avalanches, snow, glaciers, and flooding by the Arve and other streams flowing down from the glaciers." The text goes on to say that the threat still remains in 1640: "And the little (land) that remains is menaced with total ruin."[77]

The danger was real, not just invoked to arouse the pity of the authorities when it came to taxation. In 1640 there was a glacial barrier and floods at the Ruitor glacier, which shows this other Mont Blanc glacier was still advancing.

The advance can be seen most clearly at Grindelwald, where for the first time we have clear and precise visual evidence.

It was not the first picture on the subject. Breughel the Elder, Leonardo da Vinci, Conrad Witz, and certain primitives had been interested in mountains, and even, according to Baudelaire[78] and Proust, in glaciers. Between 1585 and 1635—i.e., right in a period of strong glacial development—Josse de Momper (1564–1635) painted a picture of an ambush, *Der Hinterhalt*, in which he shows a glacier advancing into the middle of a green valley.[79] But I have not been able to discover whether the glacier he painted was in Switzerland, the Tyrol, or Val d'Aosta.

In Mérian's engraving of 1640, a generation later, all is clear and precise; everything is placed, named and dated. The engraving is to be found in *Topographie helvétique*, the first edition of which was published in 1642. It shows the lower glacier of Grindelwald, bristling with séracs and reaching as far as the plain, from which it is now at least 600 meters away. The White Luschina flows out of the front, and not far off are the houses which according to the caption "had to be moved to make way for the glacier" (this confirms what the "Cronegg" says about the events of 1600–1601). Travelers and horsemen are depicted admiring the scene. In the background are the lower heights of the Eiger, the Fischerhorn and the Mettenberg.

Mérian's engraving (Plate 3) is an idyllic version of a threatening reality. As the Chamonix documents show, the glacial advance in the Alps in the 1640s was full of danger.

In 1641 the fronts of the Mont Blanc glaciers were close to the villages of Les Bois, Tour, Argentière, and Les Bossons; in 1644 they were there still. In the summer, waterspouts would descend from them every so often: in June 1641 they wiped out an entire hamlet.[80] In 1648 the people of Chamonix requested Hélias de Champrond, auditor at the Chambre des Comptes, to proceed to take note of the "other losses, damage, and floods recently caused in the said parish." They begged he would "consider and see that the said glaciers threaten the said parish with total destruction, having descended from the top of the mountains to close by the houses and lands of the same parish, in course of time overrunning the said lands, which makes the petitioners live every hour in extreme dread for fear they should perish."

An "arbitration report" of May 28, 1642, is even more explicit: "Moreover the said glacier, called the glacier Des Bois . . . advances by over a musket shot every day, even in the month of August, toward the said land, and if it should go on doing so for four years more it is like to destroy the said glebe entirely. We have also heard it said that there are evil spells at work among the said glaciers, and that the people, last Rogation-tide, went in procession to implore God's help to preserve and guarantee them against the said peril. As for the glebe of La Rosière, we have seen that at the village of Le Tour there was an avalanche of snow and ice about the month of January 1642, which carried away two houses and four cows and eight sheep among which there was a girl who came to no harm although she was buried a day and a half under the snow, this we have heard, though all that is left now of the said two houses are the ruined stones, though about two years since we saw them still standing . . . The said village of Le Tour is also much threatened by the glacier called Du Tour out of which comes the said river Arve, which is spreading and advancing over the land; and the said place of La Rosière (is threatened by) the glacier of Argentière which is the biggest of all, and which is greatly advancing, in danger of carrying away the said village, the avalanches which descend and fall from the said glacier drawing nearer each day to the said land and carrying away the cultivable fields and meadows; the people sow only oats and a little barley, which throughout most of the seasons of the year is under snow, so that they do not get one full harvest in three years, and then the grain rots soon after, and only a few poor people eat it, and for sowing they generally have to go and buy other seed, and we observed that the people there are so badly fed they are dark and wretched and seem only half alive."[81]

The glacier of Les Bossons also came crashing down. The "petition" of March 11, 1643, which announces this, places the most dramatic developments between 1641 and 1643: "In the glebe of Montquart *about three weeks since* the glacier of Les Bossons is said to have burst its bounds so impetuously that it carried away a third part of the land of the said village . . . *About two years since,* the glaciers, streams, and torrents flooded, ruined, and spoiled several lands and possessions, houses, barns, and lofts and caused other damage about the said parish." Two months later, in May 1643, the situation was no better. The inhabitants talk of "quantities of property flooded and ruined by the overflowing of the said glacier Des Bossons which on the sixth of the present month of May carried away part of the glebe of Montcard and the ninth day following the same glacier overflowed again, and had it not been for the efforts of the petitioners who hastened thither, the said overflowing would have entirely destroyed the said glebe."

As for the glacier Des Bois, it came down so low toward Les Praz and over the Côte du Piget[82] that people wondered whether it was going to block the course of the Arve! The coadjutor of Geneva, Charles de Sales (nephew of Saint Francis de Sales), was visited on May 29, 1644, by the syndics of Chamonix, who reminded him that "their parish is situated in a mountainous valley, steep and narrow, at the foot of great glaciers which break off and descend on the said place causing such ravages that they are threatened with the entire destruction of their houses and property, and wonder whether this is happening to them by divine permission as punishment for their sins."

The bishop promised to help, and at the beginning of June 1644 led a procession of about three hundred people "to the place called Les Bois above the village where hangs, threatening it with total ruin, a great and terrible glacier come down from the top of the mountain." The bishop ritually blessed it, then went and did the same to "a long glacier near the village called Largentière," then "another horrible glacier above the village called Le Tour," and finally, two days later, "a fourth glacier at Les Bossons."

These and the previous texts show the glaciers to have been much closer to the villages than they are today. The hamlets or localities of Le Tour, Argentière, La Rosière, Le Châtelard, Les Bois, Les Praz, and Les Bossons were literally hemmed in by glacial fronts, from which they are now, as we have said, separated by a good kilometer of woods, moraines, gorges, and rocks very difficult of access. But the procession of 1644 and its bishop, no matter how old or gouty, had no difficulty at all in coming right up close to the glaciers. The most impressive advance was that of the glacier Des Bois, which in 1644 apparently barred the valley of Chamonix itself and threatened to transform it into a lake.

Fortunately, the bishop's blessing worked. The glaciers slowly retired until 1663, as a text of that year observes: "Twenty-five years since, the glaciers descended and did much damage and that called Des Bois came so close to the river Arve that the people, for fear it should block its course and flood all the land by forming a lake or pool over it, called in Monseigneur of Geneva to exorcise the said glaciers, which have since gradually retired: but they have left the land they occupied so barren and burned that neither grass nor anything else has grown there."[83]

As in 1600, the maximum of 1640–1650 seems to have been fairly general in the Alps. At the Allalin glacier, the previous catastrophe was repeated, with the damned-up water accumulating and bursting through the barrier. Four years later the Allalin glacier still barred the Saaser Visp: Pierre du Val d'Abbeville's map of 1644 shows the Mattmarksee in full force and held in check by the dike of the glacier, which is thus in its advanced position.[84] The same thing with the Ruitor glacier, where a barrier lake formed and burst through once in 1640 and again in 1646.[85] The Aletsch glacier, which registered an advance in 1639, had its maximum later than Chamonix, in 1653.

"That year," says an old Jesuit text preserved in the Sitten archives, "the Aletsch glacier, which had been progressing for many years, reached an extraordinary height; it had grown very long and already threatened the fields near the Naterser. In order to avert the imminent disaster, the people of the village of Nater sent messengers to Siders, where the Jesuits had a house; and they asked for counsel, saying they were ready to appease the divine wrath, by doing penance and other good Christian works . . . Finally the Reverend Fathers Charpentier and Thomas went to the people of Nater: for a week they preached profitably to this little flock, then a procession went to the glacier, which it reached after four hours' walk. All the way, bareheaded in the rain, the people took it in turn to sing in unison. Arrived at the fatal spot, the people heard holy mass and a short sermon. A blessing was given in the name of all the saints in Paradise, in order that the snake-shaped glacier should be held in check, and in order to stop it advancing further. The most important exorcisms were used. The terminal point of the glacier was sprinkled with holy water in the name of St. Ignatius. On that very spot, just in front of the glacier, they set up a column bearing his effigy: it looked like an image of Jupiter, ordering an armistice not to his routed troops, but to the hungry glacier itself. And this trust in the merit of the saint did not go unrewarded: Ignatius forced the glacier to be still, so that since then it has ceased advancing. And all this happened in the month of September 1653."[86]

So, between 1644 and 1653, Loyola was invoked at Aletsch, and Charles de Sales was called in at Chamonix, to halt the terrifying progress of the Alpine glaciers. What was the reason for this progress? The main factor was the general climate of the "seventeenth century," during which the glaciers were always large. It was therefore not at all difficult for them to become, from time to time, very large, and dangerous to the land around. But it was also specifically the fault of the 1640s. This decade, characterized by extremely cool and damp summers, late wine harvests and famines, and by a very low rate of ablation, gave a great impulse to the growth of the Alpine glaciers, already voluminous enough. In these circumstances it is not surprising that in both Savoie and Switzerland one of the great maxima of the Little Ice Age occurred around 1644–1653.[87]

In the 1650s there are some signs of withdrawal. But the retreat is still only a moderate one. At Chamonix in 1660 the grass had still not grown again on the farthest moraines just abandoned by the Mer de Glace, and the valley glaciers were still quite close to the fields. Commenting on the abundant harvest of 1660, a witness says, "We observed that the corn sown near the glacier was the finest."[88]

And even this moderate retreat was brief. It was followed by a new advance, announced in 1644 by a deliberation of the syndics of Chamonix on the subject of the glaciers, "which have been seen gradually to withdraw ever since the benediction (by Charles de Sales), but which now begin to augment again."[89] In October 1664 bishop Jean d'Arenthon came and blessed the glaciers again at the syndics' request.[90] Yet again, "On the sixth of August 1669, we visited Chamonix . . . where after having blessed the glaciers we preached and carried out confirmation."[91]

In the same year the valley had another distinguished visitor, René Le Pays, intendant for the salt tax in the Dauphiné. Now and then, on his journey through Savoie, he would write a letter to a lady. Here is one of them:[92] "From Chamonix in Faucigny, 16 May 1669. . . . Madame, I see here three mountains which are just like you yourself . . . five mountains of pure ice from head to foot, whose coldness is unchanging." He goes on to tell the usual glacier stories about looking for crystals, precious stones, and corpses buried in the ice.

As C. E. Engel points out in his recent new edition of this text, the five glaciers in the letter are the five well-known main glaciers of Chamonix: Tour, Argentière, the Mer de Glace, Bossons, and Tacconaz. From this it probably follows that the Mer de Glace was still visible from the valley of the Arve, and therefore still in its classic position of secular advance. This is corroborated by bishop Jean d'Arenthon's visit, also in 1669. But it was only after 1680 that his benedictions produced a slight withdrawal of the glaciers.

There was an interesting event in the Alps a year after these two tests of 1669. In about 1670 the Aletsch glacier bore "on its back," as far as its terminus, a huge blue stone, an enormous block of serpentine measuring 244,000 cubic feet, which in about 1680–1700 it left stranded on its farthest terminal moraines.[93]

The 1670s show no sign of any wane. On the contrary, they bring a new maximum which in some parts of the Alps was perhaps the greatest in modern times. This maximum is clearly visible in the eastern Alps, where it equals and perhaps surpasses the records of 1600–1601. There are also signs of it, though less marked here, in the Mont Blanc area.

In the eastern Alps it is again the Vernagt glacier which provides the most decisive evidence. The documents are abundant, and have been intelligently and scrupulously edited by E. Richter.[94] They consist in the first place of the letters, drawings, and sketch maps of a Capuchin monk sent there to implore divine mercy. Secondly, there are the detailed chronicles, with sketches attached, of Johann Kuen and his son Benedikt, inhabitants of the valley and direct witnesses of what happened. Lastly, there are the official reports of a government commission which made a close study of the episode in an attempt, a vain one, to find a solution. From these three sources one can reconstruct the course of events.

In 1676 the terminal tongue of the glacier began to stretch out into the valley of the Rofen, which by 1891 would be two kilometers away, and by 1951 even farther. In the autumn, according to one source, and according to another at Christmas of 1677, it reached the rocky wall of the Zwerschwand, on the opposite side of the Rofenthal. The valley was blocked, and a vagrant suspected of practicing magic was held responsible and burned. But his potions continued to work. In May 1678 a lake was formed and filled. At half past eight on the evening of May 24 the barrier broke and in four hours the lake was emptied; this time however it caused no damage.

At the end of June 1678 the barrier had closed again and the lake had re-formed. On July 16 the dike broke again, this time with catastrophic consequences.

In 1679 the same sequence was repeated, but with little damage.

On June 14, 1680, there was another burst, and a great deal of damage.

There were repetitions in 1681 and 1682, but the barrier was then about a third lower than in 1678 and the flooding less serious.

Nevertheless, the terminal tongue, in spite of a certain withdrawal, was still much longer than it is today, and went right down into the Rofenthal valley. It did not leave it, and then only temporarily, until 1712,[95] after having been there for thirty years.

Benedikt Kuen made a shrewd and important observation about the glacier's long occupation of the valley. The text dates from 1712–1713. "This is what I was told by Mathias Gerstgrasser, a former priest there, born in Partsschin, that he read in a book [a manuscript?] at the house of a certain Thumbherr, that the glacier began in the thirteenth century [*in dreyzehnten saeculo*] because of a succession of very cold years. I note this for what it is worth; but one thing is certain: it is only in the last century [the seventeenth] that the glacier has become so big."[96]

These lines have a twofold interest. They emphasize the secular uniqueness of the two great thrusts of the Vernagt glacier in the seventeenth century, in 1600 and 1680, which Kuen describes and which were unprecedented in the past. Also, Kuen's reference to a "beginning" or *Anfang* in the thirteenth century suggests the point of departure for a first glacial advance; I shall come back to this.

But let us concentrate for the moment on the thrust of 1680. It is also attested in the eastern Alps[97] in a rather original manner, by lichenometry (dating through known annual growth rate of lichen), which gives a date of 1680 for the most advanced recent moraine of the Fernau glacier in the Tyrol.[98]

At the same time, in the Aostan Alps, the Ruitor glacier also formed a barrier in 1679. It burst shortly before September 25, re-formed, and burst again in 1680, breaking down the bridges of Villeneuve and Equiliva.[99]

As so often, the Saas glaciers parallel the Ruitor glacier. During the same few years, the Mattmarksee, barred by the growing Allalin glacier, mounted right up to the level of the huts on the mountain pastures of Mattmark. In July 1680 the great heat weakened the barrier and the lake broke through and ravaged the valley, as in 1633. The people vowed, to mollify God's wrath, to abstain for forty years from banquets, festivities, balls, and cards.

The chronicler of Zurbrüggen notes caustically: "People are always very clever and provident after the event—after the disaster has happened, after the horse has bolted, after the glacier has broken its bounds.[100]

Antoine Lambien's map of 1682 shows the Allalin crossing and barring the valley of the Saas a long way below its present position. On his scale,[101] the Mattmarksee is about 1200 meters long instead of 500 in 1915, and less today. The 1682 dimensions of the Mattmarksee are those it had in the Little Ice Age; so that about 1680 the phenomena in the Tyrol, the Mont Blanc area, and Swiss Alps were synchronous.

True, the Chamonix texts about 1680 do not say much about the positions of the glaciers. It is likely that the glacier Des Bossons was very large, because contrary to the usual practice it is this glacier which is mentioned in 1679 as marking the limit of the *mas* (plural) of Chamonix (a *mas* was sort of seignorial circumscription or messuage used for the calculating of agricultural dues.)[102] But this gives no more than a reasonable presumption.

Similarly, though the Arve flooded on 28 and 29 December 1680 near Chamonix, the texts give no indication of the role played by the local glaciers.[103]

But there is another kind of information which we have to take into account—demographic information, which the Chamonix archives supply precisely in 1679. As with other local demographies in the seventeenth century, it does not paint a very brilliant picture. But it seems that the population of the upper valley of the Arve was particularly hard hit.[104]

It should be recalled that in general, and in spite of the serious crises in the seventeenth century, populations during the age of Louis XIV were still well above what they were in the same regions during the worst period of the fourteenth and fifteenth centuries, say 1350–1450. Thus for the southern half of France, Baratier has shown that this is so for Alpine Haute-Provence, which is quite near Haute-Savoie. And in Languedoc, under Colbert, the "peoples" were at least twice as numerous as during the sad times of Charles VII and Louis XI. In Holland the demographic increase of the sixteenth century was scarcely improved on, but it was maintained, and that is the important thing. More generally, population remains stationary in the seventeenth century, and often even drops a little. But with the possible exception of Germany during the Thirty Years' War, there is no question of its dropping for any length of time to, or below, the low level of 1348–1450, the time of the Black Death and the English wars.

No question—except at Chamonix. There, incredibly, the usual pattern was reversed, and there seems to have been a demographic situation of a kind hitherto unknown.

Between 1458 and 1680, judging by the number of copyholders and taxpayers—not an absolute but a significant indication—the local population decreased by a half or more. Thus the glebe of Chavents which, taking all its mas together, had 223 taxpayers in 1458, had no more than 80 in 1679. The case of those mas nearest the glacier Des Bois is also very significant: those of Joppers and Landrieux fall from 33 tenants in 1458 to 19 in 1679. Those of Le Gerdil, which includes Le Lavancher and the vulnerable Le Châtelard, fall from 40 or 41 taxpayers in 1458 to 18 in 1559, 18 still in 1679, 21 in 1706, and 22 in 1733. It is to be noted that the fall was already considerable in 1559, which elsewhere is generally considered a period of demographic expansion.

The peak of population in the fifteenth century, in contrast with the troughs in the centuries that followed, is underlined by the fact that in 1471 the bishop of Geneva, Jean-Louis de Savoie, confirmed more than 1070 children from Chamonix,[105] a huge figure which I think it would be difficult to equal there a hundred or two hundred years later.

Of course, this selective depopulation of the high mountain area may well be explained by purely human considerations. Both in periods of economic expansion, like the sixteenth century, and in periods of depression, various factors may induce the population of ultra-marginal areas, as Chamonix was then, to move away, and thus, demographically speaking, to regress. But the case of the mas nearest the Mer de Glace (Gerdil, Joppers, and Landrieux) is a special one: it was not "climate" in general, but the encroachment of huge masses of ice on the inhabited valley which harassed an already severely tried agriculture and an already precarious population. This is shown in the texts already cited from the end of the sixteenth century and from 1630. There is also a heart-rending document from the people of Chamonix in 1730, during a period which as we shall see was one of long glacial advance. The Chamoniards say "their district is surrounded by glaciers which cause a climate and a cold which make themselves felt almost every year, and cause frosts which destroy their harvests, not to mention that the land, being almost vertical, is subject to collapses of snow, torrents and water, falls of ice, and floods and erosion caused by the river Arve and a number of violent streams flowing out of the same glaciers, and such strong winds that they sometimes carry away part of their hay and grain after it has been cut; and the icy air is so sharp that it makes the land arid and barren.[106]

The documents show that the Chamonix glaciers probably advanced in the sixteenth century (after 1540), this advance occurring between a phase of relative withdrawal before 1500 and the maxima of the seventeenth century. In these circumstances it is not surprising that local agriculture and demography were in a much less flourishing condition in the seventeenth than in the fifteenth century.

Both the demography and the economy of the highest valleys nearest the glaciers show a marked decline which seems to be the result of a special set of circumstances. We have already seen how the Chamonix tithes fell in real value between 1550 and 1625; and those of the glebes immediately under the glaciers fell much more than the others.

Also very interesting are the dues paid in respect of the mountain pastures, as reflected in deliveries of cheese to the priory of Chamonix. These fell from 52 pounds (weight) in 1540 to 45–47 pounds in 1550–1577, and 39 pounds in 1622. The mountain pastures of Blaitière (under the glacier of the same name, above the priory of Chamonix) and those of La Pendant (under the glacier of the same name and to the east of Le Lavancher) were particularly hard hit. Each paid 5 to 7 pounds of cheese between 1540 and 1580. They only paid 3 and 2 pounds respectively between 1540 and 1622.[107]

The human decline in this glacier region in the seventeenth century is appreciable even in comparison with the period around 1450, which was one of general depression everywhere. It is as if the economic and demographic renaissance of the sixteenth century had left this area completely untouched. One possible reason for this is clearly the growth of the glaciers, which had such a baneful influence on the most exposed fields and villages.

To return to the movements of the glaciers themselves: the pronounced maximum of 1680 seems to have been followed, as one might expect, by a phase of withdrawal; though this was moderate and much less spectacular than the retreat of our own period.

Some interesting information about this slight withdrawal is to be found in the biography of Jean d'Arenthon, bishop of Geneva from 1660 until his death in 1695: "The inhabitants of a parish called Chamonix showed in a singular fashion the confidence they had in their bishop's blessing. Chamonix has big mountains covered with ice . . . (which) constantly threaten with destruction the surrounding places; and every time the bishop went on a visit there the people would ask him to go and exorcise and bless these mountains of ice. About five years before our bishop's death, the people sent a deputation asking him to go and see them again, for they feared that as he was daily getting older his age might deprive them of that happiness . . . *they said that since his last visit the glaciers had retired more than eighty paces.* The prelate, charmed by their faith, answered, 'Yes, my friends, I will come, even if I have to be carried . . .' And he went, and . . . did as they had asked. I have a sworn statement made by the most important local authorities saying that since the benediction given by Jean d'Arenthon the glaciers have withdrawn an eighth of a league from where they were before, and that they have ceased to cause the havoc they used to do."[108]

This text clearly relies on a report drawn up by the syndics of Chamonix after 1690 and before 1697. The conclusions I draw from it are more cautious and less categorical than those of Mougin. From this fragment of hagiography, together with other data,[109] I merely infer that between 1664 and the 1680s the advance, or simply the proportions of the glaciers, caused the people of Chamonix great anxiety; that probably, in the 1680s, after the last but one visit of the bishop, the exact date of which is not known, they began to withdraw slowly, by about sixty meters; and that this retreat greatly accelerated between 1689–1690 and 1695–1697, until the tongues had eventually retreated by an eighth of a league, or 500 meters.

The Vernagt glacier also beat a slow retreat after 1681. But this was only a tiny episode, in no way comparable with the great retreat at the end of the nineteenth century or that of the twentieth. Kuen states categorically that up till 1712 the front of the glacier reached, if not to the thalweg itself, at least to the narrow, flat, winding groove which constitutes the valley or *thal* of the Rofen (124). Between 1860 and 1960 there were two kilometers and more between the front of the glacier and the Rofenthal.

So the retreat of the Alpine glaciers at the end of the seventeenth and very beginning of the eighteenth centuries was a limited one: 500 meters, instead of the one or two kilometers of the twentieth century. Nor was there any worldwide retreat around 1690–1700 comparable to that of today. So the Alpine withdrawal of 1700 was only a modest and purely regional episode in a long-term universal phase of glacial advance. In other words it was only an oscillation, a little local trough in the great secular high tide.[110]

This can be proved by referring to the Nordic glaciers. These, instead of decreasing as they did in the twentieth century more or less in phase with those of the Alps and the rest of the world, showed from 1695 onwards their first well-attested maximum of modern times.

In the case of Iceland the texts are precise and of one accord: between 1694 and 1698, or at the latest before 1705, the great ice sheets of the island, the Drangajökull in the northwest and above all the enormous Vatnajökull in the southeast, advanced so far that they encircled or destroyed the nearby farms, and ravaged and often covered their land.

To take Vatnajökull first, a land register of 1708–1709 notes of the abandoned farm of Fjall "that fourteen years since, its ruined buildings were still visible, but all that is now under the glacier." Similarly, the farm of Breidarmok "has been entirely deserted for four years, and like several other farms it is still ravaged by flood, moraines, and the annual passage of the glaciers, so that almost all the grass has disappeared, save on a little hill on which the houses stand." (This is from a deposition made before the Thing, June 1, 1702.) "The farm of Skaftafell has the right of summer pasturage on part of the farm of Freynes . . . But this right cannot now be exercised because everything is covered by the glacier." (Land register of 1708–1709.)[111] These texts, especially the second, are quite explicit. It is clear that by 1702 the glacier had almost reached, or was at least threatening, Breidarmok. In 1904 this site was quite free of ice, and almost a kilometer away from the front of the glacier.[112]

The documents concerning Drangajökull were collected by J. Eythorsson in an article published in 1935. The most important is by the learned Arni Magnusson who visited the region in 1710. He notes the farms which have been destroyed by the glacier, among them that of Oldugil, close to the Drangajökull: "It is said that the buildings were destroyed by thrusts from the glacier and floods; and what is now visible of the ruins is right on the edge of the glacier; according to the assertions of people still living, the glacier covered all the land belonging to the farm." In 1935, during the long retreat of the twentieth century, the site of Oldugil was two kilometers away from the glacier.[113]

As at Chamonix, so in Iceland, the advances of modern times contrast with a long period of safety beforehand. The farms of Brennholar and Breidarmok, destroyed by the Vatnajökull about

1700, are mentioned in records dating from before 1200[114]; Oldugil occurs in a document of 1397.[115] To this may be added a slightly different consideration: the plateau of Glama was occupied by the Drangajökull in the seventeenth century, whereas now, and since the end of the nineteenth century, it has been free of ice. But according to the sagas and early texts of which the latest dates from 1394, the plateau was free of ice in the Middle Ages, when troops of horsemen often used to ride across it, though Mercator's map shows it occupied by glaciers again in 1570.[116]

Similarly, in Norway, from 1695 on, the Jotunheim glacier advances across the valley of Abrekke, gradually crushing the forests and the pastures.[117] It is certain that the great glacier systems of the Nordic countries were much more advanced at the beginning of the seventeenth century than they are today.

Liestol's recent work on the Svartisen complex of glaciers (the second largest ice sheet in Norway) confirms all these separate data and makes it possible to fit them into a multi-secular chronology of advance, maximum, and retreat. In 1951 a whole series of fossil tree trunks was found below one of the Svartisen glaciers, on a site left uncovered by the retreat which began in 1940. Radiocarbon dating gives the death of these trees as occurring 350±100 years before the 1950s, i.e., some time between A.D. 1500 and 1700. So it must have been about then that the expanding glacier reached that point—from which it withdrew in 1940. This chronology in no way contradicts that which we have for the Alps.

A descriptive poem of the end of the seventeenth century presents Svartisen glacier as "very close to the sea." This can only refer to the terminal tongues of the glacier, Engabreen and Fondalsbreen, which were in fact quite close to the Holansfjord during the period of secular advance. (In 1805, according to a map and very precise documents, Engabreen was reached by the highest tides, whereas today it is two kilometers farther back.) In 1720 these same glaciers destroyed a farm and harvests. A really strong retreat (minus 2 kilometers) was not registered, measured, and finally translated into graphs[118] until much later —1865 to 1955. So the trend here is the same as in the Alps.

At all events, these various Scandinavian episodes contradict those authors[119] who, following Mougin and Kinzl, and because of the temporary silence of the documents relating to the Alps, have talked in figurative terms and, by way of parallel with the Little Ice Age, of an "interglacial" epoch at the end of the seventeenth century and the beginning of the eighteenth. In fact there is nothing around 1700–1720 to compare with the present worldwide withdrawal.

Even in the Alps, in spite of the modest retreats already referred to, the glaciers were much bigger around 1700 than they are today. Drygalski and Machatschek, in their *Gletscherkunde*, say, unfortunately without giving any references, that the highly sensitive Schwarzenberg glacier, near the Allalin, had a *Hochstand* or advance from 1690 to 1698.[120]

More certain, but still difficult to interpret, is a note I found by the prior of Chamonix, at the bottom of a report on a pastoral visitation: "In the month of July (1700) the glacier Des Bois overflowed and flooded part of the plain and carried off about two or three perches of our land at Le Soubeyron,[121] and came into Le Bouchet, with all the path that belonged to us."[122] Was this only a flood, or did the glacier itself burst its bounds as in 1643? However it may be, we should note that between 1700 and 1850 the Chamonix texts are likely to be less helpful than between 1600 and 1643. The destroyed or threatened areas were no longer cultivated or lived in. The secular advance of the glacier was now regarded as normal and caused no more cries of astonishment and indignation.

In 1703 the Grindelwald glacier also increased, and "covered, according to the documents, the meadows of the prior, which are still entered in the shepherds' books but are in fact buried under the ice."[123]

It may well be objected that all these texts are insufficient: they may suggest either a short or a long advance, but they do not give its limits or provide any fixed points of reference.

But thanks to two pieces of pictorial evidence, such points of reference do exist. The first dates from the end of the seventeenth century—"about 1686," according to Gruner, who published or republished it in 1760. It is a sort of cross between a map and a bird's-eye view of the Grindelwald glaciers (Plate 4). Gruner and Richter both comment on it, and agree that it shows the glaciers very swollen and only about fifty paces less advanced than they are in Mérian's engraving of 1640. Richter rightly adds, however, that the 1686 map is "very imperfect"—but he does not say what its imperfections are.

I have also attempted to analyze this map, which was drawn by someone called Herbord. The first question is, what scale was it drawn on? The scale actually indicated on the map, in *Stunde*, or Swiss leagues of 4.81 kilometers, would in theory give a scale of 1/93,000. But this is completely wild, and may have been added afterwards by the engraver, Zingg. Comparing Herbord with the modern map of 1956, one sees that he used three different scales, in order to get in all the information he wanted to convey.

Scale 1 concerns the left part of the sketch running from west to east along the valley of the Schwarze Lutschine, between the two main glacial fronts, *Ober* and *Unter*; in other words, between the church of Grindelwald on the one hand, and on the other the entry of the valley of the upper glacier into the main valley of the Schwarze Lutschine ("y" on Herbord's map). This is a scale of 1/32,000: i.e., 65 millimeters to 2 kilometers.[124]

Scale 2, the largest, applies to the whole central zone including, from north to south, the church and the whole valley of the lower glacier. Herbord lavished particular attention on this part of his sketch, and, following the tradition which prescribed a certain perspective and vagueness in the distance for this kind of view, used a larger scale of 1/61,000. This can be checked with reference to four points which existed then as now, and enclose the lower glacier: Grindelwald church (A); the point at which the path from the church to the lower glacier crosses the Schwarze Lutschine (B); the confluence of the latter with the Weisse Lutschine, which flows out of the lower glacier (C); and, farther up this glacier, the Heisse Platte rock (D).

EMPIRICAL SCALE USED BY HERBORD

—to represent the distance AB	1/58,000
—to represent the distance AD	1/66,000
—to represent the distance AC	1/59,000
—to represent the distance CD	1/60,000
—Mean	1/61,000

It is clear that these scales for the main and central part of the sketch converge very well, and vary between very narrow limits (1/58,000 and 1/66,000).

Scale 3, which is used for the west or right-hand part of the sketch, is quite different. The artist is not interested in this part of the landscape for its own sake, since it does not contain any glaciers. But in order to place his sketch against the more general background of the region and routes of Oberland as a whole, he crowds in all the country along the Schwartze Lutschine to where it joins the Lauterbrunnen Lutschine at Zwei-Lutschinen ("I" on Herbord's map). This area, 11.25 kilometers long from east to west, is represented by 67 millimeters (!), scarcely more than for the kilometers of the main zone, between the two glaciers. This time the scale is 1/168,000.

It is understandable that A. Zingg, making the engraving seventy years after Herbord did the drawing, was somewhat put out by these changes of scale, so that he put forward an average scale which is really the average of scales 1, 2, and 3: in other words, approximately 1/93,000 (though the true mean would in fact be 1/87,000). But Zingg's average scale is meaningless. Our analysis of the map shows that only scales 1 and 2 are relevant.

As to the glacial fronts, only that of the lower glacier is drawn by Herbord in detail: on the 1956 map this front is now 1770 meters from Grindelwald church. If we measure the distance as shown in 1686, using either scale 1 or scale 2, we find that in each case the front of the glacier is more advanced than it is today. Using scale 2, which is the more probable one, we find the distance between the church and the front of the glacier to be 1070 meters; using scale 1560 meters. In fact scale 2 is certainly the closer to reality. So one may reasonably conclude from the map of 1686:

1. That the lower glacier was more advanced then than it is today, and so nearer to the church and village of Grindelwald.

2. That it was most probably about 700 meters beyond its present front.

For the same year, 1686, we also have the firsthand evidence already referred to on the state of the Froidière of Chaux in the Jura. We have seen that the behavior of the Froidière in the 1680s was also characteristic of the conditions generally prevailing in the Little Ice Age.

Five years later, in 1691, an important but little-known document throws some light on the conditions then prevailing on what is now the Italian side of the northern Alps. In that year Philibert Amédée Arnod, a judge in the Val d'Aosta, wrote his "Account of the Passes and Cols of the Alps," preserved in the state archives in Turin. As well as being a magistrate, the author was a military engineer and a climber, and visited all the high fortifications which protected the mountain frontiers of Savoy. In 1689 he attempted to verify a popular tradition (perhaps dating back before the Little Ice Age to the medieval optimum) according to which there was an old pass, once free of glaciers, leading from Courmayeur to Chamonix over the top of the mountain and the Vallée Blanche. The tradition must have been greatly exaggerated, and in any case the pass could not be found. Arnod and his team, with climbing irons on their feet and hooks and axes in their hands, soon had to turn back.

But Arnod has left us a description of another, easier journey which took him to the Val Veni, coming down from the Vallée Blanche and Lake Combal. Leaving behind him on his left the huge moraines of the Miage glacier, of which he gives an accurate description, he entered the Val Veni: "Below this place, one comes on a great plain of meadows called Veyni and Fresnoz (Freiney), bordered on the left by great inaccessible mountains (l'Aiguille noire de Peuterey) *and ending in the La Brenva glaciers;* and to the right (these meadows) are bordered by great forests of black wood and end in the forest of the gran plana, and finally in a very narrow pass called the Cross of Le Beriex." (This latter became, in the nineteenth century, the well-known chapel of Notre-Dame de la Guérison.)

Now, as L. Vaccarone pointed out in 1881, Arnod's description implies a typical situation of advanced development in the great glacier of La Brenva; what we should nowadays call a situation typical of the Little Ice Age. In such circumstances the Brenva glacier descends into the Val Veni[125] just as the Mer de Glace descends into the Chamonix valley, or the Rhône glacier descends into the vale of Gletsch. So it is only to be expected that, descending the Val Veni in 1691, Arnod should see in the distance, barring the valley to the northwest, the white mass of the Brenva glacier. He would have seen it doing the same, and even blocking the course of the Doire, in about 1818, 1842, or 1851. But thirty years after that, in 1881, or in 1969, if Arnod had made the same journey, all he would have seen if he looked in that direction would have been the great lateral moraine, abandoned in the glacier's retreat, the glacier itself so shrunk and dwindled that it was no longer visible from the southwest.

In 1881 or 1969 Arnod would have had to go almost as far as the chapel of Notre-Dame de la Guérison, right out of the meadows of the Val Veni, in order to see at last, gripped by Le Peuterey and the rocks of Brenva, and hidden behind a kilometer of rocks and frontal moraines, the white mass of the retreating glacier, withdrawn to a position a kilometer behind its advanced positions of 1691 and 1818–1851.

Similarly with Arnod's account of what he calls the Praborna glacier, now the Gorner glacier, near Zermatt. In 1691, according to Arnod, this glacier was "an hour's walk" from the village of Praborna (the Italian name for Zermatt). Perhaps it was only one hour in 1691, when the front of the glacier reached almost as far as Furi, where it still was around 1830 according to engravings of the Romantic age. But it was two hours' walk away in 1969, as I can testify.

At the beginning of the eighteenth century, in 1705, new data present themselves which are at once pictorial and descriptive.

On August 11, 1705,[126] the Swiss naturalist Johan Scheuchzer, perhaps without knowing it, carried out Sebastian Münster's journey in reverse. "Moved by curiosity," he came down from the Furka pass to look at the Rhône glacier. His account, and the accompanying engraving to which he refers explicitly, and the slightly later engraving made by the painter Felix Meyer between 1710 and 1720, give a fairly clear idea of the *Rhonegletscher* at the beginning of the eighteenth century.

"Descending from La Furka," writes Scheuchzer, "we soon saw to the left a mass of ice, but it was small in comparison with the lower mass which we shall shortly describe. From this first mass there issued a permanent watercourse, with which the little streams from the surrounding mountain slopes mingled. These waters, flowing westward, joined, after half an hour, beneath a cave, under the mountain of ice,[127] with other glacial waters which flow abundantly: and so these waters become the true source of the Rhône. What happens is that under these mountains of ice there assemble the waters liquefied (from ice) by the heat which emanates both from above and from below the earth; and this gathering of the waters in fact occurs in two places; but soon they flow into a single bed, as can be seen in Plate XII [Plate 9 in this volume]." And Scheuchzer adds that one may also talk of a fork (an allusion to the neighboring pass of La Furka) when one sees this "two-pronged first source" of the Rhône (*Rhodani ipsius primam scaturiginem bicornem*).

The text becomes clear as soon as it is compared with the engraved sketch which illustrates it. The text mentions two masses of ice: one is a little glacier, visible on the left as one descends from La Furka, and which can only be the Muttgletscher; the other is the Rhonegletscher, Mercanton's *pecten*, which appears clearly in the engravings of 1705 and 1720. This *pecten*, spread out over the vale of Gletsch, characterizes the whole phase of multisecular advance. It did not disappear (before our own time) until after 1874, or more exactly between 1874 and 1899. Between these two dates the glacier withdrew from the vale of Gletsch and no longer barred the lateral valley of the Muttbach. It shrank back in a narrow tongue onto the rocky verrou dominated to the east by the Belvedere and itself looking down on the vale of Gletsch.

But in 1705–1720 the Rhonegletscher, as drawn by Scheuchzer and Meyer (Plates 9 and 10), is the same huge and magnificent "overturned cockleshell" as appears in the drawings of Besson (1777), Bourrit (1778), Escher (1794), Lardy (1817), Meuron (1825), and Bantli and Ritz (1834)[128]; in Hogard's watercolor (1848); in Dolfuss-Ausset's daguerreotype (1849); and in the photographs of Charnaux and Nicolas (1874)—in short, in all the early pictorial evidence, as distinct from that of our own time, which from 1899 to 1965 shows the Rhonegletscher shrunk back onto its verrou upstream.[129] (See Figs. 12A to 12K, Plates 9 to 12.)

In this connection Scheuchzer's text and his and Meyer's drawings provide information of the first importance. All three indicate the presence of streams which are visible to the left as one comes down from La Furka, and which flow westward from the mountain slopes, mainly from the "mini-glacier" of Mutt (Muttgletscher), at the entrance of the Val de Muttbach. Scheuchzer and the authors of the 1705–1710 drawings refer to the Muttgletscher rather grandly as a "little mountain of ice."

These streams form the Muttbach torrent, the behavior of which, as Mercantor has shown, reflects the position of the Rhône glacier. There are two possibilities:

1. If the Rhonegletscher is not advanced and is sufficiently far back, as it has been from 1899 to the present, the Muttbach flows directly into the beginnings of the Rhône, without any hindrance from the glacier.

2. If the Rhonegletscher is sufficiently advanced, as in the eighteenth century and the nineteenth up till 1898,[130] its terminal tongue bars the entry of the Muttbach into the vale of Gletsch. In which case:

a. Either the waters of the Muttbach penetrate through the glacier and emerge a little lower down as a distinct stream from under the front itself. If this happens, it is as Rey described it in 1835: "The waters that have come down from La Furka pass under the glacier.[131]

b. Or, a second possibility which does not necessarily exclude the first is that the Muttbach flows along beside the glacier and joins the real sub-glacial stream below the glacial front.

In the case of both a and b of situation 2, the stream flowing out of the Rhône glacier is, as Scheuchzer said, two-pronged. So it is confirmed by the 1705 text and the accompanying engraving, the engraving of 1720, and all the succeeding pictorial evidence, particularly Mercanton's scientific maps: the Rhonegletscher was in the second position at the beginning of the seventeenth century and up to the end of the nineteenth. It was in a phase of long advance, forming a *pecten* and barring the thalweg of the Muttbach, which either flowed round the glacier or under it to emerge as a separate stream. This stream finally joined the sub-glacial torrent to form "the source of the Rhône itself," at first a forked and then a single stream.

So at the end of the age of Louis XIV the glaciers were still very big, much bigger than they are today; and in the years that followed they were to become aggressive again. In 1716 the people of Chamonix begin to complain once more: a wildly spelt petition of that year says that "the parish is becoming more and more neglected because of the glaciers which advance over the land . . . causing great floods when they void their lake, and there are even some villages in great danger of perishing, which obliges the poor petitioners to appeal for help from Your Majesty's common fund . . ."[132]

In the autumn of 1716 there was also a strong thrust by the Gurgler glacier, in the Otzthal region. This system of glaciers, which bars the lateral valley of Lang, normally caused a lake to form, which became very small, and even disappeared during the summer, in the recent period of long withdrawal which the glacier has been going through since the end of the nineteenth century. But in the early period of advance the barrier of ice was much thicker than it is today, and the lake was bigger and might be there all the year round. At certain times it became dangerous.

Thus, at the end of autumn 1716 and in spring 1717, the Gurgler glacier "keeps advancing lower and lower,"[133] and the barrier lake grew to be a thousand paces long and five hundred wide. The people's fears grew likewise, and in 1718 there was a procession to the glacier. A supplicatory mass was celebrated on its left bank, on the stone table, a rocky platform where a hundred and sixty years later Richter was to see the date "1718" still carved. The lake continued to threaten until 1724, after which the documents say no more about it for half a century.[134]

It was also in 1717 that the catastrophe of Le Pré-du-Bar occurred in the Val d'Aosta. On September 12 the village of that name was buried and destroyed in a landslide so sudden it was said the birds did not have time to fly away and were killed in their nests. But what was the cause? The collapse of the Triolet glacier? Or just a fall of rocks or of part of the mountainside? Virgilio, after a close study of the original account by Saussure, who interviewed an eyewitness, concluded that the disaster was caused by the glacier.[135] The original texts support this view.

At Grindelwald, at all events, the advance at the end of the second decade of the eighteenth century is perfectly clear, and is corroborated by several pieces of evidence. In 1719 public prayers were ordered in the parish for the withdrawal of the glacier.[136] According to the texts these prayers were answered. But in 1720 the upper Grindelwald glacier was still very advanced: in 1802, when the glaciers were still much bigger than they are today, travelers and tourists were shown the extreme point reached by this glacier in 1720.

Felix Meyer's drawing, done in about 1719, confirms this: it shows the two Grindelwald glaciers, especially the lower one, completely filling their respective valleys and contiguous with the main valley, which is at right angles to them and contains the thalweg of the Schwarze Lutschine. So the lower glacier was then much more developed than it is today: its position is comparable to what it was in 1640,[137] as shown in the Merian engraving. Both pictures show the same details, the same trees, rocks, and houses on the left.

To complete this panorama of the Alpine glaciers, in 1719 the Mattmarksee burst its barrier, which indicates that the Allalin glacier had advanced.[138]

So about 1720 there was a maximum in the Alps, against a background of constant advance which had persisted for decades in spite of minor negative oscillations.

There was also a maximum in Norway. I have already mentioned the exploits of the Engabreen, in the Svartisen group, which buried a farm in that year. But the Svartisen was not the only group involved. It was also in 1720 that the Jotunheim glacier, having crossed and filled all the hanging valley of Abrekke, ran over onto the slopes of the main valley which the Abrekke joined. The farms of Tungoyane and Abrekke, the pastures of which had already been overrun, were threatened.[139]

In spite of a slight withdrawal after 1720, the Alpine glaciers maintained very advanced positions during the decade that followed. In this respect the 1730s are noteworthy. It was not a period of maximum; there were no complaints of incursions by the glaciers in the Alpine valleys; it was a sort of "normal" or "routine" period. This makes the Sardinian survey, made between 1728 and 1732, all the more interesting. This extensive document covers all the chief glaciers of Savoie, and shows them to be slightly smaller than at the maxima of the seventeenth century or of 1720, but much more developed than in 1910–1920, and therefore even more so than today.

The survey of 1730, known as the "old map,"[140] approaches the problems of the limits of the glaciers in two ways:

1. The first way, which is scientifically the better of the two, is to indicate the position of the glacial front itself (a), the position of the subglacial torrent (b), the point at which the latter emerges (c), and the position of the *graviers* or *glières* (d). The latter are the recent moraines over which trees have not yet grown, and they mark the most recent maximum extension of the glacier prior to the drawing up of the map (probably the extension of 1720).[141]

2. The second approach usually restricts itself to giving information of types (b), (c), and (d): torrent, outflow point, and recent moraines. Type (c) is of course very important, because the out-

flow point of the subglacial torrent coincides with the most advanced point of the glacial front, or at least a very significant part of the front.

But the first approach is the most valuable for the historian-cum-glaciologist, and it is this approach which was used in the part of the survey which concerns the glaciers of Le Tour, Argentière, Les Bois, Les Bossons, and of the sources of the Isère:

POSITION OF GLACIAL FRONTS IN 1730

	Compared to most recent maximum before 1730	*Compared to 1911*	*References:* MOUGIN, 1912
Glacier of Le Tour	−474 m	plus 700 m	
Glacier of Argentière	−257 m	plus 675 m	
Glacier Des Bois	−150 m	plus 1330 m	
Glacier Des Bossons		plus 25 m	
		Compared to 1918	MOUGIN, 1925 pp. 53–55
Glacier of sources of Isère		plus 810 m	

The second approach, by which the position of the fronts is derived from the outflow point of the subglacial torrent, gives the following table:

POSITION OF GLACIAL FRONTS IN 1730

	Compared to most recent maximum before 1730	*Compared to reference year in twentieth century*	*References:* MOUGIN
Gl. of Bionnassay		(1911) plus 125 m	1912
Gl. of Gébroulaz	−320 m	(1911) plus 1514 m	1912
Gl. of sources of Arc		(1905) plus 580 m	1927, pp. 140–41
Gl. of Arnès		(1919) plus 572 m	1925, p. 63
Gl. of Tré-la-Tête		(1919) plus over 1 km	1925, p. 29

Two tables covering eleven major glaciers give ten quantified parallels, the affirmation of ten trends. The 1730 map leaves no doubt at all that the glaciers of the age of Louis XV were clearly and significantly bigger than in the twentieth century. Nor was it a question, in 1730, of a temporary state of violent paroxysm: in contrast with the long retreat of the twentieth century, it was simply a long, steady state of expansion attested since the second half of the sixteenth century, with, standing out against it, the various paroxysms of 1600–1610, 1628, 1640–1650, 1676–1680, and 1710–1720.

Further evidence of this permanent state of expansion, already established around 1730, is given by events in two other places far from Chamonix but still in the Alps: the Mattmarksee broke through its barrier in 1724 and 1733; and in 1736 the Randa glacier first overhung and then collapsed.[142] More evidence still, this time from outside the Alps, comes from an Icelandic map of 1732 which shows the Vatnajökull crushing the ruins of the farm of Breidarmok, which it had reached in its great advance at the beginning of the century.[143] So here again there was not much sign of regression in 1730; the state of expansion persisted.

A decade went by. In 1741, Windham,[144] the first English nobleman and tourist to visit Chamonix, went to see the Mer de Glace, which the guides assured him got bigger every year. Windham's description of the Chamonix valley and Le Montenvers is also very significant. After referring to Chamonix as a village with a priory on the banks of the Arve, he tells how he and his companion saw from the village "the ends of the glaciers which extended into the valley," and which from where they saw them "looked like white rocks or rather enormous blocks of ice." Windham then goes on to give exact information about these "ends of the glaciers." He climbed Le Montenvers and described what he saw from it: "The glacier consists of three large valleys forming a Y, the stem of which goes as far as the Val d'Aosta and the two arms come as far as the valley of Chamonix. The place to which we climbed was between these two arms, and we had a full view of one of the valleys which formed one of them." He adds that this valley, and the glacier it contains, runs almost from north to south.

Windham thus recognized the Y of which the point of juncture is the Vallée Blanche; the upright the Brenva glacier; and the two arms the glacier Des Bossons and the Mer de Glace, that part of the latter which is visible from Le Montenvers running in fact from north to south. His reference to the "ends" of the glaciers, the two "arms" coming into the valley which are visible from the village of Chamonix, show that not only the glacier Des Bossons, as today, was visible from the priory, but also the Mer de Glace itself, or its terminal tongue. This is the classical advanced position observed between 1580 to about 1870, as distinct from the present situation, when only one of the "arms," the glacier Des Bossons, extends to and is visible from the Chamonix valley.

There is immediate confirmation of Windham in the succeeding year, 1742, when Pierre Martel[145] also followed the Arve up to the source of the Arveyron, which flows out of the glacier Des Bois. Martel writes: "Of the five gorges opening into the Chamonix valley, the one called the glacier Des Bois[146] is the most important, not only for beauty and size, but also because it contains the source of the Arveyron. It issues from beneath the ice through two icy caves, like the crystal grottoes where fairies are supposed to live . . . The irregularities of the roof, over eighty feet high, make a marvelous sight . . . You can walk underneath, but there is danger from the fragments of ice which sometimes fall off." Allowing for the fact that the shape of the cave often changed, Martel's description is very like the one given by Saussure some time later, at least as far as the dimensions are concerned (twenty to thirty meters high). Saussure saw the cave several times, notably in 1777, "by a forest of larches with a floor of beautiful white sand patched with willow herb." He describes it as "like a deep cavern entered through an icy vault over a hundred feet high, and of proportionate width. This cavern is carved by the hand of nature out of an enormous rock of ice, which because of the play of light seems in some place white and opaque, in others transparent and green as aquamarine." He goes on to say how its shape changes, and how in winter it collapses. "Sometimes it disappears completely, but it soon forms again. Martel and Saussure's descriptions are illustrated and corroborated by Bourrit and Bacler d'Albe's drawings.[147]

But the Arveyron cave is a sign as well as a spectacle: it reflects the state of expansion of the Mer de Glace when it goes beyond the rocks of Les Mottets and spreads out into the present thalweg of the Arveyron, upstream from the hamlet of Les Bois. One after the other Mougin, Ferrand, and Payot[148] have regretted that the present withdrawal of the glacier, over the last hundred years, has caused the marvelous cave to disappear. Since the end of the nineteenth century the shrunken Mer de Glace has ended prosaically, not in a fairy cavern, but in a mere "thin and puny tongue crawling over the rock of Les Mottets." In 1912 Ferrand also noted that "the cave at the source of the Arveyron, so oft described . . . has disappeared as a result of the withdrawal of the glaciers." Payot, the excellent historian of Chamonix, gives an exact date. According to him the last mention of the cave was in 1873: because of the "retreat" since that date, "the terminal tongue of the glacier (Des Bois) has hidden itself behind the rocks (Les Mottets) and been reduced to a blue strip, above the torrents, between two cliffs worn smooth by the ice."

In short, the disappearance of the Arveyron cave around 1880, after about twenty years (1860–1878) or rapid retreat, is one of the many signs marking the transition from the long phase of glacial advance to the present phase of glacial ebb. The fact that the cave was there in 1742, on the other hand, is one sign that the 1740s were right in the period of glacial expansion.

In 1742 Martel made both a map and a sketch of the Chamonix area. Both are simplifications, but their evidence chimes with the texts of Martel and Windham in 1741–1742.

According to the very rough map, the glacier Des Bossons comes a long way down into the plain, and the terminal point of the glacier Des Bois is situated 2.25 kilometers from the confluence of the Arve and the Arveyron, as against 3.5 kilometers in 1960.

The sketch is equally explicit: though the drawing itself is undistinguished, it has sound points of reference (Le Dru, Le Montenvers, and the church of the priory), and shows the Mer de

Glace definitely overrunning the rocks of Les Mottets. Its terminal tongue is visible from the priory church, beyond which Martel stood to make his sketch.

So the evidence from Windham and Martel converges to show that in 1741–1742 the terminus of the Mer de Glace still came down "almost as far as the plain," and both its positions and its configuration were characteristic of the phase of glacial expansion.

During these same years, 1741–1743, Altmann and Richter found the Grindelwald and Unteraar glaciers also in a state of advance.[149] In 1748, however, the lower Grindelwald glacier made a sudden intense withdrawal as far as the Marmorbruch, 500 meters below its present front.[150]

In and about the 1740s the traditional indicators—the Vernagt, Ruitor, and Allalin glaciers—continue to exhibit the symptoms of expansion. About 1748 the Vernagt approached dangerously close to the Rofenthal, though its advance was not sufficiently marked to produce a barrier lake.[151] The Mattmarksee did flood, however, in 1740, 1752, and 1755.[152] The glacier-lake complex at Ruitor caused trouble in 1748 and 1751, when bursting point was reached and passed several times. There is specific mention in 1751 of one "rapid and unforeseen rupture of the Ruitor lake." The engineer Carelli, who was called in as an expert, left a sketch showing how the glacier, when in a state of expansion, blocked all issue from the lake, which rushed through in force when the summer heat melted a subglacial channel through the barrier. A text of 1752 says, "Last year and in other recent years the flooding of the lake caused terrible damage."[153]

In general, about 1740–1750 the glaciers seem to have been advancing and "in majesty" in all the regions of the northern hemisphere where their limits and activities can be dated. It is difficult, in historical times, to find another period when the positions of the glaciers were so universally different from what they are today.

One example: in 1742–1745 in Norway there was a maximum[154] which has never been surpassed since, even during the first half of the nineteenth century. It caused the glacial fronts to advance several kilometers beyond the limits reached in the twentieth century,[155] and from 1740 on it buried the farms which had been spared by the lesser advances of 1695–1720.[156] About 1740 the farm of Tungoyane, which in 1667 supported two farmers, thirty-eight head of cattle and three horses, and produced 110 bushels of wheat, was completely destroyed. The farm at Abrekke was badly damaged.

Maps and texts both show a maximum in Iceland too, which this time synchronizes perfectly with the maxima in Norway and the Alps. The farm of Lonholl was destroyed by the Drangajökull in 1741, and according to what the farmer at Mofellstadir told the traveler Eggert Olaffson in 1752, the same glacier approached the buildings there during the 1740s. As for the Vatnajökull, the second of the two big Icelandic glaciers, a map of 1732 and a text of 1746 show it still close to the farm of Breidarmok, which it had reached and ravaged at the end of the seventeenth century.[157]

Two recent studies establish a similar chronology in Alaska. The oldest trees on the extreme moraine of the Lemon Creek glacier were 189 years old in 1958. Allowing for the time, which has been measured experimentally, that it takes for trees to start growing on a newly formed moraine, this moraine must have been laid down, according to Heusser and Marcus, in 1759.[158] There is a close correspondence here both in the long and in the short term with the great eighteenth-century glacial thrusts in the Nordic countries and the Alps.

The decade 1750–1760 shows only slight changes, and the long state of expansion continues.

Thus after its comparative retreat in 1748 the lower Grindelwald glacier expanded a little—*"Seither ist er wenig angewachsen,"* said Gruner, who visited the site around 1756.[159]

The Mattmarksee overflowed in 1752 and 1755; such floods, as we have seen, were always a sign of glacial expansion.[160]

Finally, just before 1760, A. Zingg drew and engraved the Mer de Glace, seen from a point in the Arve valley downstream from the church of Chamonix. Zingg's may be considered the first real picture of the glacier Des Bois—Martel's sketch of 1742, though useful for its points of reference, can hardly be called a drawing. I hasten to add that Zingg's is not very distinguished either, but as well as giving, like Martel's, sound points of reference (Chamonix church, Le Dru, the road to Vallorsine), it clearly shows the Mer de Glace in its classic position for the Little Ice Age. Zingg shows the glacier descending almost to the head of the Arve valley, from the floor of which it is perfectly visible.[161]

In Iceland, in 1754, the Drangajökull still covered the meadows where cattle grazed twenty years before, and near to the fjord of Tharalatur one of its terminal tongues was observed over a kilometer beyond its limit of 1935.[162]

The year 1760 is important in the history of the Alps for it was then that Horace-Bénédict de Saussure visited the Chamonix valley for the first time. He returned in 1761 and 1764, and soon afterward described it in his *Voyages dans les Alpes*.

Coming from Servoz on his way to the priory of Chamonix, Saussure[163] entered the Chamonix valley from Les Montées, through a small, rough pass facing south-south-east above the Arve: "Leaving this wild and narrow pass one turns left to enter the valley of Chamonix . . . The upper part of the valley is covered with meadows, through which the road passes . . . One comes in turn on each of the glaciers which descend from the valley. At first the only one to be seen is that of Taconay (Tacconaz), which almost overhangs the steep slope of a little ravine whose lower end it occupies. But soon the eyes fall on the glacier Des Buissons (Bossons). Its ice is of dazzling whiteness, heaped up in the form of pyramids, and creates an astonishing effect among the pine forests which it moves through and overtops. Lastly one sees the great glacier Des Bois which as it descends curves toward the valley of Chamonix: its walls of ice can be seen towering from the steep brown rocks."

I shall not dwell here on such details as that of the glacier Des Bossons overtopping the forest or the Tacconaz occupying the lower end of its moraine or ravine. The main thing is that for anyone who makes the same journey today, even by car, from Les Houches to Chamonix, the difference is immediately striking. Saussure saw what we have never seen and perhaps never shall see: the Mer de Glace in all its glory, curving toward the Arve valley and plainly visible to the traveler coming from the south even before he reaches Chamonix. Thus Saussure saw it, and thus Linck painted it some time about 1800 to 1820 (Plate 15). To-day the same portion of landscape—the thalweg of the Arveyron from Les Mottets to Les Bois—does not present a particle of glacier visible from below.

The rest of Saussure's itinerary and text leave no room for doubt. The next chapter, which covers his journey from the priory to Vallorsine, shows him going up the Arve beyond Chamonix toward Les Praz, Les Bois, and Les Tines. "Half a league from the priory one crosses the Arve by a wooden bridge and comes to the hamlet of Les Prés (Les Praz), where my old guide, Pierre Simon, lives . . . ; a quarter of a league from there we left on our right the lower part of the glacier Des Bois, which ends in a great arch of ice out of which the Arveyron flows . . ." Thus the approach to the glacier Des Bois is perfectly visible, easy and near; its terminal point is almost within reach at Les Praz.[164] Saussure goes on to give directions which would now be quite inappropriate. He advises against going via Le Montenvers because the way is too rough, and suggests going direct from Chamonix to the Mer de Glace by a path across the plain and bordering the Arveyron: "The journey thence from the prior is a charming one of just under an hour, all on the level, which can even be made in a carriage, through fine meadows and a superb forest."[165]

Saussure's career as climber, traveler, and writer lasted a good thirty years, and during all this time he took it for granted that the Mer de Glace was level with Chamonix and only a drive away. But what coach nowadays would dare to attempt the rocky gorges of Les Mottets, the last refuge of the retreating glacier?

Similarly, the icy cavern, the "Arveyron cave," existed in 1760, 1764, 1780, and 1790, as is shown in pre-Romantic descriptions and drawings (Appendix 10).

As for the Argentière glacier, Saussure saw it "zigzag right down into the valley"—a picture that is also given by the first daguerreotypes in 1850. The present retreat has only left behind the upper branch of the zigzag.[166] (See Plates 21 and 22.)

The same idea recurs throughout the years 1760–1790, of the *normality* of the long-term glacial expansion, in comparison with which the long-term withdrawal of the Middle Ages was only a vague memory among the peasants. This old memory going back to before 1600 was very tenacious. Gruner in 1760 and Bourrit in 1780 both found traces of it among many mountain communities in Switzerland and Savoie.[167] (See Appendix 7.)

Pictorial evidence, more and more copious after 1760, confirms this idea of the normality of the long expansion. Mougin has analyzed the data given by the watercolors and engravings, often very clear, by Bourrit, Bacler d'Albe, Hackert, and Chrétien de Méchel. They represent Argentière, the Mer de Glace, and Les Bossons between 1770 and 1790, and all show the glacial fronts much more advanced than in Mougin's time (1912), thus even more advanced in comparison with today. The difference between then and now is from 700 to 1200 meters in the eighteenth century's favor.[168]

Once again it is against a background of persistent glacial advance that the secondary episode of the Alpine maximum of 1770–1780 has to be considered.

The maximum was clearly marked at Chamonix: in 1774 the glacier Des Bossons overran the fields. In his visit there in 1778 (recorded in 1784), Saussure noted the greatest advance he had seen in the glacier Des Bois. According to Hackert's very accurate watercolor, the coordinates of which Mougin established on the spot, the front of the Argentière glacier was 1700 m beyond its position in 1911.[169]

The retreat which followed, and for which there is clear evidence at Chamonix between 1784 and 1790, is also merely a secondary episode, involving at the most a withdrawal of 200 to 300 meters measured horizontally[170]: in 1784, after five or six years of withdrawal, the Mer de Glace was still five hundred paces as the crow flies from the moraine that bordered the road from the priory to

Les Tines—in other words, according to a text and a map by Saussure, 800 meters beyond its present position.

Near Chamonix and in all the rest of the Alps also, a state of expansion is reported between 1760 and 1790.[171] The glaciers mentioned are those of the Alleé Blanche, Hufi (1760), Tacconaz, Gléret (near the St. Bernard), Grindelwald, Vernagt, Gurgler, Allalin, Le Triolet, Giétroz, Macugnana, Tschingel in the Gasterntal, and Biès.

The usual indicators are faithful as ever: the Vernagt glacier, observed and measured by the Jesuit Father Walcher, professor of mechanics at Vienna University, set up a strong and lasting barrier across the Rofenthal in 1770–1772. The Mattmarksee first formed, then flooded, in 1764, 1766, and 1772. In 1774 it was still very swollen and menacing, and the local people demanded public prayers to obtain God's mercy.[172] The Biès glacier collapsed on Randa in 1786. As in 1595, the Giétroz glacier barred the thalweg of the Dranse. The Grindelwald glaciers descended toward the valley of the Schwarze Lutschine (1768–1777); and in 1777 Besson saw "no sign of any wall" below them, which means they had overrun their previous maxima, marked by the terminal moraines. The expansion of the Grindelwald glaciers was so great in the 1770s that the parishioners called in a monk to exorcise them; but the monk, perhaps a Jansenist, not knowing whether the advance was called by God or the Devil, declined to act.[173]

Reports of expansion are very numerous during these three decades, partly because interest in the subject had become more general. With Rousseau and Saussure the Alps made their entry into western sensibility and culture, and many travelers discovered the peculiar beauty of mountains.

In some cases the expansion referred to is only short term, as in the instance of the advance of the Triolet glacier, noted by Saussure about 1778. But very often it is a question of more than a brief oscillation or momentary episode, and then it indicates a trend, a state of expansion which may be compared to its opposite at the end of the nineteenth or beginning of the twentieth century. Such is the evidence[174] given by the *Tableaux de la Suisse* (Pictures of Switzerland) by Zurlauben (Paris, 1780) and the

Merkwüdigen Prospekte aus den Schweizergebirgen (Remarkable Views from the Swiss Mountains) (Berne, 1776), "which show all the glaciers depicted as much farther advanced than in 1885."[175] The year 1885 was in the early decades of the long retreat.

A typical example is Bourrit's fine drawing of the Rhône glacier,[176] confirmed by Besson's comments and drawing (Fig. 12c). The huge *pecten* covers all the upper part of the Gletsch valley, and entirely bars the adjacent valley of the Muttbach, where the Muttglatscher, Tällistock, Blauberg, and Furka streams all join. When Besson visited the *Rhonegletscher* opinions were divided about the short oscillation then taking place: the local shepherds declared that the glacier had been retreating for twenty years, but from his own observations Besson thought that the front was advancing. But what matters to us is the long term, the trend, and here there is no doubt. Measurement confirms the pictorial evidence: Besson[177] found the glacier 300 fathoms (585 meters) from the warm spring at Gletsch, and 120 fathoms (235 meters) from the extreme terminal moraine. The same glacier, in 1870, at the end of a continuous retreat going back to 1856, was 660 meters from the warm spring; in 1874 the distance was 950 meters, in 1914 1900 meters, and today even more. The overall pattern is undeniable: the long state of expansion of the eighteenth century contrasts with the long retreat of 1860–1960.

Between 1760 and 1790, with the aid of Saussure, Besson, Bourrit, and many others, the Europeans discovered their glaciers, and they found them much larger than we do, who have the advantage of being able to make comparisons. And what was true of the Alps was also true of Iceland: in 1754 the Drangajökull was half a mile from the head of the Tharalatur fjord, according to Olaffson, and this was 1 kilometer beyond the front of 1935. In 1775, according to Olavius, the same glacier was even more advanced, and reached the end of the fjord, and the sea.[178]

The following period, that of the French Revolution and the Empire, provides rather less copious evidence (but see Appendix 10 and also Plates 27, 28, 29, and Fig. 23). It seems probable that traveling in the Alps was restricted by international events, and

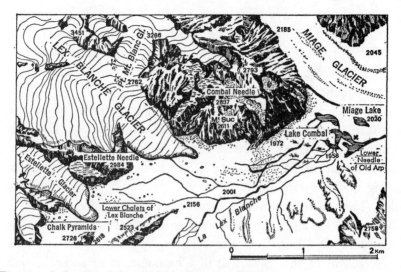

Fig. 23 *Lake Combal and Allée Blanche Glacier*
    This map shows the contemporary position of the Allée Blanche
or Lex Blanche glacier. It is withdrawn and no longer in contact with Lake
Combal, which itself is much smaller than it was in the nineteenth
century. The cross to the right indicates the point from which the drawing
and photograph of Plates XXVII and XXVIII were made. (Source: Carte
Vallot, 1960 ed., from observations of 1930.)

that centers of tourist interest shifted. Pictorial evidence is much
sparser around 1805 than it was around 1780, the first golden age
of early glaciology.

Nevertheless, Bacler d'Albe, later a general under the Empire,
lived in Sallanches from 1786 to 1793,[179] and he frequently visited
the neighboring glaciers of Chamonix. He painted the Arveyron
cave when it was in a full state of development and close to the
forest and the small rocky spur which overlooks Les Bois to the
northwest. Toward 1788–1789 Chrétien de Méchel drew a perspective view of the Chamonix valley, in which the front of the
Argentière glacier is 900 meters beyond its 1911 position.[180]

So with the Allalin glacier: in 1795, as during all the first half
of the nineteenth century, it continued majestically to bar the
Mattmarksee (see Fig. 12D).[181] This was an infallible sign of the
Little Ice Age.

In 1794 Conrad Escher von der Linth did a watercolor of the
Rhône glacier[182] seen from the top of the Maienwang (see Plates

15 and 16): as always from 1705 to 1870, the *pecten* stretches over the *Gletscherboden* into the vale of Gletsch, completely barring the Muttbach valley. All the views of the Rhône glacier at this period (1787, 1791, 1794, and the beginning of the nineteenth century) paint the same picture.

At the beginning of the nineteenth century crowds of visitors continued to make pilgrimages to the Chamonix valley. Their accounts are analyzed in Appendix 10. In the First Empire, as before, tourists found the Mer de Glace in its classic Little Ice Age state, curving down toward the plain of the Arve and visible from the valley. Their observations are corroborated pictorially.

Between 1800 and 1810 the Swiss painter Linck executed a view of the Mer de Glace from the Flégère, opposite.[183] His picture, preserved still in the town hall of Chamonix, is so accurate that one can identify on it the hamlets of Les Tines and Les Bois, the paths of avalanches, "and all the gaps torn in the common forest of Le Montenvers." The glacier Des Bois, in its typical advanced form, curves gracefully beyond Les Mottets and back toward the southwest in the direction of the Arve valley. As the crow flies, its terminus is 920 meters beyond its 1911 position; 1200 meters beyond its position in 1955. So the Mer de Glace then was still very close to the plain, which explains how Byron, in 1816, having observed the glacier Des Bossons coming right down to the wheat fields, was able to follow Saussure's advice and go up in a horse charabanc to the source of the Arveyron.[184]

One may, in fact, generalize: between 1770 and 1850, according to the abundant evidence of drawings, paintings, maps, and descriptions, the glacier Des Bois never withdrew more than 100 or 200 meters from the positions it occupied in Linck's painting, though it was often as much as 200 or 300 meters longer. So Linck shows the *norm* of the period from 1770 to 1850. Similarly, Mougin's snapshot of 1911, taken from La Flégère,[185] which shows the same panorama entirely free of ice, represents the long-term norm of the twentieth century. The photograph I had taken in 1966 exactly parallels that of Mougin.

Around 1800 the Allalin provides corroboration of the glacial advance: the Mattmarksee was duly barred and flooded in 1790, 1793, 1798, and 1808.[186]

Lastly there are the first historical data on the Oisans glaciers. In 1807 the prefect of the Hautes-Alpes who was touring the area was informed, as if it were nothing out of the ordinary, that the Arcine pass was occupied by the glacier of that name. This, as Rabot points out, is a sign that the Arcine glacier was "enormously advanced." In 1807 the Rip de l'Alp glacier, above the sources of the Romanche, was still progressing.[187]

In general, there is no sign of real retreat among the glacier systems at the turn of that century. In Iceland about 1793–1794 the Vatnajökull group were very advanced—much more so than in 1940—and very near to their terminal moraines. About 1789, and again in 1807–1812, the Norwegian glaciers advanced as far as the farthest limits they had reached in the 1740s.[188] A Russian map of Alaska in 1802 shows glaciers filling the Copper valley, now occupied by a railway:[189] at Lemon Creek, between 1769 and 1819, the glacier, although about 200 meters behind its 1759 position, was still 1 kilometer beyond where it was in 1900–1930, and 2 kilometers beyond where it was in 1958. All this is known because the terminal moraines have been dated by means of trees.

In the Alps between 1816 and 1825 there was a new offensive by the already very advanced glacial fronts just referred to. On 23 July, 1816, Mary Shelley,[190] visiting the source of the Arveyron, at the foot of its icy cavern, noted that the glacier Des Bois "is every day increasing a foot, closing up the valley." She thus expresses, dramatically perhaps but significantly, the same apprehensions as the inhabitants of Chamonix in 1644, when they feared the Mer de Glace would block the Arve valley and turn it into a lake. Mary Shelley was so impressed by the glacier that she makes her hero climb Le Montenvers (Montanvert, as she calls it) to contemplate it, and it is while he is surveying the "wonderful and stupendous scene" that he encounters the awful being he has created, who has taken refuge among the caves of ice.[191] The novel gives a detailed description of the whole appearance and atmosphere of the Mer de Glace and its surroundings in 1816. By about 1820 the Mer de Glace was 1350 meters beyond its 1911 position,

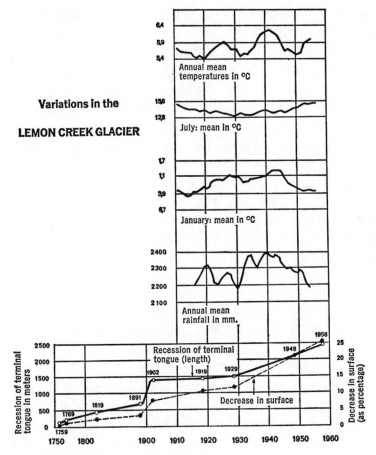

Fig. 24   *Lemon Creek Glacier (Alaska)*
Diagram brings out clearly the secular recession of the Lemon Creek glacier in terms of length and surface since its maxima of the eighteenth–nineteenth centuries. Local data on rain and temperature are also shown. For reasons of convenience the chronological scale has been modified after 1900. (Source: Diagram taken from Heusser and Marcus, 1964, p. 81.)

only 20 meters from the nearest house in the hamlet of Les Bois[192]; it curved toward the plain as Linck depicted it, but with even more amplitude and majesty. Hugo visited it in 1825 and wrote: "Beyond the glacier Des Bossons, opposite the priory of Chamonix . . . there is first of all a forest of gigantic larches . . . Above this forest, the extremity of the Mer de Glace, stretching past the Montenvers like a bent arm, overhangs and hurls down great blocks of marble whiteness . . . and opens onto the plain the terrifying mouth from which the Arveyron issues like a river."[193] The Mer de Glace, as Hugo saw it, was the same as that depicted by Combet, Saussure, and Linck.

The 1818 maximum of the Argentière glacier brought its terminal point 1050 meters beyond its 1911 position and 1300 meters beyond that of 1955—but only 300 or 400 meters at the most beyond the secular and customary norm for this glacier's expansion between 1730 and 1860. As for the glacier Des Bossons, in 1818 it advanced and spread over fields and a forest, and threatened the hamlet of Montquart. The terrified inhabitants followed the same reflexes as in 1643: they went in procession to the glacier and put up a cross at the farthest extreme of the terminal moraine. The cross survived till the beginning of the twentieth century, enabling Mougin to measure the position of the ice in the early years of the Restoration: he found that the glacier Des Bossons in 1818 was 590 meters beyond its 1911 front (it was thus more than 1200 meters beyond its 1952 position) (see also Plates 23 and 24).[194]

The glacial maximum that occurred about 1818 (in fact, from 1814 to 1825) fell right in a period of scientific, touristic, and even journalistic activity, and was therefore widely known about. In 1891 Richter[195] gave a year by year, glacier by glacier, account of it, which I summarize in the following table:

GLACIERS OR GROUPS OF GLACIERS AFFECTED BY:

Year	Advance	Maximum	Beginning of Retreat
1814	Grindelwald upper and lower		
	Lys		
1815	Sulden		
	Giétroz		
	Langtauferer		
	Langen		
1816	Palue		
	Hintereis		
	Unteraar		

Year	Advance	Maximum	Beginning of Retreat
1817	Le Tour	Zigiorenove	
	Gurgler	Ferpècle	
	Upper Grindelwald	Montminé	
1818		Hintereis	
		Rhône	
		Giétroz	
		Bossons	
		All the Valais and Mont Blanc glaciers	
1819	Renften	Sulden	Rhône
	Tschingel (?)	Argentière	
	Lötschen (?)	Lower Grindelwald	
	Gastern	Biès	
	Gygli (or Gauli?)		
	Engstlen		
	Wenden		
	Rufenstein		
	Schmadri		
	Junfgrau		
	Steinen (Steinglatscher?)		
	Schildhorn		
1820	Vernagt	Brenva	Sulden
	Leiter	Miage	Lys
		Le Tour	Allée Blanche
		Allalin	Fribouge
			Triolet
			Brenney
			Hautemma
			Tzanrion
			Chalen
			Val Ferret
			Bagne
			All the Saas glaciers
1821		Zesetta	Prafloray
			Durant-en-Tzina
1822	Allalin	Upper Grindelwald	
	Gorner	Vernagt	
		Schwarberg (Schwarzenberg?)	
		Brenney	
1824		Rosenlavi	
		Fee	
		Glaciers in the canton of Uri	

This table, in which I have corrected some of Richter's data in accordance with more recent findings, is interesting not only in itself, but also because it enables the historian to picture what the maxima said to be of "1600–1601," "1643–1653," "1680," "1720," and "1770" must really have been like.

In such paroxysms, all the Alpine glaciers do not have their maximum in the same year: while some are beginning to advance, others are still stationary or even in retreat. The maxima themselves may be spread out so that some glaciers begin their retreat in the same year as others are still in full spate. So the idea of a glacial maximum in a large mountain system like the Alps should conjure up a general but not a uniform picture.[196]

Kinzl has done on-the-spot research into the moraines of 1818 to 1820[197] and dated them without much difficulty through the age of the trees growing on them. He was able to distinguish the 1818–1820 moraines from those of the most recent maximum of "1840–1850," and, by means of trees and of geomorphological data, from maximal moraines going back to the eighteenth and, even more probably, the seventeenth centuries. He tried to discern the cases where the 1820 moraine went beyond that of the seventeenth century, and those where it remained within—it was always *just* within—it.

After years of travel and study, Kinzl concluded that out of the sixty-seven glaciers he had examined, twenty-three had their most marked recent historical maximum in the "seventeenth century," about 1600–1610, or about 1643–1644 (or, I would add, about 1680, according to the recent work of Heuberger and Mayr). Nineteen Alpine glaciers had what might be called this "maximal maximum" about 1820, and the rest had theirs about 1850. From the middle of the nineteenth century until 1930 (in fact until 1960), there were no glacial thrusts comparable with these three series of *maxima maximorum*.

In any case, the maxima of 1600, etc., 1820, 1850, are very close to one another and entirely comparable. They are simply the extremes reached by the strongest temporary *waves* during a lasting period of glacial *high tide* which continued for several centuries.

After 1820–1825 the *wave* fell back a little—in other words the glaciers withdrew a few dozen or a few hundred meters. But the *high tide* persisted.

We see proof of this in a series of excellent engravings of 1830, which show the great expansion, by comparison with the Twentieth century, of the Chamonix glaciers.[198] In the year of the three "glorious days" (of the 1830 revolution), the terminal tongue of Le Tour glacier was still very far down, much lower than in 1907 or 1960. In the same year the Argentière glacier formed the classic zigzag down to the plain which has now disappeared. The same year again, according to a drawing by Jean Dubois "on squared paper," the Mer de Glace had withdrawn only 110 meters from its 1820–1825 maximum; it was still 1240 meters beyond its 1911 position, and 1600 meters beyond where it is today.[199]

The year 1830 was also the time of the first modern surveys in the Dauphiné. The glaciers here are often smaller than those of Savoie, and their movements smaller in scope than those of Chamonix. Nevertheless in 1830 the survey shows their dimensions to be more or less comparable with what they were to be on the staff map of 1835—much superior to what they became at the end of the nineteenth and in the twentieth century.

DAUPHINÉ GLACIERS*

A	B	C
Groups and Glaciers	Position of front** in 1830 survey in relation to reference year in twentieth century	Reference year in twentieth century
Grandes-Rousses group		
Sarennes glacier	plus 530 m	1927
Grand-Sablat glacier	plus 600 m	1928
Quirlies glacier	plus 225 m	1928
Les Malâtres glacier	plus 130 m	1928
Les Rousses glacier	plus 180 m	1928
Pelvoux group		
La Selle glacier	plus 900 m	1929
Le Chardon glacier	plus 449 m	1930
La Pilatte glacier	plus 700 m	1930

*Mougin, VII, 1934
**Defined by the same methods as for the 1730 survey.

In 1834 a French traveler called Rey visited the Rhône glacier and the source of the Rhône. The description he gives is fully corroborated by Ritz's engraving of the same year, which shows the *pecten*, very swollen, quite close to the inn at Gletsch. Rey himself says that the *pecten* barred the Muttbach valley.[200] From the figures he gives it emerges that the glacier was 485 meters from the warm spring. This is only an approximate figure, but it falls within the secular norm for the period, in which this distance varies between 300 and 600 meters. In 1914, on the other hand, the distance between the front and the spring was 1900 meters; today it is even more.

In 1842, after a careful triangulation, Forbes produced the first scientific map of the Mer de Glace (Plate 18).[201] This confirms that there had been a slight regression of about 370 meters in relation to the maxima of 1820–1825. But the contrast with the twentieth century still remains: the front of the Mer de Glace as mapped by Forbes is 980 meters beyond the 1911 front and about 1300 meters beyond that of 1950–1960. Moreover, Forbes wrote that the terminus of the glacier and the Arveyron grotto could be seen facing one from the village of Les Praz.

In the 1850s pictorial evidence in a modern form makes its appearance. There are photographs of the Rhône glacier[202] from 1848 on, showing the same structure as all the prephotographic engravings: the Muttbach barred and a giant *pecten* in the vale of Gletsch. (See Figs. 12F, etc., for sketches made from these photographs.)

In 1857 the Mer de Glace was photographed by Buisson, who became official photographer to Napoleon III. A huge mass of ice is shown filling and choking the present Arveyron valley right down to where it ends near the hamlet of Les Bois—the same thalweg where today there are forests and moraines, rocks, young trees, and rushing streams. This photograph "shows conclusively that neither Bourrit nor Bacler d'Albe [nor, one might add, the Savoyard bishops of the seventeenth century] exaggerated the strength of the masses of ice piled up at the foot of the slopes of Le Montenvers."[203]

Thus a last episode occurs against the background of the long plurisecular advance, an episode which lasts from the texts by Combet in 1850 to the triangulations of Forbes and the daguerreotypes of Buisson and Dollfuss-Ausset.

This is the maximum of 1845–1858, or more precisely, of 1850–1855. Kinzl's studies of moraines,[204] and particularly the hundred or so references made by Richter, make it possible to narrow down the episode chronologically. From 1835 to 1844 the glacial advance seems fairly general in the Alps. From 1845 to 1858 maximal situations predominate, culminating in 1850 (when six glaciers or systems of glaciers were in a state of maximum) and again in 1855 (when nine glaciers or glacial systems were in a state of maximum). Signs of withdrawal, up to then scattered, become constant, regular and practically universal from 1861 on.[205]

Once again Chamonix provides good evidence. As in 1600 and in 1777 (according to the very accurate engraving by Jalabert), in 1850 the Mer de Glace, swollen up, lengthens out considerably and overruns the first, southern foothills of the Côte du Piget— thus following the same path as when it destroyed Le Châtelard and Bonanay in 1600—and flows out toward Les Tines. J. Vallot wrote,[206] reproducing what he had been told by Michel Couttet of Les Bois: "About 1850–1851, the Mer de Glace was about fifty meters away from the village of Les Bois. It had destroyed a larch wood that was in the flat part . . . the glacier filled the moraine of Le Piget right up, and in the middle of the slope ejected blocks of ice toward Les Tines . . . in 1855 the glacier filled up its moraines almost completely."

Michel Couttet was telling the truth. Buisson's photograph of 1857 shows that in that year the glacier Des Bois still thrust out a characteristic tongue over the first slopes and col of Le Piget, toward the Arve plain and Les Tines.

This 1850–1855 maximum sometimes caused panic. The atmosphere is re-created by an interesting text published thirty years ago by a Savoyard scholar,[207] from the record book of Pierre Devouassaz, a peasant from the hamlet of Le Tour.

In September 1852 the glaciers were enormous, and there were

Fig. 25 *The Glaciers of the Chamonix Valley*
This map (Le Roy Ladurie, 1960) was drawn up by Jacques
Bertin on the basis of:
1. The 1/50,000 I.G.N. map on the present state of the
glaciers (1949).
2. A 1/80,000 map on the approximate positions of the glaciers
after the 1850 maximum.
The latter was taken from J. Vallot: *Évolution de la cartog-
raphie de la Savoie et du Mont-Blanc, Barrère*, Paris, 1922, atlas, Plate
XXIV (reproduction of a map by G. H. Dufour, "Topographische
Karte der Schweiz, Martigny, 1861," observed and drawn up on the spot
between 1854 and 1861).
In Dufour's map, made just after the 1850 maximum, and
when a slight withdrawal was already perceptible, the Mont Blanc glaciers
are much bigger than on the map of 1949. They still have almost their
average dimensions for the seventeenth and eighteenth centuries (cf.
the glacier Des Bois and the Argentière glaciers near the villages of the
same name; and compare them with 1949). The 1400-meter contour
outlines the valley. The 1800-meter contour corresponds to the upper
vegetation limit.

hot winds and heavy rains.[208] The terminal masses of the glaciers, swollen and overhanging, began as a result of the melting to break off and hurtle down into the valleys. "The glaciers are exploding," as Devouassaz said. The catastrophe came between the fifteenth and nineteenth of September 1852. During those days, wrote Devouassaz, "the rains were heavy . . . the glaciers blew up at the same time . . . so that there was a great ravine covering part of the lands of Le Tour. [Farther on Devouassaz tells of the rupture of the moraine of the Le Tour glacier.] At the same time the Argentière glacier blew up too, and covered [various pieces of land] between Argentière and Les Chosalets . . . And during all these days, the glaciers of Lognan[209] burst and came down below Le Pierret which was entirely covered. And I who write this happened to be at Lognan as proxy to divide up the cheese, and we were obliged to go and hide in the cellar. There was a terrific noise, as if all the glaciers were falling . . . From the hamlet of Les Praz to Les Tines, all the highway was swept away by the waters from the glacier Des Bois."

But these disasters did not last long. From 1855 on, the first signs of ebb appeared around Chamonix. In 1861 there is evidence all over the Alps of a glacial withdrawal, stimulated by the warm springs and summers that occurred in France and Switzerland from 1857 on, producing early wine harvests. The size of the withdrawal was considerable, and for the first time for three centuries a point of no return was quite quickly reached.

In one year (1867–1868) the Mer de Glace is supposed to have retreated 150 meters. Then, in the ten years from November 4, 1868, to September 27, 1878, it lost 757 meters: 76 meters per year. By this time the Arveyron grotto had disappeared and the tongue of the glacier had withdrawn to the rocky gorges of Les Mottets, whence it has not since emerged again.[210]

Similarly, the terminal tongue of the Rhône glacier, carefully marked out year by year since 1874 by the Swiss topographical service, did not advance in any year between 1874 and 1916, the year of Mercanton's important study. Every year there was a retreat of a few dozen meters, and sometimes much more.[211]

So, from 1860–1870, the age of the great withdrawal had begun, and it affected all the Alpine glaciers.[212] It was the beginning of a new era.

The owl does not fly until dusk, says an old saw. And it often happens that a historical phenomenon does not yield up its full meaning until it is over and can be grasped in its entirety. Richter and Mougin, the great scholars and pioneers of the history of the glaciers, had scarcely emerged from the multisecular phase of glacial advance, and were not far enough away from it to see the full significance in the long term of the episodes they described. For Richter in particular, writing in 1880–1890, the retreat was so recent that he could not see its secular significance, and explained it in terms of the thirty-five-year cycle of glacial fluctuation he had adopted from Brückner. This cyclomania was characteristic of Richter's time, but glaciologists since, up to Lliboutry and Hoinkes, have dropped this element while retaining Richter's real contribution.[213]

Since 1900–1925 and up to 1955–1960, the recession of the glaciers has been such, and the contrast with the nineteenth century so strong, that we are in a better position to see the glacial phenomena between 1590 and 1860, if not as a whole, at least as a single "synchronic" structure. Two texts and a few important studies help us to do this.

The two texts concern the Ruitor complex of glacier, barrier, and lake, which during the period from 1590 to 1860 was one of the best indicators for the long advance of the glaciers. From 1864 and up to the present, this complex, as a result of the withdrawal of the glaciers and the disappearance of the barrier, has ceased, for the first time after 270 years, to spread havoc in the valleys below. To put it more simply, it has ceased to exist.

The Italian glaciologists, Baretti and Sacco, in 1880 and 1917 respectively, both realized and demonstrated, with texts and maps, the novelty of this situation. But as early as 1867 G. Carrel, a canon of Aosta, saw that something new had happened and a multisecular situation had been reversed. He wrote: "They now tell us that the Ruitor lake has disappeared completely.[214] If so, it is an extraordinary and entirely new thing. History makes no mention of it in past centuries.[215]

So in 1867 they were emerging from the multisecular phase of advance. But 271 years earlier, in 1596, again near Ruitor, the opposite was happening: they were entering it. An Italian engineer, Soldati, called in by the Val d'Aosta authorities, learned of it as he listened to the laments of the peasants. "I found," he wrote on October 10, 1596,[216] "that the lake [the Ruitor lake] was closed in on all sides by high mountains and living rock. Except in one place: in the lower part of the lake there was a shallow and fairly wide opening, through which the water of the said lake perpetually flowed out, without doing any harm in the said valley or else-where[217]; but now there has formed such a great mass of con-gealed snow, which the peasants call 'rosa,'[218] that it has closed up the opening in question to a great height." And Soldati goes on to describe the disasters that occur when the waters which have accumulated behind the barrier break through and rush down on the land below.

For the inhabitants of the Val d'Aosta at the end of the six-teenth century the advance of the Ruitor glacier, which became a recurrent phenomenon from 1596 to 1864, was indeed then a new thing, unprecedented in popular memory. (There are indica-tions, to which I shall return [see also Appendix 13], which suggest that there had been no such glacial advance at Ruitor since the thirteenth century, or at the latest since the end of the fourteenth century.)

With Soldati (1596) we enter the little glacial advance of the Val d'Aosta. With Saurel (1867), we emerge from it. Between these two dates and texts comes a period of glacial imperialism, unique and massive.

Kinzl, in 1932, was the first to call attention emphatically to the originality and massiveness of this period, after his on the spot study of the moraines of sixty-seven Alpine glaciers. He showed that the glacial thrusts of 1600–1610, 1643–1644, 1818–1820, and

1850–1855, were without precedent in the previous period, the later Middle Ages; also that they had no equivalent in the period that followed, the contemporary era. To this extent, Kinzl's admirable account does involve intersecular comparison. But it remains centered on the critical dates of the four great maxima mentioned,[219] and tends to neglect the intermediate periods of the seventeenth, eighteenth, and early nineteenth centuries. This being so, Kinzl does not yet really bring out the main features of a long-term glacial chronology.

It was only in 1942 that this decisive step was made, simultaneously, by German and American researchers apparently working quite independently of each other.

In Germany, Drygalski and Machatschek,[220] in their *Gletscherkunde* of 1942, reviewed the data on the glacial thrusts of the seventeenth, eighteenth, and the first half of the nineteenth centuries. This information was less abundant then than now, and their inventory was in any case not complete. But their arguments were clear and sound. It was not only in "1600," they said,[221] or in "1643–1644," etc., etc., that it was relevant to speak of glacial thrust. If one compares data within each group, and between the various groups (western and eastern, northern and southern Alps, and so on), one sees that it is "the whole seventeenth century" which constitutes a continuous period of glacial advance.[222] Against this background, the temporary oscillations, whether positive and maximal as in 1643, or negative, as about 1690, are only secondary and sometimes purely local episodes. In the seventeenth century there is no question of any large-scale retreat—deep, secular, pan-Alpine or worldwide—comparable to what for the authors was the "present" one, of 1855–1942, and what we should date 1855–1955. What is more, no such retreat was even possible in the seventeenth century: the few retreats recorded were much too brief to stand comparison with that of the present. As Drygalski and Machatschek put it, "there was not time"[223] for a withdrawal such as that of the present day. (For diagrams of glacial evolution see Plate XXXII.)

They go on to say that the same is true of the period 1700–1850.

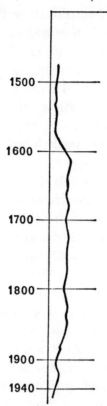

Fig. 26   *Simplified Diagram of Glacial Fluctuations in the Alps*
(Medieval low tide; modern high tide; contemporary ebb)
(Source: Drygalski, etc., 1942, p. 216, Fig. 35.)

For these hundred and fifty years, and for practically every decade
of them, there are sound data relating to the whole of the Alps
showing glaciers considerably and continually more advanced than
in the recent secular period of retreat.

Drygalski and Machatschek also proposed abandoning[224] the
outdated method of Brückner and Richter, who attempted to apply a theory of interdecennary cycles (of thirty-five years, for instance) to the pulsations of the glaciers. Instead they put forward
a multisecular schema, which they set out in the form of a simple
graph[225]:

1. On the one hand, between 1590 and 1850, a glacial "high
tide" which was never interrupted but only varied from time to

time by waves or swells. The most important of these short oscillations were the "maxima" of 1600, 1643–1644, 1820, and 1850. They constituted a kind of extreme attack from already advanced positions.

2. On the other, a long state of ebb or low water, probably at the end of the Middle Ages, and in any case "before 1600"; and certainly from 1860 up to the present.

Drygalski's and Machatschek's ideas, published in Germany during the war, did not create many immediate repercussions: they were simply added to the general store of glaciology in German. In 1949 R. von Klebelsberg incorporated them into his manual of glaciology and glacial geology, without modifying their chronology or basic assumptions.[226] Franz Mayr,[227] in an important article published in 1964, also took up the idea of a "period of glacial advance from 1590 to 1850." He listed it, using Kinzl's terminology, under the name of "Fernau stage,"[228] and numbered it "Xf" in the list of oscillations in the Alpine glaciers since the climatic optimum.

At the same time a similar question was being raised in the United States. In 1942 F. E. Matthes contributed an important chapter on glaciers to *Hydrology*, a collection of essays by various hands edited by O. E. Meinzer.[229] Under somewhat different conditions from those in Germany, Matthes too set out to inquire into the glaciological history of the last four centuries, particularly in the Alps, Norway, and Iceland. He cites the villages scarred and damaged by the ice; the fact that the glacial thrusts are much more numerous and close together than can be explained by the theory of thirty-five-year cycles; and the general state of advance in the Alps between roughly 1600 and 1850. But though we can scarcely blame him for it, Matthes' documentation is far from exhaustive. He does not know Lütschg, and only knows Mougin and Letonnelier at second hand through Kinzl and Charles Rabot. He gives 1600 as the date when Saint-Jean-de-Pertuis was destroyed by the Brenva glacier, although this episode is only known from oral tradition and it has never been possible to date it.[230]

But these are only minor criticisms. In spite of a few gaps, Matthes' vast knowledge of glaciology and geography enabled him to reach some very interesting conclusions, the chief of which are as follows[231]:

1. The Alpine, Scandinavian, and Icelandic glaciers[232] went through a phase of "moderate but persistent expansion" during the whole of the modern period, from a date which can be fixed roughly around 1600. This phase precedes the "present glacial recession," since 1850.

2. The long period of glacial advance itself followed a period of lesser expansion, "in the Middle Ages," when inhabited places covered by the fronts of the seventeenth and eighteenth centuries, in Iceland, Scandinavia, and the Alps, were still flourishing.

3. These three successive secular or intersecular episodes from the Middle Ages to the present (to put it simply, we may call them for the moment medieval ebb, modern advance, and contemporary retreat) themselves fall within a much longer intermillenary or multimillenary period, of 3500 to 4000 years. This period, which came after the climatic optimum, coincided with a general cooling down of climate and restoration of the glaciers, much diminished during the *Wärmezeit*. Matthes was the first to suggest the name "Little Ice Age" for these three or four recent millennia of cool climate and slight reglaciation (themselves containing, as we have seen, secular or multisecular oscillations more or less cool or more or less warm). It may be noted in passing that the Little Ice Age in fact equals the "subatlantic" of the palynologists.

Both Matthes' ideas and even his terminology have had a lasting influence. Richard Foster Flint[233] and his colleagues have incorporated them into their general view of the Pleistocene Age. And the idea now seems generally admitted of a big multimillenary post-optimal cool fluctuation (covering the historical period), with briefer warm or cool subfluctuations, secular or intersecular.[234]

The phrase "Little Ice Age," invented by Matthes, has had rather a strange history since. The term itself is probably somewhat too strong, conjuring up as it does, for comparatively limited phenomena, the real ice ages which covered parts of America and Eurasia with huge ice sheets. Moreover, the term, instead of being applied, as Matthes intended, to the thousands of years of the subatlantic period, has come to be applied just to the recent bisecular phase of glacial advance of 1600–1850. Lamb, Schove, and many others refer to these two and a half centuries as the Little Ice Age.[235] And since custom makes law, this usage might as well be accepted. It should be noted, however, that German authors, such as Kinzl and Mayr, refer to the recent phase, the last multisecular glacial advance of 1570–1850, not as the Little Ice Age but as the "Fernau stage" or even just as "Fernau."

But more important than questions of vocabulary is the fact that since 1942, since Drygalski and Machatschek and Matthes and Ahlmann, the long term beloved of historians has become part not only of the *geological* history of glaciers—it had been that for a long time, since Agassiz[236]—but also of their recent history, their *historical* history. Over the last thirty years, since the secular movement of the glaciers became accepted as self-evident, theories of eleven-year or thirty-five-year cycles have been replaced by a conception based on the long term and on trends, a theory which deals in decades and even years, but also in centuries and multicenturies. After all, geologists of the great ice ages deal not only in centuries but in millennia and tens of millennia.

If this was all historically proved more than twenty years ago, why do I find it necessary to go over in detail the facts of the long modern glacial fluctuation? Is not the brief account, incisive, well substantiated and brilliant, given by Louis Lliboutry in his monumental *Traité de glaciologie* (Treatise on Glaciology), sufficient?[237]

From the point of view of the glaciologist, of course it is sufficient. But it is not quite enough for the historian. The reason is quite simple: Kinzl, Drygalski, Matthes, Ahlmann, and Flint are all specialists in one of the natural sciences. They are geologists, glaciologists, or geographers, but not actual historians. They had an intuitive vision of the long modern glacial advance from

1590 to 1850, but they did not attempt to set it out in detail, year by year, or even decade by decade. What are two and a half centuries to experts used to the thousands or millions of years of geology?—a brief moment of evolution, two or three pages in a textbook, a rapid and sometimes incomplete review of events, proofs, and references.

This brevity has the usual advantages of rapid intuition and striking summarization, but it has its disadvantages too. First among these is the tendency toward too hasty or vague "historical" placing, which may entail all kinds of chronological distortions. Many scholars, including some of the best, have taken up what is too quickly termed the "Little Ice Age" of the seventeenth and eighteenth centuries, and without checking with the original documents have arranged it to suit themselves. Some make it begin in the fourteenth century or 1430; others in 1750. Writers as reputable and competent as Brooks have fictionalized their "Little Ice Age" and introduced into it a purely imaginary "interglacial" epoch" of prolonged retreat, which some place between 1680 and 1740, some between 1700 and 1750, and others again between 1760 and 1790, or even "around 1815." Sometimes they have simply extrapolated climatic series into glacial series, without reference to documents, or the Alps to Iceland or vice versa. Ignoring the original documents almost completely, although they have long been published in German, Italian, French, and Latin, and copying, inaccurately, from one another, they have each constructed their own "Little Ice Age" according to their own convenience. C. E. P. Brooks, coming from a country without glaciers and therefore without documents concerning them, too frequently relies on secondary sources.[238]

The mistakes in these secondary sources would not matter so much if it were not that, having been transmitted by a scholar of Brooks's reputation, they have acquired prestige and become widely implanted in the specialist literature of England and America, which increasingly sets the tone for the rest.

So something had to be done. To begin with, and as an indispensable preliminary to my own research, I went over the whole question in detail, returning to the sources which Drygalski and Matthes considered it either impracticable or unnecessary to consult. I have reexamined all the documentation which in one way or another concerns the history of modern glaciology: all the texts since the sixteenth century; all the pictorial evidence from 1640 to 1850; all the mpas and surveys since 1730.

One researcher, in one book, cannot cover all the glacial systems of Europe, and I have only referred to Iceland and Scandinavia in passing. But I have tried to be to a certain degree exhaustive as far as the Alps are concerned. Here, disregarding political frontiers, it is possible to reconcile the minutiae of scholarship with the preoccupations of long-term structural history.

The insights of the 1942 theorists are in fact confirmed by the documentary test: in the seventeenth, eighteenth, and ninteenth centuries there were not only isolated glacial thrusts more spectacular than lasting, but also a multisecular phase of glacial expansion which was in full force from 1590 and did not end, in the Alps, until after 1850. And the symptoms of advance during this two hundred and fifty years were so frequent, close together, continuous, and intense that they do not leave time for any secular retreat such as we have experienced during the last hundred years. The period 1590–1850 had its own internal fluctuations, its own irregular ebbs and flows of glaciers.[239] But these oscillations were secondary as against the primary fact of the long "modern" advance (1590–1850), contrasting with the lesser "medieval" expansion and the "contemporary" (after 1850) retreat.

So here we are on solid ground, and distortions and inventions will henceforth, we hope, be more difficult to get away with. The advocates of a "fixed state," also, can be more easily countered. So the historian can turn to other periods and problems and ask new questions.

1. First, what was the climatic context in which the Fernau stage or oscillation occurred? In other words, the intersecular phase of advance in the Alpine glaciers from 1590 to 1850.

2. Another question inseparable from the first: what type of phase came before the Fernau stage itself? And were there phases identical with or analogous to the Fernau stage before, either in the historical or the protohistorical period?

3. The third question presupposes that the first is resolved: is it possible to talk about the human effects of a glacioclimatic oscillation of the Fernau type?

4. To go from effects to causes, instead of as in question 3 from causes to effects, what is the climatological as distinct from climatic context in which all these phenomena occurred?

I shall try briefly to deal with these points, though it might be mentioned in passing that the historian of climate does not only ask himself questions that he can answer.

CHAPTER V

WORKING HYPOTHESES

There seem to be two types of answer to our first question, on the climatic context of the Fernau stage.

First there is the theoretical answer. Let B be the most recent secular phase of retreat (1855–1955). We know that there is a climatic fluctuation (A) which acts as its cause. It was in fact the secular warming up, recorded all over the world and in particular in Europe, which caused a lasting retreat of the glaciers. A thus determined B.

Conversely, between 1590 and 1850 there is no recorded episode of secular deglaciation comparable in magnitude to phenomenon B. From this logically follows the absence of phenomenon A: therefore, between 1590 and 1850, there was no marked period of warming up, or at least any that was comparable in magnitude and length to the contemporary one. In other words, in spite of various fluctuations, and the possibility of short or medium-length warmer periods, the secular trend of mean temperatures from 1590 to 1850 must generally have been a few tenths of a degree Centigrade below the contemporary level: probably between 0.3 and 1.0 degrees lower.

This assertion seems logically unassailable. All I have done is invert the relevant and solidly established model for the correlation between warming up and deglaciation in the twentieth century. But it goes without saying that at this stage of the argument the inverted model remains purely theoretical.

But there are means of backing it up in beginnings of experimental proof contained in early series of meteorological observations. The tables and temperature curves for the eighteenth and the first half of the nineteenth centuries by Labrijn in Holland,

Manley in England, and others in Sweden, Denmark, Germany, and Austria, tend to show that for certain seasons, in particular the winter, the mean temperatures were definitely lower (by 1 degree C or even more) than those of the twentieth century. The annual temperatures seem to have been a few tenths of a degree Centigrade below those of today.[1]

Very cogent, also, in this connection, are the early phenological series for winters in Europe, America, and the Far East. Year by year we know the date of the first snow at Annecy, from 1773 to 1910[2]; the dates when the Neva at Leningrad froze and thawed, from 1711 to 1951; and the dates for the icing up and thawing of Lake Kavallesi in Finland from 1834 to 1943,[3] Lake Champlain in the United States (1816–1935),[4] and Lake Suwa near Tokyo, where the series quoted goes from 1444 to 1954. H. Arakawa, who has done so much for the climatic history of the Far East, has also published the date of the first snowfall in Tokyo from 1632 to 1950 (on that day the daïmyos presented their respects to the Togukawa shoguns).[5] All these series converge, and their convergence is significant, at least for Europe: snow and frosts were earlier and thaws later in the seventeenth and eighteenth centuries than after 1840–1850; the winter was therefore longer (by nearly three weeks in Russia and Finland) and more severe. Both in Europe and in the Far East these hard winters seem to have begun about 1540–1560. They were particularly cold in the old world in the seventeenth century.

It may reasonably be objected that, Vienna apart, the lakes and meteorological stations just referred to (Tokyo, Utrecht, Edinburgh, Stockholm, and Berlin) are a long way from the Alps and the glaciers concerned.

What are needed are observations for the seventeenth century made at stations near the glaciers themselves and giving snow accumulation and temperature (the main factor in ablation). But such documents scarcely exist.

But if the glacial or "periglacial" area is taken in a very wide sense, there are some early temperature observations relating to the Alps. Annecy is almost the same latitude as the Mont Blanc glaciers, and in a nearby region, and Annecy has one of the earliest meteorological series in France. It was kept first by Despines, a doctor, from 1773 on, then by canon Vaullet, then by the meteorological commission of Haute-Savoie up till the 1914–1918 war. The series is confirmed from its very beginning by phenological observations on the first and last snow of the year, and the secular warming up that the Annecy series shows beginning in the nineteenth century is corroborated by the record of late first snows and early last snows, in other words by a shortening of the winter.

For the nineteenth century the Annecy series also corresponds satisfactorily with the neighboring series for Chambéry. So it seems to be reliable.

Mougin published the Annecy figures month by month, year by year, decade by decade.[6] What do they show?

In the first place they show a warming up of all the seasons, starting with the decade of 1843–1852. There is a contrast between the two halves of the series, the first half, 1773–1842, being slightly cooler, and the second half, 1843–1913, being slightly warmer.

This development is set out in the following table. The plus sign indicates the decades when mean temperature for the season or year is higher than the mean for the period 1773–1913.

ANNECY: TEMPERATURE ABOVE THE MEAN FOR 1773–1913

	Winter	Spring	Summer	Autumn	Year
1773–1782	+				
1783–1792					
1793–1802					
1803–1812					
1813–1822				+	
1823–1832		+	+		
1833–1843					
1843–1852	+	+	+	+	+
1853–1862	+	+	+	+	+
1863–1872	+	+			+
1873–1882	+	+	+	+	+
1883–1893*		+	+		
1894–1903		+	+	+	+
1904–1913	+			+	+

* The year 1892 is missing.

The chronology of this warming up from 1840–1850 corresponds to that shown at other European stations—for example, Copenhagen, published by Lysgaard.

Going into detail one sees that from 1773 to 1913 all the months of the year except May, August, and December warmed up. To put it more generally, all the seasons at Annecy grew milder. This does not apply so much to the winter, the influence of which on glacial ablation is in any case practically nil. But the warming up of spring, summer,[7] and autumn in the Savoie area is enough in itself to account for the increased melting of the Alpine glaciers from 1850 to 1914. The main period of glacial ablation comes in summer, at the end of spring, and the beginning of autumn. If these critical months grow warmer, even only slightly, ablation increases, the volume of the glacier shrinks and the fronts retire, as happened at this period around Mont Blanc.

There is no need for great differences in temperature. If the Annecy figures showing mean temperatures for 1773–1842 are compared with those for 1843–1914, the second group[8] are only about 0.5 to 1 degree C above the first. But that was enough.

So the Mont Blanc glaciers are at one and the same time ultrasensitive reactors and magnifying mirrors. They show the glacial fronts in retreat before the slightly but constantly higher temperatures of the second half of the nineteenth century.

Nevertheless one has to take into account the factor of inertia or hysteresis, the time lag in glacial reactions. For the changes which take place upstream in the glaciers, in their basins of accumulation and ablation, do not have immediate repercussions downstream on the position of the terminal front. There the effect is not felt for several years—two to six years, according to some glaciologists.[9]

The time lag seems particularly marked in the Annecy-Chamonix area. The temperatures there showed a lasting and secular type of warming up from the decade 1842–1852 on, but the lasting and secular retreat of the glaciers did not begin until the decade 1852–1863. In other words, meteorological cause is separated from glacial effect by a decade of inertia.

But here we have to remember that we are ignorant of many

data relevant to the budgets of the glaciers during 1840–1860, and in particular that of snow accumulation.[10] So we can only note the apparent inertia, without being able to say exactly how it worked or from what causes it arose.

However it may be, the Annecy series serves as an example of how the Alpine glaciers seem to have responded to local climatic oscillations. Haefeli has demonstrated the same thesis by comparing the very early temperature series for Bâle with the behavior of the Swiss glaciers.[11]

The Alpine glaciers' intersecular phase of advance, the Fernau oscillation, reached its terminal peak in 1773–1850, just as the first thermometric observations began to be made. These, both at Bâle and at Annecy, give mean temperatures nearly 1 degree C below the means for the succeeding period.

Conversely, it is apparently the rise in temperatures from 1843–1852 onwards which by various means (more frequent advection of warm air, longer periods of sunshine, reduced albedo) increases ablation and reduces the glaciers.[12]

Whatever the methods by which it works, the correlation between glaciers and temperatures, which is already solidly established for the twentieth century, is thus seen to be valid also for the end of the eighteenth and the nineteenth centuries. In the period 1770–1840 it takes the form: slightly lower temperatures equal glaciers markedly more developed.

Is it legitimate to extrapolate further from these data, right back to the first half of the eighteenth and seventeenth century? For these periods we lack thermometric observations throughout the Alps. But there is no doubt that the glaciers themselves, in the seventeenth century, already existed in their enlarged form of 1770–1850, and were much bigger than they are today. It seems therefore that during the classical period also the "meteorological character" of the Alpine seasons, their *Witterungscharakter*,[13] was roughly the same as in 1770–1840, differing only slightly from what it is in the twentieth century in the Alps and in Europe in general. Among these differences of detail we should perhaps include a mean difference of temperature of a few tenths of a degree C or more between now and the seventeenth century, our age being the warmer. Such at least is an admissible extrapolation, a working hypothesis based on the assumption that, other things being equal, similar effects usually arise from similar causes.

I should like to note incidentally that there is no need, in order to elucidate the climate of Europe and the Alpine glaciers during the modern period, to seek for new and strange explanations. It is enough just to use the models and trends provided by meteorological observations going furthest back into the past. The climate of the seventeenth and eighteenth centuries was scarcely any different from ours, and must have been almost exactly like that of the initial period of accurate observations (1770–1850): that is, slightly cooler than now. That at least is what we may suppose in the present state of our knowledge.

We now have plausible though very provisional data on the long-term climatic conditions which sustained the intersecular phase of advance (Fernau) in the Alpine glaciers. We also know of the slightly different conditions which after 1850–1900 put an end to this phase. So we have reached one of our first objectives.

But we have still not finished our inquiry into the past. The Fernau oscillation, as we have seen, became clearly perceptible at the end of the sixteenth century. But why did it begin? What climatic conditions caused it?

There is one explanation which has become classical among historians of climate. This argument tries to show that there was a slight increase in the severity of the winter, the converse of the contemporary warming up of the twentieth century.

This idea originated in the monumental work of Easton (1928), a conscientious Dutch researcher who compiled and analyzed texts throwing light on what the winter was like year by year from the Middle Ages to the beginning of accurate observations. The number and diversity of these texts, and a comparison with present-day quantitative observations, enabled Easton to give a coefficient for each winter and assign it to one of ten categories: very mild,

mild, warm, normal rather warm, normal, normal rather cold, cold, severe, very severe, extremely severe.

Easton was apparently not interested in the idea of climatic change. He worked without any preconceived ideas or basic assumptions; he ignored the behavior of the glaciers; he drew no particular conclusion from his exacting researches. He was quite content to publish, at the end of his book, a long sequence of yearly indices showing the severity of the winters.[14]

But from this raw material history could be made. Easton's series was taken up by Scherhag in 1939 and by A. Wagner in 1940. They drew up diagrams from his indices, from which they concluded that a secular cooling down of the winters set in from 1550.[15]

In 1949 D. J. Schove checked Easton's European series against local and regional ones. Schove made particular use of the meteorological chronicles and compilations of Corradi, Riggenbach, and Vanderlinden to draw up eight sequences, covering the years 1491 to 1610. The three main sequences concern Bâle, Italy, and Belgium, and Schove too finds that after 1540 severe winters are more frequent and mild ones generally less so.[16] This second phenomenon of "decreased mildness" is also noted after 1540 on Easton's curve.[17] This is very important: it confirms that the suggested cooling down was real, and not just due to an increased amount of information about hard winters made available as the sixteenth century went on. In 1950, Flohn, working independently of Schove, arrived at the same periodization.[18]

I myself have drawn up several series concerning the winters in the south of France in the sixteenth century. These series relate to the severity or mildness of the season; whether or not the olive trees survived the frost; and whether the Lower Rhône froze sufficiently to bear skaters and carts. All three types of evidence grow significantly more frequent from the decade 1540–1550 on (Appendix 11).[19]

The same chronology, with the curve changing in the second third of the sixteenth century, is found again in Manley's sequences of British winters.[20] So everywhere, from Italy to Switzerland, England to Languedoc, winters grew severer from 1540–1550 on, and the glaciers gathered in the Alpine valleys in preparation for the assault at the end of the century.

A quite recent sampling of the first order confirms all these other periodizations: it was published by the historian Van der Wee after long researches in the archives of the Antwerp area.[21] I have summarized his results in a table (Annex 11) showing absolute numbers and percentages decade by decade from 1500 to 1599. The main conclusion that emerges from the table is that the critical date locally appears to be around 1550. After that date, by comparison with the period 1500–1549, the number of severe winters and hard frosts increases significantly. From 1560 on, the frequency of heavy snowfalls also increases as measured by decades. This tendency is confirmed, between 1550 and 1580, by the meteorological journal of Wolfgang Haller, a citizen of Zürich[22]: accumulation of snow, of course, favors the expansion of glaciers. Similarly the number of mild and very mild winters at Antwerp shows a marked decrease after the turn of the half century.

So the accurate, original, and detailed series established by Van der Wee clearly confirms the other results.

Were people actually living between 1500 and 1600 aware of the secular fluctuation, this worsening of the winters[23] which appears in the second half of the sixteenth century?

It seems they were not actually conscious of it directly, and as such. The long-term climatic trend was too slow and slight, too much disguised by brief but marked oscillations, to be perceptible in fifty years of adult life. Only the historian is in a position to observe and elucidate it by collecting and comparing the various pieces of evidence.

Indirectly, however, the people who lived toward the end of the sixteenth century seem to have been conscious of certain effects of the cooling down of the winters. Lucien Febvre quotes the case[24] of the little mill of Verrières-de-Joux in the Jura, probably built in the craze for mills which seized enterprising inhabitants of Franche-Comté about 1500 to 1530. In June 1587, however,

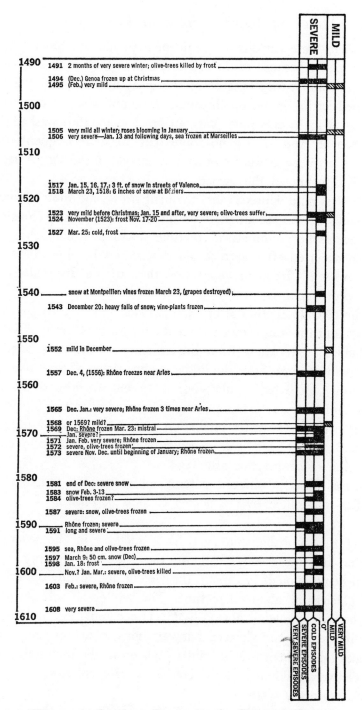

Fig. 27 *Winters in Southern France in the Sixteenth Century*
(N.B. Remember that the winter of 1494, for example, covers December 1493 and January and February 1494.)

its owners were anxious to get rid of it: it did not function regularly, "partly because of the low race and lack of water serving the said mill and partly because of the great frosts." They therefore stopped letting it out on three-year lease and allowed one tenant to take it for twenty-nine years. In other words they washed their hands of it. And there is no doubt that the more frequent severe winters after 1550 were among the causes of the "great frosts" which discouraged the owners.

The less mild winters, more frequent frosts, and more abundant snowfalls of the second half of the sixteenth century constitute an inverse but symmetrical thermic fluctuation paralleling that which occurred after 1850, when the winters became milder and the frosts less frequent and severe than in the first half of the nineteenth century.

The cold fluctuation after 1550 is made probable by the documents and the series drawn from them. But it is likely as well as probable: as we know from the evidence of the peat bogs, the climate of Europe, after the initial cooling down of the subatlantic age, only oscillated secularly or intersecularly around relatively stable millenary or intermillenary means. So it is not illogical to suppose that the winter oscillation in the direction of mildness, after 1850, may have been preceded by opposite oscillations in the direction of severity; as after 1550.

In each case the amount of temperature variation is likely to be similar—not much over 1 degree C, plus or minus.

Yet this probable and even likely increase in the severity of winter in the second half of the sixteenth century is far from being enough to explain the violent and lasting offensive of the glaciers at about that time. The winter is only a quarter of the year. Snowfall obviously affects accumulation and thereby the budgets of the glaciers, but variations in winter temperature do not affect them much: winter high up in the mountains, even if relatively mild, is still too cold to permit ablation, except on some especially warm days.

So the winter cooling down of the second half of the sixteenth century was not enough in itself to make the glaciers advance, as they did, to their historical maxima.

This advance can only be explained by an absence of ablation, among other basic factors. And lack of ablation only occurs when the main season for it—the end of spring, summer, and the beginning of autumn—also tends to cool down.

So, going by the long advance of the glaciers, it is reasonable to admit as a hypothesis, as Flohn does,[25] that the rest of the year, as well as the winter, cooled down during and after the sixteenth century.

The idea also stands up to examination from the purely climatic point of view. In the nineteenth and twentieth centuries, for example, the warming up of the winters which is the dominant factor has been accompanied, in the long term, by a more or less definite, more or less synchronized warming up of the other months and seasons of the year.[26] So, symmetrically, it may well be that the reverse happened in the second half of the sixteenth century.

It so happens that for the more clement part of the year, and in particular March–September, we have more than the circumstantial evidence with which we have to be satisfied for the winter itself. Phenological data provide annual, continuous, quantitative, and homogeneous evidence, indicating temperature trends for the spring and summer of any given year by means of wine harvest dates. A warm and sunny spring and summer mean an early wine harvest (and melting glaciers). A dull cold spring and cool damp summer mean a late wine harvest (and not much glacial ablation).

This phenological evidence is all the more relevant as the main series of wine harvest dates for modern times relate to vineyards near Switzerland and Savoie, the regions of the Alpine glaciers.

The comparison of glaciological and phenological data may be made over the whole of the modern period (sixteenth–eighteenth centuries) on two levels: "short" and medium term on the one hand, and "long" term on the other.

As far as the short and medium term are concerned, certain

correspondences seem possible between the two sets of data (Fig. 6). Three blocks of cool springs and summers stand out from the sequence of wine harvest dates for the eighteenth century: 1711–1717 (culminating in 1716); 1740–1757 (with a maximum in 1740–1743); and the years immediately before and after 1770. This series of cold years, during which the rate of ablation must certainly have been low, is paralleled (with the necessary time lag of a few years) by the most marked glacial maxima of the eighteenth century: 1716–1719; the 1740s; and 1770–1776. Manley noted in 1952 the striking correspondence between the cool British springs and summers of the 1740s and the advance of the Alpine and Scandinavian glaciers.[27] But when one has figures for both wine harvests and glaciers in the same region, the connection appears even closer.

The same juxtaposition is also suggestive applied to the Seventeenth century:

SERIES OF LATE WINE HARVESTS	BIGGEST GLACIAL MAXIMA (ALPS)
1591–1602	1601
1639–1644	1643–1644

Not all the cold years in the phenological curve correspond to glacial maxima. This is natural enough, when one remembers that spring and summer temperatures are not the only factor in glacial advance, and that in any case many glacial advances were not registered in archives or by historians. On the other hand, it is certain that known glacial advances often accompany or follow soon after a group of years which had low temperatures during the main period of ablation.

There is nothing new in this juxtaposition. As early as 1856 the geologist Favre noted that "six cool summers" had preceded the 1817–1822 outbreak of glacial activity in the Alps[28]; and the wine harvest dates, and the curve of the Paris Observatoire (Fig. 4),[29] back up his assertion. Gordon Manley, working on English and Dutch meteorological series and comparing them with glaciological observations in the Alps and in Scandinavia, noted the significant coolness of the summers preceding glacial thrusts in the decades 1691–1702, 1740–1751, 1809–1818, and 1836–1845.[30]

Finally, H. W. Ahlmann[31] has shown that between 1900 and 1940 the seventeen terminal tongues of Jotunheim advanced or retreated according to whether the mean temperature of May–September in Norway tended to diminish or increase during several consecutive years. The fluctuation of the glaciers, positive or negative, began two or three years after the start of the meteorological fluctuation, a comparatively brief latency period. As to other factors, such as winter temperatures and snowfall, they entered into the short-term movements of the Norwegian glaciers, but only to a lesser degree.

But one must not force the comparison between glaciers and wine harvest dates. Ablation, to which the phenology of the wine harvests corresponds, is not the only factor in the relatively short fluctuations of glaciers. Snow accumulation also comes into it, and on this wine harvest dates of course shed no light.

Lastly, Lliboutry has wisely recalled the often local nature of the decennary and intrasecular movements of glacial termini: "The time separating two fluctuations may vary between twenty-five and fifty years, and one may well wonder whether this time is not conditioned by the reaction time of the glaciers themselves, i.e., by their dimensions and rate of flow"[32] as much as, or rather than, by meteorological factors.

This local character is very marked, for example, in the glacier Des Bossons. It is very sensitive to climate, and in its fluctuations of climatic origin it is on the whole very like the Argentière glacier and the Mer de Glace. But because of its particular structure it always reacts several years before the two larger glaciers (Fig. 9).

Now we turn to wine harvest dates in the long term, and to the secular and multisecular aspect of the glaciers.

First, a brief methodological reminder.

Angot, Duchaussoy, and Garnier[33] established the critical and methodical study of the problem of wine harvest dates, and I have taken up and sometimes developed their arguments in earlier chapters and various other publications.[34] The dates themselves are to be found, in the form of dozens and hundreds of local series and of graphs derived from them, in the authors quoted and in my Appendix 12.

As we know from the various works cited above, wine harvest dates are good indicators of short fluctuation, which constitutes meteorological fluctuation in the real sense of the term. They may also be good indicators of secular fluctuation, when this is strictly climatic. But in the case of secular fluctuations it is necessary to recall certain particular difficulties.

At certain times the secular curve of wine harvest dates is distorted by nonclimatic, purely human factors. In the seventeenth and eighteenth centuries, the date of the wine harvest was held back because the vinegrowers wanted to get a better and stronger wine.[35] In the nineteenth century, wine harvest dates were advanced to cater for a growing popular market that was not too discriminating.[36] In the two cases, affecting three centuries, secular lateness or earliness had nothing to do with climatic fluctuation.

But in a period when viticulture and its techniques are fixed, and the demands of consumers and of the market change little over a long time, one may get stable phenological curves extending over a century or even longer. This is the case with the sixteenth century.

Let us return to the curve I have published elsewhere,[37] showing wine harvest dates in France and nearby regions from 1490 to 1610. This curve synthesizes a dozen local series whose geographical center of gravity is somewhere in the Alps of French-speaking Switzerland—not far from the glaciers.

Although it lacks the precision of the phenological curves for the seventeenth and eighteenth centuries, which are based on some hundred and fifty separate series, this diagram of sixteenth-century wine harvests has various titles to climatic fidelity. The various concordance tests give positive results: there is an internal concordance between vineyards that are close to and those which are a long way away from each other; and there is an unexpected and reassuring concordance with the dendrochronological curves.

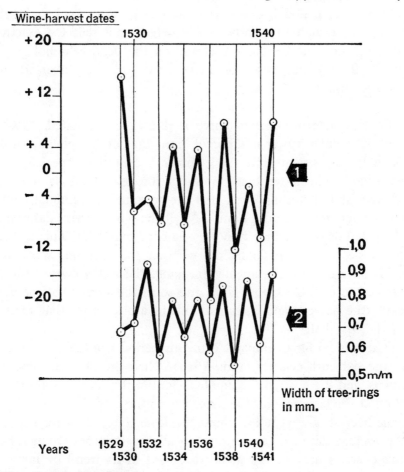

Fig. 28 *The Sägesignatur*

The *Sägesignatur* or saw-tooth curve is seen here in the 1530s dendrochronological graph for the oak in the Odenwald (curve 2) and in the corresponding phenological graph showing Franco-Swiss wine harvest dates (curve 1). Years are shown along the x-axis. On the y-axis, to the left, wine harvest dates (variations from the mean, in days; the figures are those given in Annex 12, para. B, column D, in the first (French) edition of this book, 1967). On the y-axis, to the right, width of oak tree rings in millimeters. (Curve 2 is taken from Hubert and Siebenlist, 1963, Fig. 2 and pp. 258–59.)

The strange period between 1530 and 1540 when cool and hot summers alternated regularly in a *Sägesignatur* is reflected exactly both on my wine harvest curves and on the dendrochronological graphs for the oak, based on living trees and old timbers from the Odenwald.[38]

So this sixteenth-century curve is doubly corroborated. Moreover, its secular aspect is different from that of the phenological curve for the seventeenth and eighteenth centuries, especially after 1650.[39] The latter tends upwards throughout all its length, whether it is reflecting years that are maximal or minimal, very late or very early. Under Louis XV all the wine harvest dates are about ten days later than they were under Louis XIII. This is due of course to the human factor: whether the year was warm or cold, early or late, climatically speaking, the winegrowers simply decided to pick the grapes later, and so in the second half of the curve the harvest date is regularly about a week after that found in the first half.

There is nothing comparable to this period from 1640 to 1760 in the sixteenth century (1490–1610). There are of course fluctuations, and these we shall come back to; but the curve is stabilized. Both at the beginning of the sixteenth century and at the end, late wine harvests always fall at about the same time, i.e., a week after the average date for the century, in all the localities studied. So this curve is little or not at all disturbed in its trend by human factors, and its long-term fluctuations may therefore be of climatic significance.

But what do we see? (Fig. 28). Quite clearly, during the four decades from 1561 to 1600, and more markedly from 1563 on, the number of late wine harvests (October) increases, and the number of early ones (September) declines. The same thing is repeated exactly in the quality of the wine, which is very bad between 1560 and 1600 (Appendix 12A). Conditions only return to more normal, comparable with those of the first half of the sixteenth century, during the decades 1600 and 1610.

This data shows that the number and intensity of hot summers diminished radically after 1560 and up to 1600, compared with the period before 1560. In other words, the development was the complete converse of what happened from 1880 to 1950, and there was no big series or decade of hot summers in the second part of the sixteenth century. The glaciers, comparatively uneroded by melting, had time to gather.

The winters, the wine harvest dates, and the glaciers thus all throw light on one another, and make it possible to establish the working hypothesis that during the sixteenth century, from 1540–1560 up to about 1600, there occurred a phenomenon which was the antithesis of what happened between 1850 and 1950. The decrease in annual mean temperatures determined in this way could not have exceeded, if indeed it reached, 1 degree C. But this was enough to increase accumulation and decrease ablation, thus swelling the budget of the Alpine glaciers. With incomings thus exceeding outgoings, the glaciers gradually built up until at the end of the sixteenth century they attained their maxima. After that, though there were minor fluctuations, there was no major withdrawal until after 1850.

In low altitudes like our own the slight cooling down of the "Little Ice Age" probably did not involve any significant biological or agricultural effects. But perhaps right up in the mountains this climatic deterioration, causing as it did the growth of névés and glaciers—themselves factors in local cooling down—created a kind of vicious circle and a worsening of the microclimate. That this did happen seems to be shown by the century-old larches, some of them dating back six hundred years, growing at the upper forest limit at Berchtesgaden in the Bavarian Alps.

K. Brehme, who has established a graph of their tree rings, writes that their growth "was twice as vigorous before 1600 as it became after 1700, which indicates a climatic deterioration already known from other sources, particularly from glaciological research."[40]

CHAPTER VI

HISTORICAL DATA
ON MEDIEVAL CLIMATE

The intersecular phase of advance in the Alpine glaciers, also called the Fernau oscillation, has now been defined, together with its probable climatic determinants. We have thereby been able to venture the hypothesis of a cold climatic fluctuation, a slight, almost intangible cooling down of 1 degree C or less which probably obtained, with varying degrees of intensity, over extensive regions of western Europe between 1590 and 1850. It probably began after 1540, and gradually faded out at the end of the nineteenth and in the twentieth century.

It must be stressed that this is only the least objectionable of working hypotheses, and only the research at present being carried out will tell whether it has to be discarded in favor of another.

Be that as it may, the glacial fluctuation of 1600–1850 was not the only, nor even (in spite of what Kinzl and Matthes thought) the biggest one recorded in the historic era. It is the one which has been most exhaustively studied because of the comparative wealth of documentation about it and the evidence available in quite recent and well-preserved moraines. But it was in fact only a repetition of many similar episodes during the historical era, or more exactly since the end of the optimum or *Wärmezeit* and the beginning of the subatlantic cooling down.

Following Leo Aario,[1] Franz Mayr,[2] in a remarkable study published in 1964, lists a total of five secular or multisecular episodes of the Fernau type since the beginning of the subatlantic period. Mayr was working on the Fernau glacier itself in the Tyrol, which through Kinzl[3] has given its name to the glacial thrust of 1600–

1850. By a fortunate coincidence, the maximal moraines of the Fernau glacier end in the marsh or peat bog of Bunte Moor. (The retreating front is at present 800 meters away from the bog.) The stratigraphy of Bunte Moor shows layers of peat corresponding to glacial minima (comparable to the present minimum), which left the peat bog uncovered by ice and free to produce peat. These layers of peat alternate with layers of morainic sand, corresponding in turn to times of pronounced glacial advance, when the glacier reached as far as Bunte Moor. The dates here are given by the pollen layers, which have been worked out by Firbas and Aario[4]; by various geomorphological methods; by radiocarbon dating; and by the growth rate of the peat, which has been accurately measured. Mayr's final diagram shows the alternating strata of peat and glacial sand against a time scale.

The long-term historian is bound to be interested in these five glacial episodes, though of course there can be no question of applying the chronology worked out for the Fernau glacier alone to the glaciers throughout the Alps. As far as the period covered by this book is concerned, the Fernau peat bog merely offers some relevant examples which converge with many others, both in the Little Ice Age and in the "little optimum" of the Middle Ages. However that may be, Mayr's diagram (Fig. 31) shows two major and contrasting periods. The first millennium B.C. is almost entirely occupied by two long glacial thrusts, just separated by a century of withdrawal. But in the two most recent millennia the Bunte Moor peat bog was covered much less often by the terminal ice of the Fernau glacier: the warm or mild intervals of retreat add up to a considerably greater total than the cool periods of advance.

Here, in chronological order, are the five major episodes of the last thirty-five hundred years:

(*a*) The "maximum of the Alpine glaciers between 1400 and 1300 B.C.," when the terminal tongue (750 meters beyond its 1850 maximum) "reached its greatest extension since the Ice Age."

(*b*) The "maxima of the glaciers between 900 and 300 B.C." Here we have two successive glacial thrusts, each lasting two or three centuries, and separated by only a century and a half of withdrawal. These two thrusts left their mark on almost all the glacial systems of the first millennium B.C., and can be traced in many other Alpine glaciers besides Bunte Moor. Heuberger and Beschel, using the growth rate of morainic lichens as a chronological indicator, also give 600 B.C. as the date of a big advance in the Alpine glaciers.[5] The long cold episode which began about 900 B.C., and lasted with internal fluctuations for about six centuries, introduced the cooler conditions of the subatlantic era, and thus offers many analogies[6] with the Little Ice Age of A.D. 1550–1800.

(c) After an interval of retreat corresponding to the Roman era, there was another glacial maximum between 400 and 750 B.C.

(d) The brief medieval thrust, from about 1200 (perhaps 1150) to 1300 (perhaps 1350).

(e) Finally, the 1550–1850 maximum with which this book has been largely concerned.

The spectacular thing about the peat bog evidence is that whereas it was necessary, using documentary and pictorial evidence, to build up the picture of the modern maximum very slowly and bit by bit, the stratigraphy of Bunte Moor reveals it in a flash, in the form of a layer of morainic sand between two layers of peat representing respectively the deglaciation of the Middle Ages and the not yet ended contemporary deglaciation.

It makes one think of the astronauts, seeing at a glance the contours of a continent which cartographers took many centuries to map. And in terms of millennia, the 1590–1850 intersecular phase of advance in the Alpine glaciers appears as just another example of a recurrent phenomenon.

The same applies to the present, so far secular phase of withdrawal among the glaciers of the Alps. It has been going on since 1860, and we still do not know whether it will end in a few decades or continue for one or more centuries. But one thing we are certain of is that it too is a particular example of what has happened five times before and been reflected by a simple layer of peat at Bunte Moor.

While it does not resolve the problem, the stratigraphy of Bunte

Moor also throws light on the difficult question of the secular and intersecular movement of western climate during the historical era.

Up till now, recent meteorological observations made it possible to establish this movement for two centuries, and documentary and morainic evidence about the glaciers enables us to extend our knowledge over four or five centuries.

Now all this painfully assembled data is suddenly placed in a much wider context and given a much more general significance. The two main types of secular or intersecular oscillation known to us, the cool fluctuation of 1590–1850, and the warm fluctuation of 1860–1960, are suddenly each reproduced five times. And the characteristics of intersecular movement suggested by the climatologists—particularly by Pierre Pédelaborde in 1957[7]—are authenticated and verified experimentally. There is no doubt that this movement involves long oscillations, alternations, periodicities if you like; but this periodicity is not regular, nor is it cyclic in the strict sense of the word. The "phases," whether mild or cool, are never of the same length as each other. Periods of glacial advance may last about two and a half centuries, like the Fernau oscillation of 1590–1850; or hardly more than one century, like that from 1150–1200 to 1300; or over three centuries, like that of the first millennium B.C. Similarly, phases of glacial withdrawal and warm climate may last scarcely a century, or four or five hundred years and more.

Another thing which emerges is the homogeneity of the overall pattern. Many climatologists think that in fact all climatic oscillations are of the same type, whether they are, individually, annual, decennary, secular, millenary, or have to be measured in geological time.[8] At all events, the intersecular oscillations shown by Bunte Moor combine to give a picture which is millenary and intermillenary, which, falling as it does within the subatlantic period, forms part of the fundamental chronological divisions of the postglacial age and the Quaternary. There is thus a continuity between historical and geological time.

These glacial thrusts, which though they vary in length all last over a century, present a great homogeneity in space as well

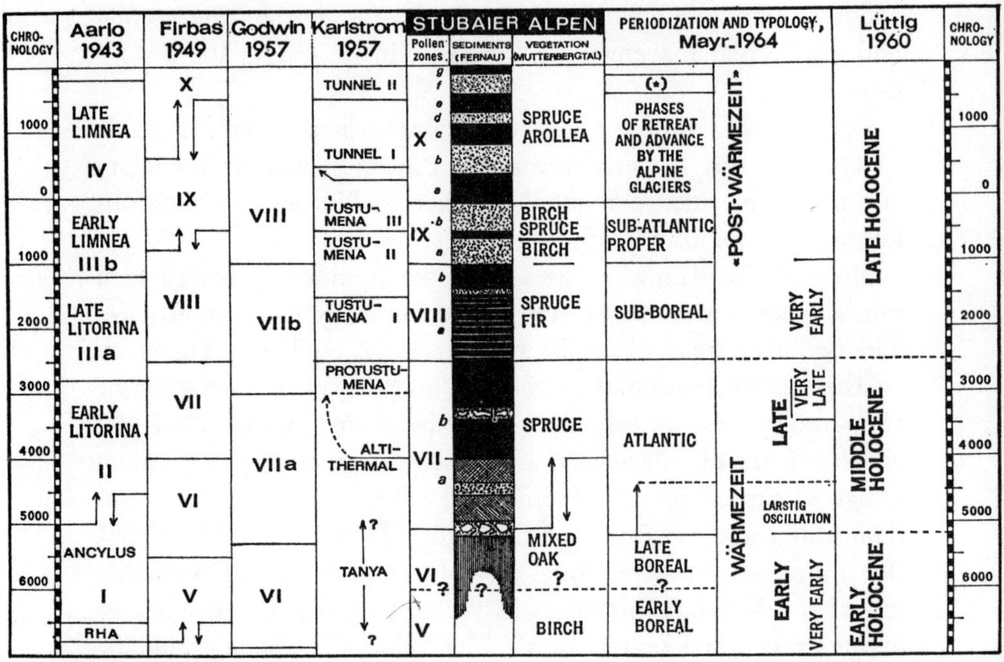

CHRO-NOLOGY	Aarlo 1943	Firbas 1949	Godwin 1957	Karlström 1957	STUBAIER ALPEN Pollen zones	SEDIMENTS (FERNAU)	VEGETATION (MUTTERBERGTAL)	PERIODIZATION AND TYPOLOGY, Mayr 1964			Lüttig 1960		CHRO-NOLOGY
1000	LATE LIMNEA IV	X		TUNNEL II	g f e d		SPRUCE AROLLEA	(•) PHASES OF RETREAT AND ADVANCE BY THE ALPINE GLACIERS	"POST-WÄRMEZEIT"		LATE HOLOCENE		1000
				X c b								
0	IV		VIII	TUNNEL I	a								0
1000	EARLY LIMNEA IIIb	IX	VIII	TUSTU-MENA III / TUSTU-MENA II	IX b a		BIRCH SPRUCE / BIRCH	SUB-ATLANTIC PROPER					1000
2000	LATE LITORINA IIIa	VIII	VIIb	TUSTU-MENA I	VIII		SPRUCE FIR	SUB-BOREAL	VERY EARLY				2000
3000	EARLY LITORINA II	VII		PROTUSTU-MENA	VII b a		SPRUCE	ATLANTIC	LATE VERY LATE / LATE	MIDDLE HOLOCENE			3000
4000			VIIa	ALTI-THERMAL					WÄRMEZEIT				4000
5000		VI						LARSTIG OSCILLATION					5000
6000	ANCYLUS I / RHA	V	VI	TANYA	VI ? / V		MIXED OAK ? / LATE BOREAL ? / EARLY BOREAL / BIRCH		EARLY VERY EARLY	EARLY HOLOCENE VERY EARLY			6000

(•) "FERNAU" GLACIAL THRUST

Fig. 29 *Recent Glacioclimatic Periodization Derived from a Subglacial Swamp*
In the middle, in the column headed *Stubaier Alpen, Sediments, Fernau,* is represented schematically the stratigraphy of the subglacial peat bog of Bunte Moor-Fernau. The black portions show the stages at which peat was laid down (glacial withdrawal). The speckled portions show the levels of morainic sand (signs of glacial advance). To left and right of this column are time scales and reminders of the postglacial and postoptimal periodizations proposed by various authors, including that of Aario for the Baltic, those of Firbas and Godwin for West European pollens, that of Karlström for the Alaskan glaciers, etc. (Source: Table is derived from Mayr, 1964, pp. 384–85, Table 6.)

as in time. In none of them has the Fernau glacier ever gone more than between 200 and 800 meters beyond the position of the terminal moraines of 1850, which mark the extreme point of the latest thrust.

The medieval offensive of the Alpine glaciers (about 1215–1350)

Two of these advances stand out from the rest in respect to the methods by which they may be investigated. The modern advance of 1590–1850 is, as we have seen, amply described in various documents. But the last advance but one, the advance which took place in the Middle Ages, although like its predecessors it is undocumented in the ordinary sense of the term, is still accessible to the historian.

In other glaciers besides Fernau, documentary evidence and geomorphological methods can be brought together very fruitfully. The Vernagt, Aletsch, and Grindelwald glaciers are all very good witnesses, and prevent our chronology for the Middle Ages from resting on the evidence of Fernau alone.

I have already cited the curious text by Benedikt Kuen,[9] who in 1712, on the strength of an old register, stated that the initial thrust of the Vernagt glacier began "in the thirteenth century" (*Anfang . . . im dreyzehnten saeculo*). Mayr's geomorphological researches confirm this.

At Aletsch, a fundamental souce of medieval datings, archaeological data, documentary evidence, and radiocarbon datings all converge. It was when he was tracing the *Oberriederin*, the early irrigation course near Aletsch, that Kinzl found that the glacier itself was smaller in the Middle Ages than in the seventeenth century, and even smaller in fact than it is today.[10] The *Oberriederin*, the course of which is indicated by various remains of walls,[11] started in a subglacial stream or one of its tributaries, at a point which is today still covered and obliterated by the glacier. According to a text in the archives, it was already abandoned in 1385. And recent radiocarbon datings have pushed the date back even further.[12]

In its retreat of the past few decades (1940), the Aletsch glacier has uncovered the remains of humus deposits and a fossilized forest of larches and deciduous trees rooted in the rock. Various samples which have been radiocarbon tested at Berne show that this forest lived for about two centuries and then was crushed by the advancing glacier. The dates are:

Sample 1 (larch): 720±100 years before the present era.

Sample 2 (deciduous): 800±100 years before the present era, or about A.D. 1190.[13]

Since the starting point of the *Oberriederin* was still covered by the glacier in 1961, it follows that this irrigation conduit must already have been unusable in 1190, by which time the glacier had already gone beyond this point to a slightly more advanced position from which it began to withdraw in 1940.

So at Aletsch as at Fernau, the thirteenth century was a period of glacial advance, in contrast to the ninth–eleventh centuries, when there was a marked retreat during which the forest was able to grow and the *Oberriederin* to function.

The local chronicle of Grindelwald, in a later text of the sixteenth or seventeenth century,[14] says that the church was moved in 1096 from the "Burgbiel" to its present position "because of the glacier and the danger of flood." This text is a doubtful one: the old church at the Burgbiel only seems to have dated back to 1140; should "1096," then, be read as "1196"? Or should the text just be disregarded?

Much sounder and more valuable is the evidence of the fossil forest at Grindelwald. According to Gruner,[15] a reliable eighteenth-century witness (1760): "In the Grindelwald glacier, near to the place called *Hot Leaves* or *Black Shelf* (*Heisse Platte* or *Schwarze Brett*), you can see larches which are still quite fresh, although they have been a long time in the ice . . . These trees clearly show that this was once fertile land." He goes on to say later: "At Grindelwald, on the sides of the Fischerhorn and the Eiger, there are several larch trunks in the middle of the ice, which have been there perhaps several centuries. People who have been there say it is impossible to prise out the smallest scrap of them with the sharpest knife." Gruner compares these fossil trunks with the petrified wood found in the Danube under the piles of a Roman bridge.

I visited this site in 1960. In the right lateral moraine (Fig. 20), near Stieregg, there were still magnificent trunks of arolla (cembro pine) emerging from the morainic gravel. They have been scraped and smoothed by the ice, and would certainly yield dendrochronological evidence. The most striking thing about the place is the complete absence of living trees on the slopes all round the glacier and its moraines, at this altitude and even higher.

So it seems as if, "once," in this bare and partly ice-covered area, there was a forest which was able to grow up because of a fairly prolonged glacial retreat, and probably also because of a partial or complete absence of sheep grazing over the land. It was probably later destroyed by the advance of the glacier, which preserved relics of the dead trunks in its moraines.

One of these relics, collected by Pastor Nil of Grindelwald,[16] was dated some ten years ago by the radiocarbon method, which gave the death of the forest as occurring 680±150 years before the present era, i.e., A.D. 1270±150. This is still the glacial thrust of the thirteenth century, and very near the Aletsch datings (A.D. 1190±100 years). So there seems to be a sort of *terminus a quo* about A.D. 1200.

This was the chronology at my disposal for the first edition of this book. Since then, Stuiver and Suess[17] have put forward correction tables rectifying and bringing up to date the carbon datings. For the twelfth century these corrections are very small, and such as they are tend to narrow the bracket of possible variation. On the whole they confirm or bring slightly forward the dates cited above.

Site	Previous radiocarbon datings	Rectified radiocarbon dating
Aletsch 1	A.D. 1230	A.D. 1230
Aletsch 2	A.D. 1150	circa A.D. 1200
Grindelwald	A.D. 1270	A.D. 1280

The Aletsch advance is the most relevant, because it is attested by rooted trees that were killed where they stood. This advance is now dated 1215 (1200–1230), while that at Grindelwald is now 1280. So the *terminus a quo* (the end of the little optimum of the year 1000, and the beginning of the Little Ice Age of the Middle Ages) is now placed at about 1215 or even slightly later, and it is then that we have to think of the Alpine glaciers as seriously embarking on their movement toward a temporary maximum.

The complex formed by the Fernau glacier and the Bunte Moor peat bog provides the *terminus ad quem* of this medieval thrust.[18]

The two layers of barren morainic sand, clay, and pebbles which correspond to the two glacial advances (one the "Fernau stage" of 1590–1850—Xf in Mayr's terminology—and the other the medieval thrust, or Xd) are separated by a band of peat from 8 to 10 centimeters thick, which corresponds to the intermediary phase of glacial withdrawal (Xe). At this altitude the rate of formation of the peat is estimated at 3 or 4 centimeters per century.[19] There are thus two hundred and fifty or three hundred years between Xf and Xd. All things considered, the medieval glacial thrust must have begun about 1215, and ended about 1350. It was a comparatively brief episode, but a violent one: the erosion, denudation, and soil disturbance accompanying Xd were more intense than at the beginning of Xf.

This chronology agrees very well with the data we have for the year 1300 concerning the Allalin glacier, in the Saaser Visp valley in Switzerland.[20] Our information comes from a text dated April 13, 1300, written by two stock farmers who were tenants of the mountain pasture of Distelalp in the upper valley of the Visp, above the Allalin glacier. The two men were making an agreement with Jocelin de Blandrate, mayor of Visp, and wrote: "We ask you to grant us tenure of the pasture in question, from the glacier upwards (*a glacio superius*), according to the custom of the men of the Saaser Visp valley, so that the said men do not prevent our herds from grazing as far down as the glacier."[21]

It appears clearly from this text that in 1300 the Allalin glacier (and perhaps also that of Schwarzenberg upstream) *barred* the upper valley of the Visp; thus the upper mountain pastures on the right bank of the Visp—the pastures of Distelalp—were also separated by the glacier from the lower valley of the Visp. So it seems that in 1300 there was a situation typical of the Little Ice Age, analogous to that which recurred during the great advance of the Allalin glacier between 1589 and 1850. This situation is illustrated by the watercolor by Maximilien de Meuron, entitled "Distelalp and Allalin, 1822."[22] In our own time, however, the front of the Allalin glacier has withdrawn from the bed of the Visp valley, and only overhangs, from a considerable height and

above steep slopes, the *left* bank of the valley. Between the retreating front and the thalweg of the Visp there are thick moraines marking the withdrawal of the last hundred years. Thus the
Allalin glacier, and still more the Schwarzenberg, has ceased to be
the natural barrier between the high mountain pasture of Distelalp and the meadows by the Visp farther downstream, as it was
in 1300. Another sign of glacial thrust at the end of the thirteenth
century and the beginning of the fourteenth may perhaps be found
in the text adduced by Christillin, suggesting that the Ruitor
glacier in the Val d'Aosta was in a state of advance in 1284. But
we do not know how much reliance to place in this text (Appendix 13).

At any rate there is enough archival and radiocarbon evidence
now for us to say that there was a glacial advance in the Alps
during a substantial period between 1215 and 1350.

Did this medieval advance, like Xf at the end of the seventeenth century, have a parallel in Scandinavia? To answer this we
should have to reexamine the whole record of Viking colonization
in Greenland, and reinterpret the conclusions that have been
drawn from P. Norlund's excavations.[23] We should also have to review the data established by T. Longstaff[24] on the positions of
Norman farms in Greenland in the eleventh century; these positions later became almost inaccessible because of the ice barring
the fjords farther downstream. The well-known text by Ivar Baardson, which is unfortunately unique, and perhaps somewhat late in
relation to our Alpine chronology, is one of the most suggestive
references.[25] Ivar Baardson was a Norwegian priest who lived in
Greenland between 1341 and 1364 as steward to the bishop of
Gardar. He wrote: "From Snefelness in Iceland, to Greenland,
the shortest way: two days and three nights. Sailing due west. In
the middle of the sea [in fact off the coast of Greenland] there
are reefs called *Gunbiernershier*. That was the old route, but now
the ice is come (*en nu er kommen is*) from the north, so close
to the reefs that none can sail by the old route without risking
his life."

Do these geographical correlations extend even farther than

Greenland, as in the case of phase Xf? It is quite likely. At any rate Y 7-1955, a sample of carbon-dated fossil wood from Glacier Bay, Alaska, indicates, according to Franz Mayr,[26] a glacial advance there contemporary with the medieval Alpine advance.

The medieval advance, in the Alps at least, seems to have been of fairly short duration—scarcely more than a century or a century and a half, certainly not more than two centuries. It was thus the shortest of the long glacial episodes of the last three millennia, the rest of which have generally lasted two or three centuries and sometimes more.

Before the glacial offensive of the thirteenth century: the "little optimum"

Can we estimate the dimensions of phases of withdrawal which preceded and followed the medieval advance? The Aletsch and Grindelwald evidence is some help here. The retreat of the early Middle ages (ninth–eleventh centuries) seems on the whole to have been slightly more marked than that of the twentieth century has been, at least up till now. The "preglacial" trees which were eventually destroyed by the 1200 offensive grew on sites where in our own time trees have not had time or the necessary conditions to grow again. And at Aletsch the starting point of the *Oberriederin* is still covered by the glacier.

All this being so, is it correct to apply the name "little climatic optimum" to the period of marked glacial retreat which occurred between A.D. 750 and 1200–1230?[27] It is probably appropriate, in so far as the word "little" differentiates this from the "big" optimum of prehistory. In the first place, the "little optimum" of the Middle Ages lasted only a few centuries, whereas the Neolithic *Wärmezeit* lasted thousands of years.

The description is justified above all in terms of temperature differences: the climate of these four centuries, from the Carolingians to the great land clearances, seems to have been quite mild—as mild as, or even a little milder than, the twentieth century. It is reasonable to think of the Vikings as unconsciously taking advantage of this to colonize the most northern and inclement of their conquests, Iceland and Greenland.

But we cannot really go beyond these cautious hypotheses. For we know from pollen analysis that vegetation groupings characteristic of the great prehistoric optimum did not recur around the year 1000. To take one of a dozen examples, the hazel did not return in the eleventh century to its "optimal" positions toward the north of Scandinavia.

So the mean annual temperature difference between this mild period and the cool one which followed is apparently of the same order of magnitude (perhaps a few tenths of a degree Centigrade more) as that which separates the cool period of 1800–1850 from the mild period of 1900–1950: i.e., about 1 degree Centigrade.[28] The special characteristics of the warm period around the year 1000 can be explained by a temperature difference of this order; there is no need to suppose a difference comparable to the "great optimum," the atlantic period of prehistory (2 degrees Centigrade or more).

In terms of more precise chronology, series of meteorological events[29] taken from medieval documents place the most clement period of this little optimum, with its mild winters and dry summers, in A.D. 1080–1180, at least as far as western Europe (England, France and Germany) are concerned. This documentary evidence coincides with other evidence of a fine twelfth century: the Alpine glaciers, which were very shrunken during this period; cores of foraminifera from the Atlantic bed,[30] which show a warm tendency culminating toward 1200; and the curve showing the extent to which the coast of Iceland was encumbered with ice, probably minimal[31] between A.D. 1020 and 1200.

In western Europe this mild period of the early Middle Ages seems to have been accompanied by quite marked droughts, the combined result of lack of rainfall and strong evaporation.

The stratigraphy of the peat bogs shows damp episodes in the form of "recurrence surfaces," which correspond to a renewal of the growth of sphagnum peat. It is difficult to interpret these layers with certainty, because they may have been caused either by a sudden temporary flooding of the peat bog or by a real phase of heavy rainfall.

Overbeck, by using many radiocarbon datings, has been able to establish the chronological "brackets" within which recurrence surfaces are met with. (Ry is the notation used for them.)

According to Overbeck, recurrence surfaces occur in the peat bogs of Germany (after the beginning of our era) between A.D. 400 and 700 (Ry II) and in 1200 (Ry I). There are none at all in the ninth–twelfth centuries, which seem to have been dry as well as mild.[32]

As to this dry period, there is an interesting text by Boulainvilliers, unfortunately somewhat late, taken from *intendants'* reports at the end of the reign of Louis XIV.[33] The text concerns the Sarthe, a typical river of a very damp region, usually well provided with water. But it appears to have dried up, just like a Mediterranean wadi, three times in history: first in some unspecified year in the reign of Charlemagne; then, similarly, under Louis le Débonnaire (814–840); and a third time in June 1168, for just under half an hour. Could it be that these three episodes are characteristic extremes of the dry or relatively dry period of the ninth–twelfth centuries?

At all events, the little optimum of the Middle Ages caused Europe to experience various gusts of warmth, and even sometimes great heat. These were responsible for the plagues of locusts which in the ninth–twelfth centuries sometimes spread over vast areas, sometimes far to the north.[34] In A.D. 873, a time of great famine, they were found from Germany to Spain; during the autumn of 1195, they reached as far as Hungary and Austria.

The little optimum, warm and probably dry as well, which extended on either side of the year 1000 and faded out in the thirteenth century, seems to have affected very large areas both in the old world and in the new. Traces of it are found on many sites in western and central Europe[35]; in the already mentioned cores from the Atlantic bed; even in the far north of Canada. A few years ago Reid A. Bryson, prospecting west of Hudson's Bay, discovered on the shores of lakes Ennadai and Dimma the remains of a fossil forest, 25, 40, and even 100 kilometers north of the present forest limit. Four radiocarbon datings done for different sites showed that this forest was living about A.D. 880, 870, 1090, and 1140. This corresponds exactly with the medieval little optimum, which both in Canada and in Europe temporarily displaced toward the north the Arctic polar front and the northern forest limit.[36]

The evidence converges from all sides to suggest a chronology in three movements for the climate of the last millennium: 1. the little optimum of the year 1000; 2. the glacial thrusts of the Little Ice Age, culminating first about 1200–1300 and then about 1580–1850; 3. the contemporary warming up.

This chronology, which had already been substantially verified, has recently had splendid confirmation from Greenland, where the activities of the Vikings have provided so much useful information of fluctuations in climate since the Middle Ages. The recent evidence comes in the main from two sources. The first, both fascinating and uncertain, is the "Vinland Map." The other is the great ice core from Camp Century, the stratigraphy of which has just been published by Danish and American scholars.

Quite recently, thanks to the researches of the librarians of Yale University,[37] an extraordinary new and controversial document was added to Nordic history. The Vinland Map is supposed to have been drawn up in the middle of the fifteenth century,[38] and seems to condense the information amassed by Scandinavian sailors in their voyages during the tenth–thirteenth centuries. It shows a fairly exact outline of Greenland, and if authentic may well confirm Ivan Baardson's previously mentioned chronology consisting of two periods[39]: first the end of the tenth, the eleventh and twelfth centuries, during which the southern half of the east coast of Greenland, at the latitude of Gunnbjørn's Skerries (Gunbiernershier),[40] was relatively free of ice, so that ships coming from Iceland could sail in a direct east-west line to Greenland. (More generally, Nordic sailors and settlers are supposed at this period to have acquired a good experimental and even cartographic knowledge of the whole coast of Greenland, a knowledge which is supposed to be distantly reflected in the Vinland Map.[41] Baardson's second period[42] includes the thirteenth–fourteenth centuries, and perhaps even the twelfth, and during this phase the ice moved southward, making the approach through Gunnbjørn's Skerries impracticable and forcing ships from Iceland and Norway to follow a much more southerly route to Greenland.

This periodization has the advantage of agreeing with the recent indisputable discoveries about medieval fluctuations in the climate of Greenland, made from ice cores by American and Danish researchers.[43] But is the Vinland Map genuine? G. R. Crone categorically denies it,[44] and his skepticism should make us cautious and at least wait for fresh evidence. For Crone, the very fact that the Vinland Map gives an outline of Greenland is an additional reason for mistrusting it! Arguing exactly the opposite from Skelton, he writes: "Another difficulty of the Vinland Map is the apparent accuracy of the outline of Greenland, which was not circumnavigated until the nineteenth century. It is generally accepted that, firstly, this great island could not have been circumnavigated at an early period, despite a somewhat milder climate; secondly, that there seems no motive for the Norsemen to have undertaken such a voyage; and thirdly that the Norsemen did not use or make charts. It is possible that the map was reconstructed in or before 1448 from oral tradition and a study of the sagas, though even this hypothesis would not explain the Greenland outline. On the cartographic side, a solution to the puzzle may result from a further study of contemporary maps and charts of the northern Atlantic. For the present it remains an enigma."[45]

Let us leave the enigma for the moment, and turn to the many certainties about Greenland recently provided by the experts on glacial fossils.

In 1966 an American research unit, the C.R.R.E.L. (Cold Region Research and Engineering Laboratory), succeeded in extracting an ice core which went vertically right through the ice at Camp Century. The sample thus obtained was 12 centimeters in diameter and 1390 meters long. The C.R.R.E.L. arrived at the approximate age of the different sections of this column of ice by means of a complex formula taking into account the rate of accumulation of

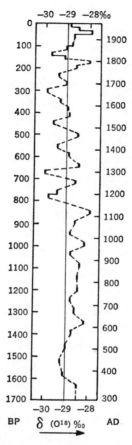

Fig. 30 Variations in $\delta(O^{18})$ in the upper 470 meters of the Camp Century ice core plotted against the calculated age of the ice. The climate in Greenland was warmer before A.D. 1130, and colder after that date. (Source: Dansgaard, et al., 1969, p. 378.)

the ice (35 centimeters per year), and the rate at which it was compressed under the weight of successive upper layers. Over a thousand centuries of ice, gradually built up right to our own day, thus became available for systematic research. Dansgaard and other authors have undertaken this study.[46]

From the varying ratios of the oxygen isotope O^{18} present in glacier ice, it is possible to explore temperature conditions of the past: the concentration of O^{18} in precipitation of rain or snow preserved in the form of fossil ice is chiefly determined by the temperature at which the precipitation condensed. "Decreasing temperature at formation leads to decreasing content of O^{18} in rain or snow." And vice versa.

The highest and most recently formed layers of the Camp Century ice sample show high concentrations of O^{18}. These correspond to the well-characterized climatic optimum of the years 1920–1930. Here it may be mentioned that climatic fluctuations in Greenland do not correspond exactly in magnitude or time with parallel fluctuations in Europe. But over all, and at the level of secular trends, there is an approximate convergence, and this is confirmed by the work of the C.R.R.E.L.[47]

Below the levels representing the recent warm years from 1900 to 1950, the Camp Century ice core presents layers corresponding to the Little Ice Age and characteristically poor in O^{18}. The Little Ice Age here extends roughly from the thirteenth to the nineteenth century, and breaks down into three basic cold waves. The first occurs between 1160 and 1300, and is followed by a not uninterrupted but moderate remission (1310–1480). The culmination of the cold, foreshadowed in the sixteenth century, duly arrives in the seventeenth century first, then again toward 1820–1850. By contrast the eighteenth century (1730–1800) seems to have been a temporary but marked period of warming up.[48]

Of course this periodization is not definitive or hard and fast. Other ice cores from other sites will be studied, and will rectify or add detail to this chronology. The important thing here is that beyond the secular fluctuations occurring roughly (very roughly!) in terms of a "cycle" of 120 years, the major period of cold indicated by the Alpine glacial advances of the thirteenth, seventeenth, and nineteenth centuries, is now, in Greenland, determined with precision, always allowing for important differences due to the geographical distance between the continent of Europe and the Greenland subcontinent.

Then as it reaches even farther down the ice core, the Camp Century diagram reflects the pleasant warmth of the little optimum of the early Middle Ages. Suddenly, in the five centuries preceding A.D. 1125 (i.e., from A.D. 610 to 1125), the curve rises and remains at a ceiling; the concentration of O^{18} increases, and for the whole period remains higher than during all the Little Ice Age (thirteenth–nineteenth centuries). This brings out the continuity of the intense warming up which lasted from the seventh to the eleventh century, half a millennium. There is no doubt that the Norsemen took advantage of normally ice-bound coasts which then became accessible in the Arctic.[49] They landed in waves on the marginal lands forming the kingdoms of Thule. The climatic amelioration may have facilitated the colonization of Iceland in the ninth century, and certainly facilitated that of Greenland in the tenth. Two thermal maxima stand out on the Camp Century curve against the background of the whole favorable period from A.D. 610 to 1125. One occurs in the last third of the tenth century, the other in the first quarter of the twelfth. Nothing similar is met with again until the fine warm periods which occurred in Greenland at the very end of the eighteenth century (1780–1800), and above all during the recent optimum (1920–1930).

The two early medieval culminations of the Greenland little optimum present interesting parallels with two essential episodes in the history of the Arctic subcontinent. From 978 to 986, first Snaebjørn Galti, then Eirik the Red, took advantage of a sea relatively free of ice to sail due west from Iceland to reach Greenland in the latitude of Gunnbjørn's Skerries; thence Eirik turned to the south of the island, where he established his great farm of Brattahlid at the same time as the eastern settlement.[50] Two and a half centuries later, at the height of the climatic and demographic fortunes of these northern settlers, a bishopric of Greenland was founded at Gardar in 1126.[51]

So the great ice core of Camp Century confirms the patient researches of the Danish archaeologists who since 1925 have first suspected, then demonstrated the existence of a little optimum in Greenland in the Middle Ages.

Going back further still, the same enormous sample throws light on and confirms many other important episodes. The Fernau peat bog suggests a maximum in the Alpine glaciers at some indeterminate date between A.D. 400 and 750; the Camp Century ice core presents a probable equivalent of this in a cold episode occurring between A.D. 340 and 620. As with the Little Ice Age of

1580–1850, this episode was probably an intercontinental one, affecting at least both Europe and America. John Mercer, in an important article on "Glacier Variations in Patagonia"[52] notes that according to radiocarbon datings the glaciers of the American continent (Alaska and Patagonia), after showing certain symptoms of advance in A.D. 250, were in a state of maximum about A.D. 450.

Going back earlier yet, beyond the relatively warm phase[53] at the beginning of our era (50 B.C.–A.D. 200), the Camp Century curve forcibly underlines the amplitude and violence of the subatlantic cooling down which affected the whole of the last millennium B.C. (the coldest period occurring between 500 and 100 B.C.). Mercer's excellent demonstration enables us to extend our conclusion: it was all the glaciers which reacted with a strong maximum to the peak of the subatlantic cold between 500 and 300 B.C.: not only those of Greenland, but also those of the Alps, Iceland, Sweden, New Zealand, and Patagonia (the latter magnificently dated).[54]

Last but not least, the Camp Century ice core definitively confirms the existence of the prehistoric climatic optimum. This phase reached its maximum in Greenland between 5200 B.C. and 2200 B.C., more exactly between 4000 B.C. and 2300 B.C. So the fourth millennium before our era (4000–3000 B.C.) really was, both in Europe and Greenland, the "sunny millennium" long suggested by the pollen diagrams for the Nordic countries.

So, thanks to O^{18}, the ice sheets "remember" climatic fluctuations right from the great Ice Ages down to the recent amelioration.[55] Dansgaard and his colleagues are extremely definite about the causes of these fluctuations, and if one accepts their analysis, which in many respects agrees with that of Suess, one must conclude that there is an indissoluble connection between:

(1) the small fluctuations in the quantity of solar radiation, which cause the world to grow warmer or cooler;

(2) secular trends of solar activity, as measured by long-term variations in the number of sunspots;

(3) the production of O^{18} in the atmosphere;

(4) the production of C^{14} in the atmosphere, the variations

in which are known through the analysis of precisely dated fossil wood.

Dansgaard writes: "The prime cause of the oscillations in oxygen-18 concentration is probably related to fluctuations in solar radiation. Solar variations are also considered to cause changes in the C^{14} concentration in the atmosphere. During periods of high sunspot activity, an increased amount of short-wave radiation reaches the earth, and at the same time the magnetic field of plasma emitted from the sun shields off the cosmic radiation to a relatively high degree, causing low production of C^{14} in warm periods associated with the high concentrations in O^{18}. This effect gives rise to temporal oscillations in the uptake of C^{14} by plants."[56]

According to this view, cold centuries like those of the Little Ice Age, and particularly the seventeenth century, are "centuries of quiet sun."[57] In a recent article, Suess has developed this idea at length. He refers again to his curve for carbon-14 production, obtained from the closely dated fossils of the Californian bristlecone pine. He puts forward the threefold chronology suggested by this curve:

1. Little optimum of the eleventh century: low production of C^{14};

2. Little Ice Age, beginning in the thirteenth century and culminating in the seventeenth: high production of C^{14};

3. Recent amelioration: a new fall in the concentration of C^{14} in the atmosphere.

Suess concludes: "In view of the complexity of the problem it appears most remarkable that the observed C^{14} values indeed show a correlation with the more general patterns of climatic change.[58] In particular, at the time of the so-called 'Little Ice Age' the carbon-14 concentration in the atmosphere was rising during the seventeenth century. During this period, central Europe had a sequence of exceptionally severe winters, and glaciers were advancing everywhere. During the same time the sun was quiet, and sunspot numbers were low. It seems that climate and carbon-14 concentrations were independently influenced by the unusually quiet state of the sun."[59]

In short, if we believe Suess and Dansgaard, there is a reassuring convergence between carbon-14 and oxygen-18: in contrast with centuries of quiet sun and climatic cooling down (for example the seventeenth century), there are the pleasant periods of the optima (the twentieth-century amelioration, and especially the longer "little optimum" of the early Middle Ages). The optima, according to this view, are characterized by slightly intenser radiation and solar warmth; by increased numbers of sunspots, in the secular long term; by low production of C^{14}; by high production of O^{18}; and by a slightly warmer and more clement climate (see Appendix 19).

After the glacial advance of the thirteenth century: the climatic uncertainties of the end of the Middle Ages

As we have seen, the climatic chronology of the Middle Ages up to the thirteenth century now rests on many converging kinds of evidence. We must now turn to the difficult problem of the climatic fluctuations of the end of the Middle Ages: to the apparent period of glacial retreat, very moderate retreat, which followed the advance of the Alpine glaciers in the thirteenth century and may be supposed to extend between about 1350 and 1500. This regression of the glaciers seems to have been less marked than that of the ninth–eleventh centuries, and their terminal tongues must have oscillated within limits fairly comparable with those of 1900–1950, or very probably within a slightly narrower range. This at least is what may be suggested, with the usual reservations, on the basis of certain indications.

The first of these is that during the period 1350–1550 the Aletsch glacier probably did not withdraw as far as its 1940 positions. If it had, the trees killed there in the twelfth century and preserved down to our own day by the ice, would not have lasted long exposed to the air.

Nor does the fossil forest of Grindelwald, killed in the thirteenth century, seem to have grown again in the fifteenth.

So apparently climate had not returned to the mildness of the year 1000. But there was a retreat, or, to put it differently, between 1350 and 1550 there was no glacial advance comparable to

XIX. THE ARGENTIERE GLACIER in 1780

This admirably precise engraving by Hackert (reproduced from the copy in the Bibliothèque Nationale, Cabinet des Estampes, "Etats sardes et Savoie," dossier V b 1) shows the Argentière glacier quite close to the church of the same name. It is 1005 meters beyond its position on October 26, 1911, as measured by Mougin with a Sauguet tacheometer (Mougin, 1912, p. 154), and even more in advance of the frontal positions of our own time.

(Photo: Ed. Flammarion.)

XX. The same view in 1966

Two centuries later the mountains and the church are still there, though the site depicted by Hackert has been invaded by forest and buildings. But the glacier has withdrawn far to the left and disappeared from the scene.

(Photo: M.L.R.L., July 1966.)

XXI. THE ARGENTIERE GLACIER about 1850–1860

This engraving, by a German or a Swiss, bears the words "*Nach Photographie arrangirt von L. Rohdock; G.M. Kurz sculps.*; Druck und Verlag von G.G. Lange." The Argentière glacier has only just begun its withdrawal; it still forms the characteristic zigzag mentioned by Saussure about 1770; its front is still close to the plain and the village.

(Photo: Ed. Flammarion.)

XXII. THE ARGENTIERE GLACIER in 1966

This photograph of the same view as the previous engraving shows the very marked secular retreat of the glacier. All the lower branch of the zigzag had disappeared, revealing huge rocks once rasped by the ice. The moraines have been covered with larches. The village has grown a little, but the church and certain houses are easily recognizable.

(Photo: M.L.R.L., 1966.)

XXIII and XXIV. THE GLACIER DES BOSSONS
(Nineteenth–twentieth centuries)

Early pictorial evidence about the Des Bossons glacier is less abundant than for the previously mentioned Swiss or Chamonix glaciers.

Plate XXIII is an engraving by Winterlin executed about 1830–1850; the view-point was a site on the right bank of the Arve. Unfortunately this view-point can no longer be used because of forests and buildings. But a photograph taken in 1966 from slightly nearer shows the withdrawal of the glacier. The strange tower of séracs, which stands out so clearly in Plate XXIII in front of the tongue of the glacier, has disappeared in Plate XXIV, and the forest has crept over the moraines and areas abandoned by the ice. The Tacconaz glacier can just be glimpsed on the right.

(Private collection, and photo by M.L.R.L.)

XXV and XXVI. THE BRENVA GLACIER (1767 and 1966)

Jalabert's drawing (1767) was made on the spot during a visit he paid there with Saussure, and was published as an engraving by Saussure, 1786, Vol. IV, p. 27, Plate III. This is the finest and most accurate early representation I know of any of the Mont Blanc glaciers. Note, lower right, "the huts of the farmers who cultivate the fields near the glaciers" (Saussure, p. 27).

Plate XXVI (photo by M.L.R.L., 1966) shows how marvelously accurate Jalabert was, and also the magnitude of the recent withdrawal: all the "piedmont" of the eighteenth century has gone, leaving a huge moraine.

The withdrawal is of the order of 750 meters, according to the I.G.N. map of 1958.

It is on the left, in the moraines at the foot of the slopes running down from the Aiguille Noire and the Mont Noir de Peuterey, that local tradition places the site of Saint-Jean-de-Perthuis, a village destroyed by a glacial advance. (According to Kinzl, 1932, and a verbal communication by Mme. Zanotti, hotelkeeper, who belongs to an old Entrèves family.)

**XXVII and XXVIII. ALLEE BLANCHE GLACIER and
LAKE COMBAL (Courmayeur region)**

In 1861 (Plate XXVII), as in the year VII of the Republic (Fig. 23 and Plate XXVIII), the Allée Blanche or Lex Blanche glacier came right up to Lake Combal. (Drawing and engraving by E. Aubert, from Aubert, 1861.)

The photograph in Plate XXVIII (M.L.R.L., 1966) was taken from exactly the same point of view as Aubert used, as is proved by the trapezoid overhanging rock in the bottom right of both pictures. In the photograph the glacier has withdrawn, and is only just visible behind the somber rocks that descend on the right from the Combal needle. The withdrawal is at least 500 meters.

The lake itself has partly dried up and been overgrown with vegetation in its upper parts. It is to be hoped that the sediment of the lake will be studied, for it would provide information about the long phases of drying up and of replenishing the lake. These phases may be related to variations in the volume of the glacier, which feeds the lake. (But it is also necessary to allow for the artificial dike which has raised the level of the lake since some unknown date before 1691. See Vaccarone, 1881, p. 183.)

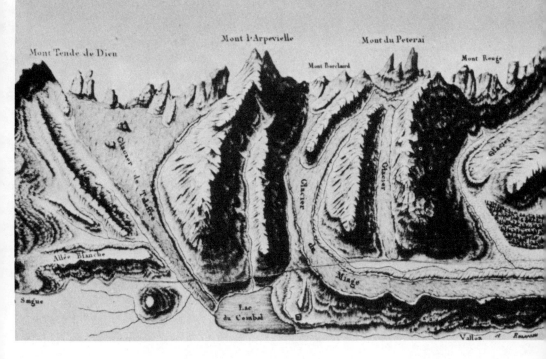

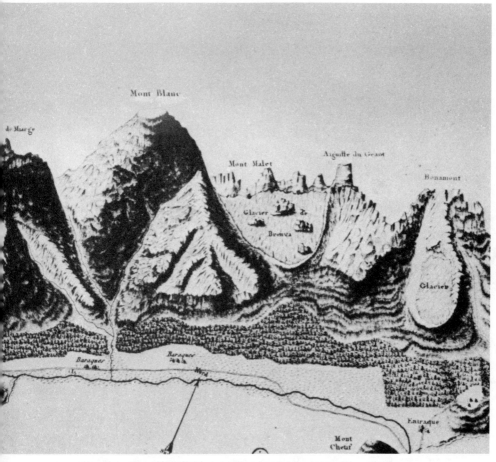

XXIX. LAKE COMBAL in the year VII (1798)

This map by citizen Bourcet, drawn during the revolutionary years when pictorial documents on the glaciers become comparatively rare, is probably one of the first maps of Lake Combal (left).

At this period the lake is shown as fully developed and in direct contact with the glacier which feeds it. This corresponds to a relatively marked glaciation, comparable with that of 1861 and radically different from the contemporary configuration (see above, notes on Plates XXVII and XXVIII). But Bourcet, though he placed the glaciers correctly, made mistakes in their names: he calls the Allée Blanche glacier the "Talèfre," and calls the little glacier De L'Estellette, adjacent on the left of the picture, the Allée Blanche. (The real Talèfre glacier is not in this region at all; it is 20 kilometers to the northwest, beyond the Mer de Glace.)

(Source: see Plate XIII.)

XXX and XXXI. On the MONTENVERS

Early pictorial evidence (of the eighteenth century) mainly provides evidence for comparing the glacial *fronts* then and now.

It is rare to find data on the thickness of the glacier upstream from the terminal tongue.

But Hackert's fine engraving (Photo: Ed Flammarion), done from the Montenvers in August 1781, enables us to make a comparison of this kind when it is set beside a contemporary photograph (M.L.R.L., August 1966). Both pictures show the rocky shelf in the middle ground on the left, covered with vegetation. It is much clearer of the ice now than in the eighteenth century. Then it only looked as high as it was wide; now it is about twice as high as it is wide.

Similarly the rocky spur on the right which extends into the middle of both pictures is much more exposed now than then. In Hackert's engraving it plunges straight into the ice, from which it is separated now by a narrow belt of moraine.

(Source: Private collection.)

Hackert's engraving has often been reproduced, for example in Mougin, 1912, and Engel, 1961. There is a copy in the Cabinet des Estampes in the Bibliothèque Nationale (reference as for Plate XIII).

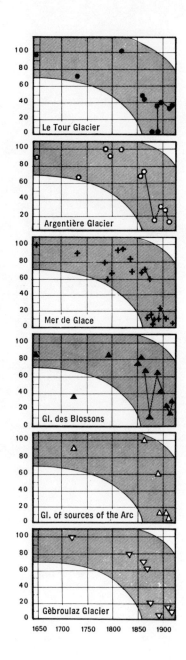

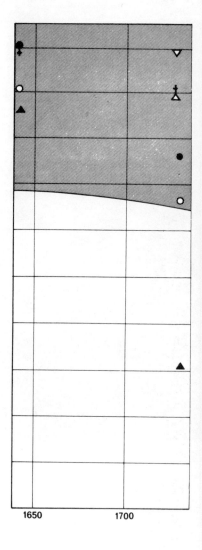

XXXII

These diagrams try to put in visual form the long evolution of the Savoie glaciers. On the left, the evolution of the separate glaciers; on the right, their general evolution. Time is shown on the x-axis, the positions of the fronts on the y-axis. The positions of the glacial fronts were expressed by Mougin in terms of horizontal distance, calculated from a fixed base or imaginary line arrived at in relation to local points of reference (for example, the "ligne du chapeau" in the case of the Mer de Glace). Each of these bases or lines was placed "behind" the position occupied by its respective glacial tongue in 1911, and at a variable distance from the tongue's extreme point in that year.

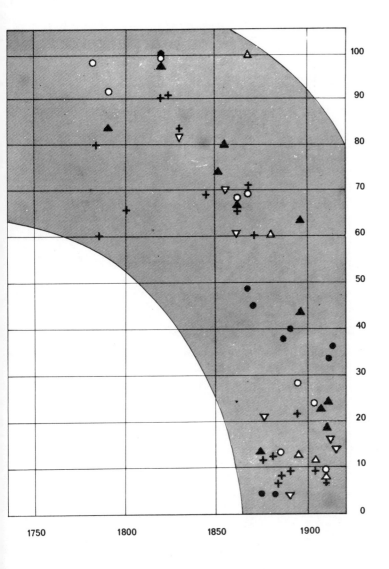

To facilitate comparison between the evolution of the different glaciers, all these distances have been expressed in terms of a single conventional line (corresponding to point o), situated 100 meters above the known historic minimum of each glacier in the period in question (1640-1911). The figure 100 represents the known historical maximum. The measurement 0–100 is 1600 meters in the case of the Mer de Glace; 1700 meters for the glacier Des Bossons; 1340 meters for the glacier of the sources of the Arc; 1800 meters for Le Tour glacier; 1400 meters for the Gébroulaz glacier; and 1150 meters for the Argentière glacier.

The period 1850–1860 marks a clear separation between two eras: high tide before, ebb afterwards (the ebb would be seen continuing after 1911 if the series were prolonged). (Source: Based on an article published by the author in the Annales E.S.C. for May–June 1960, Fig. 1.) N.B. See similar diagram for the Iceland glaciers in Thorarinsson, 1943, article cited.

that of 1150–1300, still less to that of 1600–1850. If this had not been so, how could we account for the fact that a layer of peat (Xe) was able to form at this period at Bunte Moor, between the morainic strata Xd and Xf?[60] How also should we explain that the hamlets of Châtelard and Bonanay at Chamonix, and the chapel of Saint Petronilla at Grindelwald, which were to be crushed or driven away by the neighboring glaciers in 1600–1610, flourished quite uneventfully between 1380 and 1600?

Even the data from Greenland does not necessarily conflict with this. The Norman burial places excavated at Herfoljness in 1921 were found in earth which was frozen all the year round, in spite of the warming up already noticeable at that time in Greenland as elsewhere. The excellent state of preservation in which materials and wooden objects were found could only be explained by this permanent state of frost going back a very long way. It may not have existed at the time of the actual burials, the last of which, according to the styles of the clothes, took place about 1450, because the roots of shrubs were able to bore through the biers and the skeletons and the materials they were clad in.[61] Did the Norman colony dying out in the fifteenth century do so in a phase that was slightly less severe climatically than the "modern" era (seventeenth century and after)? I merely raise the question.

Such is the main outline of what is known—in much less detail than for the modern era—of the glaciological and climatic history of the Alps in the Middle Ages and up to the sixteenth century. This history makes necessary certain revisions, and leaves certain gaps. I shall list some of these.

Modifications first. Under the influence of the Scandinavian school, which was naturally much impressed by the disappearance of settlers from Greenland in the fifteenth century, and influenced also by the catastrophes which occurred in Europe in 1348–1450, too hastily identified with a period of climatic cooling down, certain historians of climate, including some of the best of them (Utterström, Flohn, Lamb),[62] have included the fourteenth and fifteenth centuries with the intersecular phase of glacial advance called the "Little Ice Age." But in the present state of the evidence, whether historical, botanical, or geomorphological, I think this conclusion, though on the whole valid, needs to be expressed more precisely. It is indeed very probable that the Alpine glaciers of the fifteenth century[63] were more developed than in the ninth or the twentieth, i.e., than during the two optimal phases of the historical period. But they were slightly less developed than they became in the seventeenth century, leaving unharmed the localities they were to destroy about 1600.

It is true there were cold episodes in the fifteenth century. Pierre Pédelaborde tells us about the most picturesque of them, when in 1468 wine was distributed with a hatchet, and people carried bits of it away in their hats.[64]

But the cold episodes of the late Middle Ages, severe as they might be, were not lasting or general enough, nor did they affect enough of the year (including the ablation period), to cause a maximal thrust in the Alpine glaciers.

As for the catastrophes of the end of the Middle Ages (1348–1450), they had little to do with the rigors of climate. They arose from, among other things, the tragic pattern usually symbolized by the Black Death and the English wars, and mark the end of a long agrarian cycle.

The main inadequacies and gaps relating to the late medieval period concern the texts. In the seventeenth and eighteenth centuries glaciological documents are many and accurate, but the case of the Middle Ages is different. The medieval advance of 1215–1350 is mentioned in only a handful of texts, and most of those are vague and of doubtful authenticity. The decisive proof is geomorphological and botanical. It is because of this that the chronology of the medieval advance is elastic about fifty years either way. This type of chronology is quite satisfactory for a glaciologist or a botanist, who is interested in trends and the long term. But the historian has to be more precise.

I have already referred to the least doubtful of the medieval texts, and will here cite a few others, sometimes suggestive but often not entirely reliable (see Appendix 13).

In 1285 there is mention of an overflow of the lake at Ruitor. This suggests an advance of the glacier itself, but no details are given. The date coincides with the radiocarbon datings for Aletsch and Grindelwald, but all my efforts to track down the original reference in the archives and libraries of Val d'Aosta have proved vain. In other words this text must be regarded with suspicion.

Other texts, cited by Stolz and Du Tillier, suggest advances by the Ruitor and Vernagt glaciers in 1315 and between 1391 and 1430 respectively. But these texts are later, hypothetical, and entirely deductive, and the chronology they give is not at all certain. The same applies to texts of 1513 and 1531 on the beginnings of a vaguely specified advance in the Tyrol glaciers, and to another of 1540 on a marked retreat of the Grindelwald glacier. The retreat is quite probable; but the evidence of it here is not conclusive.

As for the texts discovered by Allix and quoted by him in his little book on the Oisans glaciers in the Middle Ages, they are reliable—but they are completely vague about topography, and all that can be derived from them about advances or withdrawals are hazy presumptions.

In fact, most of the medieval texts on the glaciers have yet to be discovered and interpreted. The Chamonix archives, so abundant for the fourteenth and fifteenth centuries, must hold some surprises. I have only been able to look through a small part of them, but once they are properly examined they should supply data of the first importance on the Chamonix glaciers.

Climatological texts proper, as well as glaciological ones, are also lacking for the medieval period, but here again the lack can only be temporary.

There are no wine harvest dates for before 1400–1500; but perhaps some series will be discovered in Italy.[65]

There *are* texts referring to contemporary events connected with climate. A certain number of these are to be found in ad hoc collections of the nineteenth century,[66] but these compilations, worthy as they are, are out of date and need to be replaced by new and systematic studies.

There are plenty of unpublished texts describing the winters, summers, inclemencies, droughts, floods, and so on, for this or that year of the fourteenth or fifteenth century. Such evidence abounds in all the manorial, legal, ecclesiastical, and administrative archives.

System is what is needed. These texts should be used to establish valid series, by region, year, season, or even month, by kind of climatic phenomenon (cold, heat, drought, humidity, etc.), and by intensity of climatic phenomenon. All this should be done with the aid of the most up-to-date classifying techniques, including computers. Take for example an isolated text like the one published by John Titow[67] in his article on the manorial accounts of the see of Winchester: "Hay was scarce this year [1292], because it was given to the lord's sheep on account of the severe winter (*propter maximam hiemem*)." In itself this does not tell us very much, but it takes on its full significance when it is added, as in Titow, to the enormous corpus of similar texts, known or yet to be discovered. Brought together and broken down into their main elements (kind of climatic inclemency, season, time, place), these texts provide material for accurate, long, precise, and subtle series which can be put against the decisive but vaguer evidence of the glaciers.[68]

This is what the historians present at the 1962 Aspen symposium tried to do for the eleventh and sixteenth centuries. I shall briefly describe this collective experiment as it related to the end of the Middle Ages and the sixteenth century.

Climate in the sixteenth century

In the last few years there have been many scientific conferences on the theme of the recent climatic fluctuations. In 1961 there was one held in New York under the auspices of the American Meteorological Society, on climatic changes and related geographical phenomena.[69] In September 1962 the international symposium held at Obergürgler dealt with "variations in existing glaciers" and the causes of these variations.[70] In the same line of fundamental research was the conference organized at Aspen, Colorado, in June 1962, by the Palaeoclimatology Committee of the U. S. Academy of Sciences: the Aspen conference took as its subject "climate in the eleventh and sixteenth centuries."[71]

This concentration on two particular centuries was intended to make it possible to go beyond generalities to concrete realities clearly situated in history. Another new feature was that not only natural scientists took part in the conference[72]; as well as geologists, glaciologists, meteorologists, and "dendrologists," professional historians were also invited. Although this was probably the first time such a thing had happened, the invitation was understandable enough. In "short term" chronology, such as is involved in the study of one century, or, as in this case, the study of two separate centuries, the usual methods of palaeoclimatology, based on the very long term and even on geological time, are inadequate. Finely graded observations and annual series are needed. But most of these are derived from archives, and thus are in the historian's province.

In response to the climatologists' request, the historians at Aspen, together with colleagues from other disciplines, supplied over fifty annual or decennary series.[73] With all this material, problems of method soon arose. Should each series be discussed, criticized, and interpreted separately? This stage was soon left behind. After the first discussions of the conference and the first meetings of the History Section, it was clear that the material needed to be pooled and each element considered in relation to the others.

The synthetic approach made it possible to test the various series against each other; also to bring out significances which otherwise might have remained latent; and finally to draw the beginnings of an overall picture.

Using the series and diagrams supplied by its members, the history committee of the Aspen Congress drew up two chronological tables, one for the eleventh and the other for the sixteenth century. Jacques Bertin and Janine Récurat, of the cartography laboratory of the École des Hautes Études, in 1962–1963 took on the job of converting these into graphs, which are reproduced at the end of this book, together with notes and explanations.

The object of the present chapter is to comment on these two tables, and to indicate their sources, the general methodology used, and the principles of classification followed. I end by putting forward a brief attempt at interpretation.

Naturally, in such an account it is not possible to expound or justify in detail each of the fifty diagrams involved. Here I can give only a brief account of the content and method behind each series; for more details the reader should consult the books of the individual authors or their sources (Appendix 15).

1. SOURCES AND METHODS

The sources of the diagrams fall into two groups: primary, which relate to climate itself (e.g., harvest dates, early or late); and secondary, which relate to the human effects of climatic phenomena (e.g., volume of harvests).

Phenological data are among the most important of the primary sources. A typical example is the series on Kyoto, presented by Arakawa (Diagram XVI–7). Arakawa relies on an ecological fact which has been well demonstrated[74]: the close correlation between spring temperatures and the flowering dates of plants. A warm mild spring means early flowers; a cold spring means late ones. The earliest phenological documents in Japan relate to the local variety of cherry tree (*Prunus yedoensis*): every year, when the cherry tree was in blossom, the emperor or governor gave a sort of garden party under them, and the annalists noted the date. T. Taguchi published a first chronology in 1939, which Arakawa has revised and put in the form of a graph. Diagram XVI–15 gives a fragment of his total curve, which covers—with some gaps—the ninth to the nineteenth century.[75]

After the flowers, the fruit. Since Angot[76] the ripening of the grape has been recognized as a climatic document of the first order, and recent work by experts in viticulture has provided us with details. Pierre Branas has shown the decisive role played by the "heliothermic complex" in the length of time taken by the grapes to ripen: the brighter and warmer the spring-summer in any given year, the quicker the grapes grow, and vice versa. Georges

Montarlot has done similar work on the growth of the Aramon, a common kind of vine, during the cold late year of 1932: from February to September a prolonged thermal deficit retarded first the coming into leaf, then the flowering, then the harvest, the lateness increasing as the growing season advanced. The same happened in 1957: spring frosts, persisting cold through the spring and summer, all joined to retard the harvest and the opening of the cooperative wine cellars in the south of France. More generally, the work of Godard and Nigond shows that there is a linear connection between summer temperatures and the date of the wine harvest.[77]

Let us now turn to the sixteenth century. Here we have a certain number of regional sequences of wine harvest dates. They relate to the vineyards of the north and center-east (Dijon, Salins, Bourges, Lausanne, Aubonne, Lavaux) and of the south (Comtat, Quercy, Languedoc, Gironde, and Old Castile).[78] These local curves—in constructing which allowance has to be made for the Gregorian calendar introduced in 1582—show common reactions to seasonal tendencies, from one year to another. This convergence authorizes unification of all the series, and from the means we get a single diagram (XIV–16). This diagram gives, in days, the mean difference in France,[79] for every year of the sixteenth century, from the secular mean of wine harvest dates. Thus in 1545, in the vineyards for which we have information, the harvest was an average of a fortnight earlier than the normal date for the century, calculated for each locality. In 1600 it was a week late. Thus the ups and downs of diagrams XVI–16 are largely a translation of the fluctuations of annual mean spring-summer temperature.

The application of a running mean of three years (XVI–16, ii) to this diagram softens it outlines (see Appendix 15, XVI–16, ii), but brings out the cyclic type of fluctuation, showing up groups of late or early years more clearly.

A similar attempt has been made with cereal harvest dates for the south of France (Appendix 15, XVI–12), insofar as they are known to us through certain ecclesiastical accounts. But this series, as well as being too short, rough, and intermittent, is also more complicated to interpret than those for wine harvests. The date of the cereal harvest certainly acts as a climatic integrator: in Provence, Languedoc, and Catalonia, where the harvest is gathered in June, a late harvest indicates a cold spring, and vice versa. But though spring temperature is the main factor, the harvest date is also influenced by the date of the sowing, which in turn is determined by the farmers according to various factors, including the autumn rain, the state of the soil, the amount of plowing needed, and so on. So dates for cereal harvests are a much more complex phenomenon than those for wine harvests, and their climatic significance is more variable.

Lastly, there are other phenological sources which do not concern flowering or fruition, but record the dates of frosts, icing up, and thaws which mark the beginning or end of the winter.

The series for Lake Suwa in central Japan is almost complete for the period 1444–1954. The priests of a nearby temple observed the lake and noted the day when it became completely frozen over; if it was a mild winter and the lake did not freeze, they usually noted that too. H. Arakawa, continuing the work of S. Fujiwara, turned this information into a series of which Diagram XVI–17 is an extract.[80]

From an early period the city of Riga recorded the date of the spring debâcle and the opening of the port, and from these records V. Betin has constructed a series that goes from the sixteenth century to our own time (Appendix 15, XI–6 and XVI–6).[81] This series agrees with comparable data from the Neva and Lake Kavalesi in Finland.[82] The ends of each of the three curves show a tendency toward early thaws, as a result of the amelioration from 1880–1950. Diagram XVI–6 shows the beginning of the Riga series, covering 1530–1610, with a few gaps.

Another primary source is dendrochronology. H. Fritts, J. L. Giddings, and G. Siren have supplied three groups of diagrams, covering the southwest United States, Alaska, and Finnish Lapland. The first series, on the Rockies, is by far the richest, based as it is on the records accumulated during more than fifty years by the Tree-ring Laboratory at Tucson, Arizona. Diagrams XI–8 and XVI–8 show the annual mean growth of five groups of trees,

Douglas firs or Ponderosa pines, two groups of which are in Arizona and the others in Montana, Colorado, and California. All the sites are extremely dry, and the varieties of tree chosen are sensitive to dryness. So the growth of their tree rings is correlated to rainfall, and in inverse relation to the aridity index, which covers both rain and evaporation.

The Alaska series (XI–8 and XVI–19) is derived from old trees (*Picea canadensis* Mill, *Larix alaskensis* Wight) and from excavated Indian timbers. All these samples come from sites just north of the Arctic Circle.

The Finnish Lapland series covers the whole of the period from 1181 to 1960 and gives the growth index of over two hundred old trees—Scots pines (*Pinus sylvestris*)—growing in the far north of Finland, at the northern forest limit. Diagram XVI–18 gives the part of this curve which covers the period 1490–1610.

Alaska and Finnish Lapland have certain ecological conditions in common. In both regions summer temperature is the critical factor, and rainfall plays only a minor part in the formation of tree rings.

In these very cold latitudes there is also a certain prolongation of some effects. In Lapland a warm summer stimulates tree growth not only that year but also the year after, and a cold summer inhibits it similarly. So Nordic tree rings tend to relate to two successive years, and dendrochronological curves are softened as they would be by a sliding biennial mean (Appendix 15, XI–8, XVI–18 and XVI–19).

A new primary source is documents relating to rogations asking for rain. In Spain the municipal authorities, in cases of urgency, quite often obtained Church permission for days of processions. Emili Giralt Raventos has worked out the Rogations at Barcelona in the sixteenth century in terms of "numbers of days of prayer," using the complete records kept of them by the "ancient Council of Barcelona." The days are shown divided up according to season, then by year-harvest (Diagrams XVI–9, XVI–13, XVI–20, XVI–25, and XVI–30) to give a comparison with fluctuations in cereal harvests.[83]

Emili Giralt himself emphasizes that this series is only of relative significance. Religious fervor enters into it as well as the main factor, absence of rain: and religious fervor varies. (Perhaps it increased at the end of the sixteenth century, with the Counter-Reformation.) On the whole, the series only seems to be objectively representative from 1520 on.

The Giralt series is so far, apparently, the only one of its kind, though perhaps a similar one could be made for Valencia. In Paris, for example, although the reliquary of Saint Geneviève was carried in procession to bring rain, this happened too rarely to form a series.

A fourth type of primary source is made up of meteorological observations, in the literal sense. Long before the appearance of measuring instruments, journals, diaries, and record books of various kinds—particularly those which were kept daily—provide systematic information about climate. The information may be somewhat rough, but significant facts may be drawn from it. Thus Wolfgang Haller, a citizen of Zürich, scrupulously noted down in his journal, between 1550 and 1576, all the days when it rained or snowed, and from this H. Lamb has derived the statistics from which Diagrams XVI–11 and XVI–23 are derived.

We also have genuine hydrological series for the historical period. H. Lamb, using the data published by Prince Omar Toussoun, gives the annual levels of the Nile in the eleventh century (in cubits converted into meters) (Diagrams XI–16 and XI–17). The material here goes right back to the Arab chroniclers and their Nilometers.

The glaciers too provide a whole class of information of their own, especially in the sixteenth century. This has been dealt with in a previous chapter, and the most characteristic episodes in the Alps will also be found in schematic form in Diagram XVI–34.

But whereas texts are plentiful for the Sixteenth century, especially in the Alps, there are no eleventh-century texts to give the position of the glaciers just after the year 1000. But radiocarbon dating does give a chronology for two fossil forests near the present glaciers of Grindelwald and Aletsch.

Carbon-14 dating gives the probable mean date for the death

of these forests as about A.D. 1215. Judging from the tree rings, and from the layer of humus under the roots at Aletsch, we estimate that these trees lived for at least two centuries undisturbed by the glaciers before the ice advanced and killed them.

So there is some possibility that the eleventh century formed part of the medieval phase of glacial retreat which was comparable to or perhaps even more marked than that of the twentieth century. Conversely, the sixteenth century, at least the second half of it, is already characterized by a tendency toward the glacial advance which continued, through various oscillations and a series of maxima, until about 1850. (See Appendix 15, XI–19 and XVI–34; also Chapter IV.)

The last kind of primary source through which we may approach the knowledge of early climate is the series relating to contemporary meteorological events. These series show year by year some phenomenon which struck people living at the time as a departure from normal: an exceptionally severe or an exceptionally mild winter, the freezing up of major rivers, floods, torrential rain, prolonged drought. This kind of series is often based on heterogeneous documents of varying validity, and may be inadequate and intermittent. Such series are not to be compared with those which have a phenological or dendrological basis, and which are annual, continuous, quantitative, and homogeneous. But it would be absurd and hypercritical to reject such contemporary evidence out of hand. The historian who did that would be flying in the face of texts and wantonly refusing to admit witnesses who have a right to be heard.

In fact, this kind of evidence can be used to build up quite meaningful collections of data, but of course it is material which must be handled according to very strict rules.

The first rule is never to try to derive climatic information from any but genuinely meteorological texts. The mere mention of a bad harvest is not in itself a direct source of information about climate. The bad harvest may be due to all sorts of things, according to place, time, and crop. It may be an indication of extreme drought and scorching sun; of a damp summer or wet winter; of frost killing the seeds; or exceptional winter mildness which makes them rot in the ground. Only if the text gives the cause of the bad harvest can it really be used as climatic evidence.

The second rule is that this sort of information must be broken down into seasons. In the case of vines, for example, a spring frost which kills the buds but not the plants is not the same as a winter frost which attacks the plant itself and may destroy a whole vineyard. The two phenomena belong to two different series, according to season. Similarly, winter or spring floods which may be caused by the thawing of the ice have not the same climatic or ecological character as the torrential rains of autumn. It is legitimate, for research purposes, to construct some "cumulative" annual series from this sort of evidence (see Diagram XVI–29), but as far as possible this should only be embarked on when single studies have been done on the separate seasons.

These seasonal series themselves must be based on a precise chronology. The meteorological year begins in December: so, for example, the frosts of December 1564 must be ascribed to the winter of 1565, which goes from December 1564 to January-February 1565. Unless care is taken over this, some winters will be halved and others doubled.

Good series of this kind, or rather the least bad ones, are thus built out of selected material, properly classified. Next they need to be tested, otherwise the uncertainty of the basic documentation may produce a distorting effect. For example, severe winters seem to be more numerous after 1540–1550 (Appendix 15 and diagrams XVI–1 to XVI–6). But this may well be because documentary evidence becomes ampler as the century goes on. Such suspicions can only be overcome by putting the matter to the test of contradiction. Let us suppose that the increasing number of hard winters corresponds to a real phenomenon. In that case, there should be a corresponding decrease in the number of mild winters. If, on the other hand, the increasing amount of evidence of cold winters arises merely out of an illusion, a sort of documentary mirage, the growing amount of documentary evidence ought also to produce an increased number of allusions to mild winters. In fact, the diagrams survive this test. After 1545, in spite of the ever-growing amount of documentary evidence, references to mild winters are definitely rarer than before.

The second test is the classic one of agreement and disagreement. Here the series of mild winters around 1530 may serve as illustration. Early chroniclers and historians were much struck with it.[84] And all the diagrams confirm it: to take one example, Weikinn has collected all the cases of rivers being frozen in the sixteenth century, and not one occurs between 1528 and 1532.

Another reassuring parallel is that not one of the dry summers in Barcelona (Diagram XVI–20) occurs in a year when France had a wet summer. And out of seventeen summer droughts in Belgium, only two occur when the French summer was a rainy one (XVI–21 and 22). There is another excellent convergence between the French and Belgian series of autumn precipitation (XVI–26 and 27). In the series for spring precipitation in the sixteenth century, Belgian droughts fit so neatly between French floods that data for both countries have been run together in Diagram XVI–14.

But there are many disagreements also. One discordance is the differences shown between the French and Catalan autumn and spring (XVI–9 and 10; XVI–25 and 26). These may arise either out of a real geographical difference in rainfall, or merely out of inadequate documentation.

It goes without saying that concordances themselves only apply within a comparatively limited area. Russian and Japanese winters show very little resemblance to those in the West. Meteorological conditions, and the distribution and influence of the circulation of the air, may in the same period produce different and even opposite effects according to longitude.

Even so, when discordances appear between neighboring regions we should be on our guard as to the validity of the series themselves. This applies particularly to those of the eleventh century. The vagueness of certain medieval texts makes it impossible sometimes to work out the chronology of the winters exactly to the year.[85]

In some cases, only one series exists (Diagram XI–5, on Russia), so no comparison is possible, and the series has to be treated as provisional until it calls forth new information against which its own may be tested.

A last difficulty is the classification of contemporary evidence in quantitative terms. As far as possible, some sort of gradation has to be attempted, even if it is bound to be only a rough one.[86] In the case of winters, for example, a single cold episode, lasting days or even weeks, is not the same thing as a really severe season which freezes rivers and kills the olive trees. This is the system of notation we adopted for the series concerning winter in the south of France (Diagram XVI–4). Small white rectangles denote single cold episodes (falls of snow, short frosts), which may be of little significance. Medium-sized rectangles stand for severe cold which kills the olive trees and vines, as in 1591 and 1600. Long rectangles signify exceptionally hard winters, when the Rhône is frozen as far south as Arles or Avignon, and the ice is thick enough to bear skaters, guns, and carts. This happened very rarely indeed between 1900 and 1960, but was quite frequent in the second half of the sixteenth century, as Hyacinthe Chobaut saw from the daybooks kept by country notaries.[87]

C. Easton, in his important study of the winters of western Europe, has used similar criteria to establish a slightly more complex system,[88] and Manley followed him in his own study of English winters in the sixteenth century (XVI–3). There is a certain amount of objectivity in this kind of notation: the freezing solid of a great Mediterranean river certainly shows that very low temperatures must have been reached, and thus implies an exceptionally severe winter. But there is always a subjective element in value judgments of this kind.

Our study of contemporary series can be pushed further if they are put into decennary form, which may give a glimpse into long-term trends. H. Lamb has attempted this by a very simple method[89]: for each decade he has worked out how many more mild winters there were than severe ones, and vice versa. The resulting histograms cover several centuries and give a glimpse of multisecular trends. Those dealing with the eleventh and six-teenth centuries, with others similarly constructed for precipitation, can be seen in Diagrams XI–6, XI–15, XVI–8, XVI–24, and XVI–33. They relate to western, central, and eastern Europe.

So far we have been concerned with primary and direct sources for the study of climate. We now come to secondary and derivative ones. These provide information not about climate itself but about its effects, primarily the agricultural ones: such series include appraisals of harvests, mentions of scarcity, and, to a certain extent, short-term oscillations in the prices of cereals, leaving out of account the long-term movements. In Diagram XVI–32 E. Helleiner gives data on famines, scarcities, and high cereal prices, derived from texts and statistics. David Herlihy and others give series on famines in the eleventh century derived from texts (XI–20 to 22), but because of the scarcity of documents they could not connect the famines to particular harvests. As always, the chronologies are much less accurate for the eleventh than for the sixteenth century.

Emili Giralt uses another method, perhaps to be preferred in cases where it is feasible. He simply presents the "phenomenon as it happens," and in Diagram XVI–31 translates into visual terms, year-harvest by year-harvest, from 1499–1500 to 1599–1600, the price of wheat in Barcelona. From quantitative series of harvests which have survived for the years between 1532 and 1548, he shows that fluctuations in the prices at Barcelona were related to concomitant fluctuations in the harvests in Catalonia, among other factors (Appendix 15, XI–31).

But in the long term the price of wheat in the sixteenth century was affected by an overall revolution in prices which has nothing to do with climate. To give the Barcelona figures their true ecological and climatic significance, one has to eliminate the secular movement and keep only the short-term fluctuations since they are directly influenced by meteorology and harvests. This has been done in Diagram XVI–31 by the use of a running mean of nine years. The diagram shows, for each year of the sixteenth century, the annual difference between the price of wheat and the running mean; in other words, it shows the short-term fluctuations. A series worked out in this way is all the more interesting because it can be set beside the records of prayers and processions organized in Catalonia in times of drought.

All these sources, primary and secondary, provide the historian of climate with between fifty and sixty curves. How are they to be classified? The system adopted divides them by century, season, and country.

The centuries are the eleventh and sixteenth.

As for the division into seasons, it was only possible to break down the eleventh-century data into three groups: winters, summers, and general information on the year as a whole. The better-documented sixteenth century could be broken down into winters, springs, summers, autumns, and general information. Within each group an attempt was made to distinguish between temperature and precipitation. The white parts of the diagrams indicate cold, damp, or a bad harvest; the black parts indicate heat, or dryness, or good harvests.

But an exception was made for Catalonia, where drought is more of a danger to wheat than rain. So here scarcity and high prices are represented in black, and good harvests in white; this facilitates comparison with the series on Rogations, which are also shown in black (XVI–30 and XVI–31).

There are a few problems concerning the division into seasons which may seem to have been solved somewhat arbitrarily and which I shall explain briefly. In the tree rings series, the "northern" diagrams (Lapland and Alaska), which give temperatures for the more clement months of the year, are put into the "summer" grouping; the "southern" diagrams (the arid southwestern United States), which give rainfall conditions for a whole year, are put in the "general information" grouping.

Variations in the Nile floods reflect the amount of summer rain in Abyssinia. Variations in normal water level, on the other hand, relate to annual rainfall in equatorial Africa, as regulated by the "flywheel" of the great lakes of central Africa.[90] So Diagrams XI–16 and XI–17 go into "summers" and "general information" respectively.

Neither the glaciological series (primary sources) nor the sequences of famines and scarcities (secondary ones), can be reduced to terms of the influence of one season. They are influenced by temperature and precipitation, which concern every month in the year. So they all go into "general information."

Each seasonal group is divided into regions, usually going from west to east: America, western and central Europe, then Russia; or sometimes western Europe, Riga, and Japan.

2. LINES OF INTERPRETATION

Let us begin with the better known and more extensively worked period, the sixteenth century.

Here the series for the Alpine glaciers (XVI-34) is the most suggestive, in bringing out the trend toward glacial expansion between 1546 and 1590, and the historical maximum around 1600, which equals and often surpasses those of 1643, 1820, and 1850.

These phenomena show that, in spite of the general resemblance, there were slight differences between the climate at the end of the sixteenth century and that of the twentieth. It may be that mean temperatures were lower, as toward 1850. According to recent researches the difference in the annual means may be as much as 1 degree C.

As Hoinkes showed at the Obergürgler symposium, these differences in temperature act on the glaciers in a variety of ways: cool dull summers with snow at high altitudes help to arrest melting and so to promote expansion. Lack of ablation, through a series of such summers, results, after a time lag of three or five or up to ten years, in a marked advance of the terminal tongues, which break through the morainic fronts and in extreme cases submerge forests and even houses and villages.

Our diagrams for the end of the sixteenth century do not contradict Hoinkes. The sequence of bad summers stands out clearly on the curve of wine harvest dates from 1592 to 1600, the most marked of all such groups in the sixteenth century. Diagram XVI-16 shows that it affected the whole of France, and it appears to have affected all western Europe, since a very similar chronology is also found in England and in Finnish Lapland. Just as Hoinkes suggested, the decade of cool summers from 1590 on resulted, after a time lag of about six years, in the glacial peak of 1598–1602 (XVI–34).

More generally, the whole of the second half of the sixteenth century shows a decrease in warm or hot summers. Their absence diminishes glacial ablation and favors the trend of glacial advance in the Alps.

The part played by winter is also relevant. According to Hoinkes, long winters with heavy snow which cause accumulation are essential factors in glacial advance. What were the winters of the sixteenth century like? Easton finds seventeen winters considered as mild by contemporaries between 1500 and 1550, and only six between 1550 and 1600 (Diagram XVI–1), in spite of the fact that documentation increases during the second half of the sixteenth century.

Belgium, England, and the south of France show the same radical decrease in the number of mild winters as between the first and the second half of the century (XVI–2 to 4). Particular examples suggest not only less mildness but also a marked severity in the winters of the second half of the century. Take for example the freezing solid of the Rhône in the Mediterranean south of France, which occurred in 1556, 1565, 1569, 1571, 1573, 1590, and 1595. Yet in the first half of the century, when information is already abundant, there is only one reference to its having happened, in 1506.

The death by frost of the olive trees is another indication of very low temperature. And it happened much more often in Mediterranean France in the second half of the century (1565, 1569, 1571, 1573, 1587, 1595, and 1600) than in the first half (1506 and 1523), or in the twentieth century (1914, 1929, and 1956).

Were these cold winters snowy ones too? It seems like it. Let us take the case of the decade 1565–1574, one of the bitterest of modern times, in that it saw the Rhône frozen four times for a considerable period, and the olive trees were killed by frost four times also (1565, 1569, 1571, 1573). Haller's journal shows that the decade was snowy as well as severe. The proportion of days when it snowed, as against the total number of days (Diagram XVI–11) when it either snowed or rained, was regularly less than 50 percent between 1551 and 1560. But from 1564 to 1577, when the journal comes to an end, this proportion is surpassed considerably every year. The Swiss winters were extremely snowy for almost fifteen years, during which the Alpine névés and glaciers were probably glutted. This is considered an additional explanation of the already amply determined trend toward glacial expansion which manifested itself in the Alps in 1570–1580 and culminated at the end of the century.

This group of hard winters was not an isolated phenomenon, either geographically or chronologically.

To take geography first, the winter at Riga, as reflected in the freezing up of the port, lasted an average of nine days longer from 1550 to 1600 than in the twentieth century (1880–1950). Similarly, in Japan,[91] Lake Suwa was frozen over every year from 1560 to 1680, which is far from being the case now. Moreover, the freezing over took place earlier round about 1560–1680 than in 1702–1950 (XVI–6 and 7).

Chronologically, the sequence of hard winters after 1550 has a parallel in more recent meteorological history. The period of milder winters which appears so clearly on all the meteorological series for the temperate zone between 1900 and 1950[92] was preceded, on earlier temperature curves, by a period of more severe winters in the nineteenth century (Fig. 8). As in the sixteenth century, this period of severe winters was associated with glaciers in a more advanced state than now. The oldest meteorological series also show that the winters at the end of the eighteenth and the beginning of the nineteenth were much more severe than those in the first half of the twentieth century.[93] The hard winters of 1550–1600 and those of the modern or contemporary era probably belong to the same meteorological family; but the later group are more exactly known.

So far we have dealt with those aspects of our series which are clearest and relatively simplest in interpretation. But there are many others which present many problems. Examples are the cold, wet springs around 1570 in Belgium, northern and southern France, and Catalonia (XVI-12 to 14); the marked aridity in Catalonia in 1530–1550 (XVI-30); and during the 1590s, at the time of the advance of the Alpine glaciers, the decrease in droughts (was it an increase in rainfall?) in the Barcelona area. These phenomena correspond to certain documentary data already pointed out by Fernand Braudel and Jean Delumeau,[94] and suggest possibilities which must be examined without prejudice.

As regards the "eleventh century" our results are critical rather than positive. The C^{14} chronology for the glaciers, suggestive as it is, is much too loose to provide material for research on any particular series of seasons. These series themselves, though very carefully worked out by their authors, are by their very nature debatable. The most reliable ones are based directly on sources (XI–1, XI–9, and XI–20), the others make use of still useful collections of texts often edited in the last century (XVI–2). It follows that concordances between these series may arise purely and simply from a given text being used at first hand in one instance and at second hand in another. So concordance tests, even when positive, are not necessarily conclusive.

As to the quantitative series for the eleventh century—those on the hydrology of the Nile and U.S. dendrology—the climatic zones they relate to are too far away from each other and from western Europe for cross reference to be possible.

In short, our studies of the eleventh century constitute only a beginning: the series we have need to be either confirmed or refuted, and cannot yet give rise to solid or clear conclusions. Only for the sixteenth century do some provisional glimpses of an overall vision emerge.

So far we have been talking about climatic history pure, and trying to establish a minimum of fact. Now we must turn to the influence of climate on human activity and ecology.

In a few extreme cases, such as the arid zones, this is a

fairly simple matter. Take for example the severe drought which lasted for twenty years (roughly 1570–1590) in the already dry southwest of the United States (XVI–28), and in the space of a generation turned the area into a temporary desert. The volume of rain fell to 20 percent of its twentieth-century level, and the drought strangled the primitive maize-based agriculture and perhaps led to the abandonment of certain farms and villages.[95]

But climate does not have this sharp and simple effect on the European societies of the sixteenth century, and there its influence is mainly short term, affecting the level of harvests and agricultural production.

For the Old World, let us compare what is most immediately comparable. Emili Giralt's Barcelona series provide us with two curves, one for drought and the other for prices, one ecological and the other economic. On what basis are they to be set against one another?

In a Mediterranean climate drought is a threat to wheat. If it occurs in the autumn it hinders plowing; if it occurs in the spring it kills the ear and so deprives the farmers not only of their harvest but sometimes also of seed for the next year's sowing. At whatever season, it interferes with the plowing of fallow land. Thus the droughts which occur in any one year-harvest (say 1548–1549) affect mainly the harvest of that year (June 1549) and secondarily the harvest of the following year (June 1550).

As regards prices, the effect of drought is often deferred, with a time lag of one year. To take once again the drought of the year-harvest of 1548–1549, which partly destroyed the harvest of June 1549: it thus reduced the supply of grain available during the twelve months that followed, when this supply ought normally to have dominated the market, and so sent prices up during the year-harvest 1549–1550. So, in a Mediterranean time-scheme based on the year-harvest, drought may affect either the prices of the same year, or, more probably, those of the year following.

This can be seen clearly happening in the case of Barcelona

in the sixteenth century. The great price peaks here[96] after 1525 follow the droughts with a time lag of one year (XVI–30 and 31).

These examples apply only to Mediterranean or arid ecology. In more northern regions the factors limiting cereal yields are very different. In a climate such as that of Paris or London the main danger to harvests is excessive humidity. In the Baltic, it is cold and absence of sun, producing "green years" where the wheat cannot ripen for lack of warmth and light. The Aspen diagrams give some indications here (compare XVI–16 and XVI–18 and 32). The 1590s seem to have seen several agricultural disasters in oceanic and temperate Europe. But these northern series are not so finely subdivided as the Catalan ones, and lack their regional and chronological precision. In the present state of research, no systematic conclusions can be drawn about the ecology of sixteenth-century harvests in northwestern Europe. So although the human effects may easily be guessed, they cannot really be known yet. It is not impossible, but it is not absolutely proved, that the cold years at the end of the sixteenth century caused a marked series of bad harvests in northern Europe.

It still remains to give a truly climatic interpretation to the best-established factual data. This is not the historian's business, and it is a pure meteorologist, R. Shapiro, who puts forward certain hypotheses.[97] These are based on current theories of general atmospheric circulation, on the changing distribution of air masses and jet streams.[98] By means of a number of synoptic maps, an attempt is made to pick out the most contrasting and best-known situations, such as that which existed at the end of the sixteenth century, with cold winters and advancing glaciers in western and central Europe, and very dry years and "desertification" in what is now the southwestern United States.

The collective spadework done at Aspen produced significant though limited results. The experiment should be resumed, applied to a wider time span, and mechanized. The tens of thousands of scattered climatic references, published and unpublished, should be classified and grouped and related to one another with all the thoroughness and subtlety made possible by modern electronic resources. The meteorology of the twentieth century is not the only one that can be computerized.

CHAPTER VII

HUMAN CONSEQUENCES AND CLIMATOLOGICAL CAUSES OF FLUCTUATIONS IN CLIMATE

Two points remain to be dealt with in a little more detail.

The first is the relation between climatic and human history. So far I have only touched on this question in passing, but here I want to expand on it a little, though it would take a separate book to deal with it exhaustively.

The problem is mainly that of agricultural meteorology, and I shall treat it season by season with reference to cereals, the fundamental element in a traditional economy.

In the first place, what is the incidence of winter on cereal yield? As regards France all agricultural meteorological studies agree: a cold winter, unless it is exceptionally severe, is favorable rather than dangerous. In Seine-et-Oise, "where the winter normal is 3.8°, years when the mean temperature is below 3° have good harvests and years when it is above 5° have bad ones."[1] The statistical study which gives these results covered a period of thirty years (1901–1930), and Sanson's conclusions are confirmed by experimental observation[2] and agricultural practice.[3]

In fact, in the northern half of France, a "bad" winter for cereals is not a cold one but a wet one. This is true not only of Seine-et-Oise, but even more so of a department like Loire-Atlantique, where the amount of winter rainfall often exercises a despotic influence on harvests, which can definitely be expected to be good when the winter is dry and poor when it is wet.

So, to return to the seventeenth and eighteenth centuries, their winters, though probably more severe than now, are unlikely to have harmed the harvests, except in such cases of extreme cold as in 1709. The trend toward colder winters and more extended glaciers which many authors have detected from 1540 on, a trend which persisted between 1600 and 1850, and was catastrophic from the economic point of view. In the present state of our knowledge it still seems as if the long "crisis," hypothetical or real, of the seventeenth century had some other explanation.

It does seem, however, as if the Nordic countries are different: too much winter cold there really is a danger to the cereal harvests, and a trend of severe winters has entailed serious consequences when in France it has been harmless or even beneficial.

Next, what is the impact of spring and summer, the growth period, on cereal yields? In the northern countries the impact is simple and the chief operative factor is temperature. A hot growing season, particularly a hot summer, is the best guarantee of a good harvest. This applies both to Sweden[4] and to Finland.[5] Conversely, a thermal deficit during the growing season results, in Sweden and Finland, in a cereal deficit. At the worst there are the famous "green" years, when the wheat will not ripen and rots in the ear. Scarcity years in Scandinavia include 1596–1602 and 1740–1742.[6]

Farther south, in England and France, the influence of the climate from March to August is more complex. In spring, warmth and light remain the chief factors. In Seine-et-Oise, even if winter conditions have not been unfavorable to start with, the success of the harvest can be seriously compromised if the amount of spring sunshine is insufficient and the mean temperature under 9°.[7] Conversely, a warm sunny spring bodes well both for winter and spring cereals,[8] in one case favoring the ripening and in the other the sowing.

But in summer, both in the Paris and in the London basin, rain plays a decisive role in the final yield, not so much by its absence, as a literary tradition of Mediterranean origin might lead one to expect, but by its excess. In Seine-et-Oise and Loire-Atlantique, rainfall has only to rise above the annual summer mean during the summer and up to the harvest (including the gathering in of the grain), and this is enough to make the harvest a poor one even if the conditions in winter and spring were favorable. For summer rain lodges and beats down the wheat, making it rot.

A dry summer, on the other hand, though unfavorable to stock-raising, benefits both winter and spring cereals. In England, where the harvest is much later than in France and often occurs in September, a dry summer is an important factor in high yields. It affects not only the harvests of the same year, but also those of the next, which can be sown under good conditions.[9]

So what wheat needs in Britain and the northern part of France is a dry summer, not especially a hot one.

But this is still rather vague. Slicher van Bath, in an excellent article, has drawn up a detailed "identikit" portrait of the ideal climate for wheat.[10] His study applies particularly to Holland and England, but suggests the complexity, in general, of the correlations between climate and harvests.

IDEAL CLIMATE FOR WHEAT IN HOLLAND AND ENGLAND
(from Slicher van Bath, 1965, p. 9)

1. End of September	Rather damp.
2. October, November, up to December 20	Rather dry, not too mild.
3. December 21 to end of February	Rather dry, a little snow, no frost below −10°, no strong wind.
4. March	Frost is dangerous once germination has begun.
5. April	A little regular rain, especially for spring sowing; sun.
6. May to June 15	Warm, but not a heat wave; still plenty of rain.
7. June 16 to July 10	Cool, cloudy, not too much rain.
8. End of July, August, beginning of September	Dry, warm and sunny, no heat wave.

This minute analysis is a suggestive one, and in its own way emphasizes the favorable effects which are often produced by dryness and sufficiently high temperatures between December 21 and the harvest. But on one point this model may need supplementing, for it deals mainly with contemporary cereals, with seeds which have been rigorously selected. But does it tell us all we want to know about the cereals of earlier times, which had to face the rigors of climate with much less developed technical and genetic equipment?

As regards the thirteenth century, John Titow has provided us with the elements of a good answer to this question.[11] He has drawn his climatic texts from the annual accounts of the see of Winchester, which owned manors all over the diocese. About eight hundred texts, covering every season, give information about the weather, rain or shine, between 1209 and 1350. We learn, for example, that in the summer of 1263 a certain "meadow in the manor of Pillingbere was not cut because of the great dryness" (*in prato de . . . levando et falchando nihil hoc anno propter magnam siccitatem*),[12] and that in the manor of Weregrave, in the winter of 1272, "eleven acres of a field of oats were not sown because of flood" (*propter inundacionem aque*).[13]

Titow has tabulated all this data by year and season, and set it beside the annual cereal yields, which are also given in the Winchester accounts. The results are as follows.

Good harvests (15 percent above the secular mean) result from this sequence:

Summer and autumn of previous year: very dry.

Winter: severe, or indeterminate (average?)

Summer: very dry.

Bad harvests result from two different types of sequence:

Type 1 (wet)

Autumn of previous year: wet or very wet, soaking the fields for several weeks.

Winter: wet.

Summer: wet.

Type 2 (dry)

Previous autumn: wet.

Winter: indeterminate (average?)

Summer: dry.

Titow's findings thus confirm those of Slicher van Bath: precipitation plays the most important part. Sometimes it is lack of it which causes a bad harvest in England in the thirteenth century, but generally it is too much rain that does the damage. When the soil is saturated with rain for successive seasons the seeds are drowned, the natural nitrates are washed away, weeds flourish, the wheat is lodged where it stands, and the sheaves blacken and rot.

Excessive wet in already watery England can lower grain yields so as to pave the way for the great famines typical of the Middle Ages.

Temperatures show no marked correlations with harvests. Winter cold tends to be favorable to cereals except when it is extreme: there are examples of this for Winchester in 1236, 1248, and 1328. We come back once again to the same conclusion: the great series of cold winters, such as are seen in the Fernau phase, were not necessarily harmful to food supplies.

To extend the discussion further, the limiting factors in cereal yield vary geographically according to region. Adverse climatic conditions are not the same to the north of the Baltic, on the shores of the Mediterranean, and in the temperate zone between.

In Mediterranean Europe, it is mainly dryness which lowers cereal yields.

In northern Europe, temperature, in every season of the year, is the critical factor, and a long series of cold years can do great harm to agriculture.

Between these two extremes, in oceanic and temperate Europe, the main danger comes from rainy winters, cold damp springs, and wet summers—in other words, from the recurrence of years of high humidity.

The relations between physical climate and human history can be approached by another method than through the fine distinctions of agricultural meteorology. We can attack the problem direct, and ask if there is any connection between such and such a secular fluctuation and some major episode in human history— migration, a long phase of economic depression or expansion, etc. Did the mildness of the centuries around the year 1000 stimulate the great land clearance in the west? Did the rigors of the seventeenth century contribute to the so-called "economic lassitude" of the period?

These are fascinating questions, but difficult to answer: the presuppositions they imply are not clear, and the right method of approach has not yet been found. Can a difference in secular mean temperature which is under, or at the most equal to, 1 degree C, have any influence on agriculture and other activities of human society? The question has not even been settled as regards the twentieth century which we know so well (see Chapter III). So it is all the more insoluble, at least for the moment, in the case of earlier periods about which we are much more ignorant.

As regards migrations, the influence of climate is a completely ambiguous question. The Teutons of the first millennium before Christ are supposed to have left their countries of origin because of the cold. The Scandinavians of the period before A.D. 1000 are supposed to have done the same thing for exactly the opposite reason—the mildness of the climate, stimulating agriculture and thus also population growth, is said to have led to the departure of surplus male warriors. But what is one to conclude from such contradictory and unprovable speculations?

Similarly, the Fernau oscillation (a multisecular phase of glacial advance and relative coolness, between 1590 and 1850) coincides with phases of economic depression (during some parts of the seventeenth century) and with phases of expansion (eighteenth century). In such circumstances how can one assert that there is causality?

In short, the narrowness of the range of secular temperature variations, and the autonomy of the human phenomena which coincide with them in time, make it impossible for the present to claim that there is any casual link between them. Until the subject has been gone into exhaustively, if this is possible, we can only suspend our judgment—which is not necessarily the same thing as being skeptical. I am satisfied if this book establishes certain primary phenomena of pure climatic history. The secondary question, of the impact of climate on human affairs, belongs to another province, and to researches not yet carried out.

Or rather, these researches have so far been undertaken with only relative success and in connection with a marginal area. In 1961 R. Woodbury made some interesting suggestions concerning the arid southwestern United States, and the possible relations between local fluctuations of climate from the twelfth to the sixteenth century and long-term demographic trends among the Indians.

In this region archaeology supplies information about human trends, and dendrology shows oscillations in humidity. Certain comparisons may perhaps be made between the two series.

Excavation and radiocarbon dating show that Indian agriculture in New Mexico and pre-Columbian Arizona began shortly before the beginning of our own era; flourished between A.D. 700 and 1200, when the farmers of the "desert culture," who lived on gourds, maize, and beans, grouped themselves together in large villages; and then, just before the end of the thirteenth century, started to decline.

From then on the rural communities began to shrink. People abandoned the land, by an irreversible process which continued through the fourteenth and fifteenth centuries. Vast areas turned into desert again in the basins of the little Colorado, the Gila, and the Rio Grande; in east-central Arizona; and in south and west New Mexico. In less than three hundred years (from the thirteenth to the sixteenth century) nearly two thirds of the cultivated area was abandoned: the 230,000 square miles that was being farmed in about 1250 had shrunk like the peau de chagrin until by about 1500 it was only 85,000 square miles. Christopher Columbus, Cortés, and his conquistadors, with their massacres and contagions, their measles and their smallpox, had nothing to do with these catastrophes, which happened before the coming of the Spaniards. The Conquest only confirmed the decadence of the southwest, which lasted right up until the twentieth century, in spite of the Indians' supplications to the rain-gods.[14]

What then were the initial factors in this fall in the native population? There was probably a whole cluster of variables concerned, both human and physical. And among them there stands out a climatic episode which according to Woodbury is very significant. This is the great drought which in the second half of the thirteenth century was general in what is now the southwestern United States. The trees are irrefutable witnesses: between 1246 and 1305,[15] eight groups of Douglas firs, ponderosa pines, and bristlecone pines appear in the tables by Harold C. Fritts, which cover more than fifteen centuries; and they show that droughts in the southwest were never so general and continuous as during the sixty years from 1246 to 1305, and particularly between 1276 and 1299.

Yet the Indians who lived there had tried to provide against lack of water. In their heyday, between A.D. 900 and 1100, they had invented and generally applied irrigation and the terracing of steep valleys. But according to Woodbury such meager defenses were of no avail against the arid episode of 1250–1300, which fell like a thunderbolt on a population which was already potentially above saturation level. The drought parched the land, endangering harvests, thinning out the inhabited areas, and reversing the previous demographic trend.

But it would be wrong to attribute all this to meteorological causes alone. Single explanations are always dangerous, and in this case climate cannot be the only factor, since the uprooting of the peoples of Arizona and Colorado continued after 1305 and the return of the rains, and right through the fourteenth century, which was a wet one. So climate simply contributed to the launching of a tendency which perpetuated itself through purely human causes and effects.[16]

That, very briefly, is Woodbury's historico-climatic argument. But scholars are far from agreeing over it, and recently a young archaeologist from Tucson, Jeffry Dean,[17] has attempted to reappraise the problem of the desertion of the land in the southwest United States in the thirteenth and fourteenth centuries. According to Dean the great drought of 1276–1299 (see earlier reference), despite its intensity, was not the only operative factor in the decline of population in Arizona in the Middle Ages. He specifically analyzes the case of the Kayenta Indians. In the thirteenth century the Kayentas settled in large villages in the Betatakin and Tsegi canyons)[18]; then, in 1300, just at the time of the great drought, they deserted these places to go and settle much farther south, in the region of the Hopi mesas.

According to Dean this move was only partly due to the great drought of 1276–1299. The most important factor was the erosion resulting from irresponsible land clearance, deforestation, and the mere establishment of agriculture. Primitive as the latter was, it still meant the inevitable destruction of the delicately structured soil, which when eroded formed temporary channels or *arroyos*. These gradually climbed up to the highest canyons, eating away the soil and ultimately destroying the arable land and lowering the water table. Without soil for their fields or water to irrigate it, the Indians living in the Betatakin canyon were forced by about 1300 to abandon it. So the great drought of 1276–1299 was only an additional and precipitating factor in a complex situation which drove the Kayentas to move south to the Hopi mountains.

Thus, while it is easy to decipher the human incidence of climatic fluctuation in the case of famines and the short term, it is much more difficult to measure it in the long term. Nevertheless the example of the Kayentas shows it is possible to come to a balanced conclusion, in which climatic change is seen as one factor among a number of other causes which are of specifically human origin.

Another and at present more productive question is that of climatological causality. Unfortunately here the historian has up till now been limited by the nature of his traditional training.

As long as it is only a question of unearthing texts and building up long series the historian of climate can perfectly well produce creative work. But when the time comes to explain the phenomena concerned—to expound, for example, the general circulation of the atmosphere which is responsible for climatic oscillation—then the historian becomes a mere spectator. All he can do is refer his readers to the most recent findings of the traditional natural sciences.

The first piece of data he has to hand on here is the fact that the glacial episodes observed in the Alps and in the west cannot be separated from a world context. Both in the seventeenth and the twentieth centuries the glaciers of Iceland and Alaska advanced or retreated, in the long term, more or less at the same time as these in the Alps.[19] Even in cases most remote from one another there are certain remarkable parallels: when, in the thirteenth century and at the end of the sixteenth century, the Alpine glaciers advanced, in both cases what happened in Europe was accompanied by a long and clearly marked period of drought in the arid southwestern United States: from 1210 to 1310, and from 1565 to 1595 the growth of trees sensitive to drought slowed down considerably in Arizona, Colorado, and California.[20] Shapiro's recent studies suggest that these two regions at these two periods were affected by the predominance of worldwide, analogous, and recurrent barometric conditions.[21] Thus the history of the glaciers and of the giant sequoias, as revealed in the series of data that have been compiled, ultimately leads to a synoptic history of the atmosphere.

This synopsis itself now has its own theories, new and attractive and sometimes contradictory. Just a few words about them here—their history, relations and methods—in order to encourage the reader to consult the experts.

H. H. Lamb,[22] following up the work begun by Rossby and Willet and continued, in German and French, by Flohn and Pédelaborde,[23] has provided the most extensive tables concerning climatological interpretation and historical descriptions, whether early or recent.

He sees a succession of phases of optima on the one hand and pessima on the other. There is the great "atlantic" optimum of prehistory; the brief one about the year 1000 which lasted a few centuries; and the recent amelioration. In contrast to these, the "subatlantic" pessimum of the Iron Age, about 500 B.C.; and the pessimum of the modern era, beginning about A.D. 1200 and culminating in the Little Ice Age of about 1550–1700, of which the most intense phases of cold or coolness came in the 1590s, the 1640s, and the 1690s.

The 1690s, culmination of a European pessimum, with their wet summers, severe winters, ruined harvests, and famines, belong to a more general context. According to Lamb, the alternation of optimum and pessimum in every era is fundamentally to be explained, if one is willing to undertake a dynamic, and hence historical, analysis, in terms of the general circulation of the atmosphere. The analysis attempts, in particular, to define the westerly flow of the temperate zone—the great "zonal" and circular movement which between the 40th and 70th parallels carries from west to east, parallel to the lines of latitude, particles of air which encircle the earth like a ring.

Since Rossby's definitive studies of 1947 we know that this westerly flow is subject to great variation. These variations govern the changes in the weather in North America, the Atlantic, and Britain, and on the continent of Europe, and affect all the components and characteristics of the westerly flow.

(1) In the first place, they affect the latitudes of the depression tracks at sea level, which are displaced more or less toward north or south. These tracks, which run from west to east, may, according to the year, or the century, or the type of climate prevailing, pass either over the north of Scandinavia or, much farther south, over Scotland, Denmark, and the Baltic.

(2) These variations also affect the westerly flow at higher altitudes, including what are known as the upper westerlies, which operate at a height of 4 or 5 kilometers.[24] The upper westerlies, whose west-east paths are closely linked to those of the depression tracks at sea level, may also, and in correlation with the depression tracks, pass sometimes nearer to and sometimes farther from the North Pole. Such northerly or southerly variations in the westerly flow are accompanied, in the upper westerlies, by changes in structure and intensity. When the upper westerlies are displaced toward the north they become stronger, more intense, and swifter, driving toward the Continent an increase of warmth, humidity, and oceanic influence. When, on the other hand, they are displaced southward, they slacken and weaken, and, in Europe at least, continental influences prevail over oceanic ones.

Structural changes in the upper westerlies also affect the geographical distribution of their characteristic features. The upper westerlies, through the masses of air which they set in motion, produce a system of warm ridges and cold troughs.[25] Both warm ridges and cold troughs alter the topography of the 500 millibars surface, 4 or 5 kilometers up, displacing it upwards or downwards respectively.

When the circulation becomes more northerly and faster, the wavelength separating one cold trough from its predecessor becomes longer: so cold troughs become fewer, and spread, more loosely, toward the east. The converse happens when the circulation is slower and more southerly: then troughs become more numerous and closer together, and tend (starting from the quasi-permanent disturbance located in the Rockies[26]) to be driven more toward the west.

In accordance with these different models, the temperate latitudes of the northern hemisphere may have, in summer, either a system of four cold troughs (one over the Bering Strait; one over the Atlantic coast of the United States; one over a line from Finland to the Adriatic; and one over Lake Baikal) or a system of five troughs (two over North America; one over England; one over the Aral Sea; and one over Manchuria).[27] We may note incidentally that the four-trough system tends to be representative of an optimum climate such as that of the eleventh or twentieth centuries, while the five-trough system is typical of a pessimum, such as the Little Ice Age at the end of the sixteenth and in the seventeenth century.

These, summarizing Lamb, are some of the variations which affect the westerlies in the temperate zone. What is of capital importance to the historian of climate, seeking facts from which models may be derived, is that these variations can be reduced to two main types. As Pierre Pédelaborde has written, "The essential fact here is the existence of two types of circulation . . . ; the alternation between the two enables us to explain the variation of climate on every scale and in every age."[28] For convenience we may call these two types Model I and Model II.

Model I: this type of circulation, spreading out toward the equator, predominates during periods of cooling down and "pessimum." It involves, at least in the zone which includes America and Europe,[29] a displacement southwards of the depression tracks, which in summer are situated about 57°–60°N.[30] During the famous "bad decades" of the 1590s and 1690s, eight out of ten of the summers were characterized by depression tracks passing relatively far south,[31] over Scotland and Denmark, between 56°

and 60°N. When this happens floods of cool and damp pour over western Europe during what should be the more clement months, and under early agricultural conditions might destroy the harvests and cause famine.

Correlatively, and still in a situation evolving toward a pessimum, in which the circulation tends toward the south, the winters tend to get colder. As the depression tracks move southwards, the winter exhibits "a southward shift of the zone of prevailing easterly and northerly winds in the Atlantic sector, which has been associated with more frequent snowfalls in the British Isles."[32] This dual phenomenon is closely linked to the movements of the glaciers: the summers grow cooler and so there is less ablation; the winters have more north winds and snowfall, increasing accumulation in the higher reaches of the glaciers. These factors act cumulatively to produce the classic advance of the Alpine glaciers, and more generally of those of Europe and America, during the Little Ice Age when the circulation of the atmosphere was displaced toward the south.

Model I also affects the sea. In a movement roughly parallel to that of the depression tracks of the atmosphere, marine currents of northern origin and cold tendency, such as the Labrador current, also tend to "descend" toward the south, accompanied by the oceanic isotherms. So the North Atlantic, in the latitudes of Canada and France in winter and of England in summer, cools down in the Little Ice Age, for example during the first third of the nineteenth century.[33]

Model I is also characterized by a lessening of energy and a general weakening of pressure gradients in the main windstreams. Insofar as the westerlies form part of a general system of transference of warm air from the tropics to the polar zones, this loss of force "should imply a weakening of the transport of heat and moisture toward the polar regions."[34] So we are in the presence of a cooling factor of major importance in high and temperate latitudes.

In a period of Model I type of circulation, the structure of the upper westerlies is modified: the cold troughs grow closer together and shift toward the west. Over Europe, for example, they shift from eastern Europe to the central and western part of the continent. This explains the increased frequency, during the centuries of the Little Ice Age, of incursions of cold air from the north, passing southward from the Sea of Norway down to the western Mediterranean.

So there is Model I: a type of circulation spreading out southwards, with a comparative lack of intrinsic energy, with a ridge-trough pattern crowded together toward the west in the upper westerlies. All this is correlated, by numerous interrelated factors, with periods of cooling and pessimum in North America and Europe (e.g., between 1550 and 1850).

Model II, on the other hand, is characteristic of phases of warming up and optimum, whether early or recent. In periods when Model II prevails, the whirl or "circumpolar vortex" of the westerlies, instead of spreading out toward the equator, contracts around the pole. The depression tracks become mostly northerly, and during the summer, instead of passing over Scotland and Denmark, pass farther north over the south of Greenland, Iceland, Lapland, and the Kola peninsula. This produces a warmer climate, in particular over the west of Europe. The summers become hotter and brighter, as the cyclones pass farther to the north, and western Europe comes more and more under the influence of the southern anticyclones. The winter too grows warmer with the intensified circulation of Model II: the stronger westerly flow brings oceanic warmth and moisture to western Europe. Such summers and winters finally converge to produce periods of optimum, long or short, which may be described as oceanic and summer-anticyclonic.[35] Lamb's views here are corroborated by the glaciologists. Hoinkes, for example,[36] using a systematic classification of meteorological situations (*Grosswetterlagen*) for central Europe, considers, like Lamb, that the recent regression of the Alpine glaciers is the result of a dual phenomenon affecting summer. On the one hand, anticyclonic situations, with high temperature and insolation, became much more frequent during the summers after 1930; while during the same period, cyclonic situations, with snowfall and therefore increased albedo, became correspondingly fewer.

In this connection the contrast between Models I and II can be expressed by the fact that in one the circulation is more *meridian*, and in the other more *zonal*.

Model II, in which there is a more intense west-east circulation, is characterized by a growing influence of *zonal* and oceanic factors which produce a warming up of climate.[37] Model I is distinguished, at least in Europe, by the preponderance of *meridian* factors: north-south exchanges, streams of cold air toward the south, all of which are the natural accompaniment of the weakening and fragmentation of the westerlies[38] and the formation of a cold trough over western Europe.

As regards the ocean, Model II is characterized by displacement toward the north of the warm Gulf Stream–North Atlantic Drift.[39] As regards the trough-ridge pattern of the upper atmospheric circulation, it produces changes which are the converse of those produced by Model I. In conditions of optimum and warming up, the wavelength of the upper westerlies increases correctively with their increase of strength and speed.[40] Cold troughs thus tend to space themselves out toward the east, and to spare western Europe. This disappearance of cold troughs at high altitude probably explains the remarkable diminution, at sea level, in the number of northerly outbreaks and northerly weather types in the European sector during periods of zonal circulation and warming up: for example, between 1890 and 1950, during the culmination of the recent amelioration. There is nothing surprising about this: a diminution of cold outbreaks means milder weather, by definition. Finally, all these factors together—*oceanic* and "zonal" warming up in winter, *anticyclonic* warming up in summer, the lesser frequency of cold outbreaks—result in a general rise in mean temperatures which produces intense fusion in the glaciers.

All these theories, and the two contrasting models, help climatologists understand better the main features of the synoptic situations in the temperate zone. What chiefly interests the historian is the universal validity of the theory, and the unity thus introduced into climatic and historical interpretation. This unity is, in the first place, worldwide. It is henceforth permissible to relate the glacial thrusts of the eighteenth century in the Alps and in Norway because we know these phenomena have a common origin in the persistence during that period of a certain type of atmospheric circulation (expanded from the pole toward the equator, weak, fragmented). Similarly, geologists are now entitled to do what they have been doing for a long time on a purely empirical basis, and relate the Würm glaciation to that of Wisconsin in North America. The two episodes are not only roughly contemporary, but they are also correlative, occurring as they do in a similar climatic ambiance.

An even more general unity is introduced into history by the alternation of these two types of atmospheric circulation. Variation from one week or one year to another are now seen to be of the same kind as secular oscillations, both geological and climatic.[41] These episodes of varying duration differ from one another in scale, magnitude, and frequency, but not in kind. This illustrates how universal and stimulating is the effect of the new climatology.

It accounts, in particular, for the major oscillations of the historical period which this book has been concerned with: multi-secular episodes of coolness and glacial advance, separated by intersecular phases of mildness and glacial retreat (e.g., the Little Ice Age of glacial advance in the Alps from the thirteenth to the nineteenth centuries, sandwiched between the withdrawal of the late Middle Ages and that of the twentieth century).

At the same time as the first wave of great syntheses based originally on traditional methods of calculation (Rossby, Willett, Pédelaborde, Lamb), the last decade has seen the coming of age of electronic computers. These machines have made it possible to bring together world data on atmospheric circulation on an unprecedented scale. B. L. Dzerdzeevskii, of the Moscow Academy of Sciences, is one of the masters of this new, computerized, climatology, and it is worth while to refer briefly to his theories, though he himself does not regard them as definitive.

Dzerdzeevskii[42] has undertaken to interpret in terms of dynamic climatology the fluctuations of climate, as we now know them, in the first half of the twentieth century. According to him, all is fluctuation. Consequently, everything is evidence. Daily meteorological changes are fluctuations in relation to the background of seasonal and annual changes, which in their turn are fluctuations in relation to long-term trends. These are fluctuations in relation to very long-term trends, and so on, until we reach the categories of glacial and interglacial epochs.[43]

The conditions of world atmospheric fluctuation which produce climatic fluctuation may be typified, in our own latitudes, as in our two models: zonal circulation, when cyclones and anticyclones proceed along west-east tracks in an annular movement round the earth; and meridian circulation, when the cyclonic and anticyclonic tracks tend to move at right angles to those of the zonal type, mostly along north-south axes.

Of course this simplified account only applies to those parts of the globe for which we possess early and continuous information: the "extratropical" zones of the northern hemisphere, our own temperate and Arctic regions.

The raw material for such researches is provided by the synoptic maps of the atmosphere which have been gradually built up by the meteorological services over the past sixty years. Dzerdzeevskii and his colleagues, in their researches into the first fifty-six years of the present century, have used over twenty thousand such maps.

Once the enormous mass of data had been digested by the machines, the Russian researchers were able to put forward an extremely subtle statement of the problems. They derived, from their statistics, six categories of predominant circulation[44]:

1. Meridian circulation, North.
2. Violation of zonality.
3. Zonal circulation, West.
4. Zonal, East.
5. Meridian, South.
6. No circulation ("stationary" type).

These six types alternate with one another, and their respective duration and average life have been worked out for each of the main sectors of the northern hemisphere (Atlantic, European, Siberian, Far Eastern, Pacific, American).

The main conclusion of these researches, in 1961, was that meridian circulation had markedly decreased between 1899 and 1948,[45] while zonal circulation seemed to have become much more intense.

This intensification appeared to coincide with an increase in solar activity, as measured by the total surface of sunspots. From 1900 to 1954 both curves—sunspots and west-east circulation—rise slowly and simultaneously, when rounded so as to bring out the secular trend. Is there a real correlation? Or is it coincidence pure and simple? The old problem of the influence of sunspots on the earth would find a new solution if these parallels were confirmed.

But we must leave that and pass on to the main point. We already had a *climatic* trend in the recent amelioration. Now we have a parallel *climatological* trend in the intensification of zonality during the last half century. It is tempting to compare the two phenomena and try to explain the first by the second.

All this shows how dynamic climatology makes it possible to go beyond the vague generalities of "warming up" and "cooling." It opens the way to more differentiated and complex ideas, which at the same time can be unified, as in the case of the general circulation of the atmosphere. Finally, it poses, on a slightly less hypothetical basis than before, the problem of the relations between sun and atmospheric circulation and climate and extremities of weather. And in this area, the postulate of uniformity means that what is valid for the twentieth century, as regards sun, circulation, and climate, is also valid, *mutatis mutandis*, for the twelfth century, or for the seventeenth.

So the prevalence at a certain epoch of a certain kind of general atmospheric circulation helps to explain variations in the climate. But why does the circulation itself change? This ultimate question has been put by several researchers, who have arrived at various results and suggestions by way of reply.

The work of J. M. Mitchell is among the most typical. As we have seen, Mitchell has made a worldwide study of the recent amelioration, and he considers that in the present state of our knowledge the underlying causes of the recent changes may be said to include:

(1) The accumulation of industrial CO_2 in the earth's atmosphere ("greenhouse effect").[46] (*Industrial pollution*—factor in general warming up.)

(2) The varying frequency of volcanic eruptions, which "inject a large amount of dust into the mid-stratosphere" and thus reduce, by the interposition of a temporary screen, the amount of solar radiation received by the earth. (*Volcanic pollution*—factor in climatic cooling.)

(3) The possible "microvariability" of solar radiation itself, assumed to be sufficient to warm or cool slightly the earth's atmosphere. ("Extraterrestrial" factor.)

(4) The autovariation of the ocean-atmosphere system. Mitchell writes[47]: "Instabilities of general circulation and climate may be caused by dynamic or thermodynamic feedback from the oceans, or from certain qualities of the earth's land surfaces (e.g., soil moisture, snow cover, and continental ice caps[48]), which are in turn conditioned by prior states of the atmospheric circulation . . . Owing to the vast heat storage capacity and long 'memory' span of the oceans, it is reasonable to suppose that air-sea interactions in particular are the most important class of feedback mechanisms giving rise to sustained climatic anomalies (of the order of years, decades, and perhaps over much longer periods as well)."[49]

As regards the recent amelioration, Mitchell thinks it may be explained *in the last analysis* by a combination of these various factors, acting independently of one another:

(a) In the first place, the greenhouse effect may well have played an important role. Since the beginning of this century, coal combustion has released enormous quantities of CO_2 into the atmosphere,[50] thus helping to warm it up. But in that case, Mitchell asks, how is one to explain the recent cooling, when since 1950 industrial pollution by CO_2 from coal and oil has become increasingly massive?[51]

(b) The relative inactivity of the volcanoes between 1890 and 1950 was a supplementary factor in the warming up (or more exactly in the noncooling down) of world climate. This is not to say that all the volcanoes in the world were completely inactive in those sixty years. But statistically speaking, known volcanic eruptions in the world as a whole were less frequent and less intense between 1890 and 1950, and so the pollution of the upper atmosphere by volcanic dust was less then than before or after.

(c) The superficial activity of the sun, as measured by the "Zürich numbers" concerning sunspots, slightly increased between 1925 and 1960. So it is not absurd to suppose that there was also a slight increase in the "solar constant" during that period, and in this case it would be natural that the earth's climate should accordingly warm up slightly. This at least is the view that Mitchell put forward in 1961 and 1965.

However that may be, all these possible causes,[52] together with others more difficult to detect, explain, according to Mitchell, the general warming up of the earth's atmosphere during the first half of the twentieth century. It is at this point of the argument that Lamb's propositions come in, and are adopted by Mitchell: the general warming up tends to increase the vigor of atmospheric circulation and to draw the increasingly contracted ring of the westerlies more and more toward the pole. These dynamic changes in their turn lead to a northerly redistribution of surplus heat received or retained by the earth. Thus is explained the particularly marked warming up of the temperate zones of the northern hemisphere, especially in winter, European examples of which have been described in detail earlier in this book.

The little optimum of the Middle Ages, which is comparable to, though longer and more marked than, that of the twentieth century, could be explained in terms of similar factors,[53] with the exception of course of industrial pollution.

Conversely, it is reasonable to suppose that the Little Ice Age was caused by a group of underlying causes opposite to those which produced the periods of warming up.

So the special contribution of the historian of climate links up without too much difficulty with that of other specialists. There is a fruitful mutual exchange between them. By means of working hypotheses, meteorological historiography projects into the past the discoveries of modern dynamic climatology, while the latter can set in a wider, rational context, and validate, the empirical series worked out by the historians. The intellectual probability of the one activity is joined to the factual truth of the other, and the two forms of research complement and reinforce each other. They tend toward the common goal of all science, and bear witness to the universality of knowledge.

And so, taken as a whole, the history of recent climate, i.e., of the last millennium, now possesses its own *methods*, some of which were described in Chapters I and II. It has its *models*, of which the recent amelioration is the most striking illustration (Chapter III). It also has its own *chronology*, to be read, unlike most others, backwards from the better to the lesser known and from the more recent to the earlier. From the mild twentieth century to the cool seventeenth, from the Little Ice Age to the little optimum and the Middle Ages, I have attempted to show the film of time backwards, going further and further back into distant epochs (Chapters IV, V, and VI). This history also has its *human and climatological implications* (Chapter VII). Finally, it is built around a double necessity: for whatever century it may be concerned with, there can be no good history of climate unless it is interdisciplinary and comparative.

AFTERWORD

Since 1971 I have strayed somewhat from the areas of research covered in *Times of Feast, Times of Famine*, and I have not been able to prepare a new edition of the book. Instead, I have included here a review of H. H. Lamb's most recent work, *Climate History and the Modern World* (1982). His book provides the professional climatologist's qualified view on the state of historical-meteorological questions.

As a historian of climate—though he is by training and profession a meteorologist—H. H. Lamb starts by asserting something which was long denied: the Unesco *General History*, published twenty years ago, proclaimed that the "climate" had, in the main, been stable, or become so, since the sixth millennium B.C. Lamb attacks this notion of "fixity" just as he challenges that of a "normal period" which is supposedly representative *ad aeternum* of the average meteorology of a given region: for a long time, specialists considered as "normal" the thirty-year periods from 1901 to 1930 and 1930 to 1960. In fact, these groups of three decades were the warmest experienced for many years and cannot therefore be considered to fall within the "norm."

At the same time, Lamb discreetly qualifies some of the more sweeping theories. Marxism, for example, believes in economic and material determinism; it is not, in theory, hostile to examination of climatic variations, since these affect the "ecological" base of society, "beneath" the economy. As for the ideas of Aristotle and Montesquieu, they may partially explain civilization, or its absence, by the climate of the continent in question, but need to be put into perspective: "civilized" capitalism has indeed flourished in temperate climates, but also in the equatorial city of Singapore (cooled, admittedly, by office air-conditioning).

Every historian must first tackle the problem of his sources, and Lamb readily submits to this golden rule. For a start, there are glaciers, well documented thanks to iconography, records, and Carbon-14 dating (applied to the trunks of fossil trees which give evidence of previous glacial advances). For the past 100,000 years, and up to and including the seventeenth and nineteenth centuries, glaciers are consequently a major guide to climatic change. Late wine harvests imply a cold season, and vice versa: early vintages are well documented, year by year, since the start of the fourteenth century and give a considerable amount of information. The same is true of tree rings: their annual growth in dry or semi-desert regions is proportional to the humidity occurring in a particular year. Records of events (successions of cold or mild winters, for example) are highly informative when compiled by reliable observers, though this is by no means always the case. Through them, Easton and his colleagues established the existence in the second half of the sixteenth century of the fall in temperature which preceded the great advance of the Alpine glaciers, around 1595–1600; the same evidence also allowed Christian Pfister to make a proper study of the Swiss climate in the eighteenth century. Pollen series in peat bogs are significant for prehistoric climates, but not from the neolithic onward, when they were disturbed by land clearance, which destroyed trees and replaced their pollens by those of crops, with the "invention" of agriculture. Variations in the price of corn, on the other hand, may be variously caused, so graphs illustrating them should not be relied on too much as indicators of variations in climate, except in some obvious cases: the famine of 1709, for example, was directly attributable to the celebrated cold winter in that year.

Lamb boldly starts his survey with the great ice ages. The last recorded of these began gradually 115,000 to 90,000 years ago; then, after some hesitations lasting for periods of between two and five thousand years, finally set in 70,000 years ago and lasted 50,000 years. A short cold snap, 10,800 years ago, lasted a mere six centuries, and started a few small glaciers in the Lake District. The melting of the ice over the past 10,000 years encouraged the beginnings of agriculture and stock breeding, naturally enough, but cannot be seen as the prime mover in this respect: in Mesopotamia and Palestine, where wheat was first cultivated, glaciers never had any significance at all.

On the other hand, over those ten millennia, for reasons which are easy to understand, the melting of the ice did involve a rise in sea levels: the Pas-de-Calais thus emerged around 7600 B.C. and the present outline of the French, German, and British coasts was more or less fixed around 5000 B.C. (farewell to the reindeer of Hamburg and to the Copenhagen tundra!). The post-glacial heat wave reached 2°C. above nineteenth-century temperatures during the so-called Atlantic or optimal phase, between 5000 and 3000 B.C., before another chill set in, bringing us our present-day climates. All climatic regions (Arctic, temperate, etc) are now once more shifting southward. Until around 3500 B.C., the Sahara still enjoyed the rainfall brought by a few meridional cyclones. From then on, it again dried out, by virtue of what is only apparently a paradox, and this put an end to the extraordinary vegetation seen in the Tassili rock paintings. At the same time, the glaciers once more advanced a little in the Alps. A nascent agriculture certainly took advantage of the pleasant warmth of the Atlantic phase in Europe.

Apart from this, Lamb's other historical speculations about these millennia are often hypothetical: he is well aware of this, judiciously introducing them with phrases like "It may well be that . . ." or "It is tempting to think that . . ." Among such hypotheses, one might cite the following (which he does not, of course, take too seriously): that there are stone circles in Britain might suggest that the sky was clearer than it is today; the development of the major religions in the first millennium B.C.; the appearance of Jesus himself at a "mild" period in climatic history; the emigration of the Italian vine under the Roman Empire to southern Gaul, a region receptive to grapes in any case since it had a basically sunny climate, regardless of minor climatic variations.

Thanks to the many studies which Lamb conveniently draws together, we now have some virtual certainties as far as the Middle

Ages are concerned: there was indeed a medieval "mini-optimum," with temperatures comparable to those of the "good" years 1900–1950, or even a little warmer: it is clearly shown in the texts, as in the Greenland and Alpine glaciers, and lasted from roughly A.D. 800 to 1200. It certainly favored the colonization of Greenland by the Vikings in the tenth century and did not hamper—far from it, though one cannot with certainty say more—the great land clearances in Western Europe during the eleventh century. It seems to have ended during the thirteenth century, and Lamb tends to explain the great crises of the fourteenth and fifteenth centuries by the slight drop in temperatures following this "optimum," which would have meant smaller harvests. But historians remain cautious about this: the *natural* conclusion of a cycle of economic growth which started in the eleventh century, culminating around 1300, and the appalling disasters of the late Middle Ages (Black Death and other plagues from 1348, and the Hundred Years' War) were all so overwhelming in their effects that slight variations in climate can only have been decidedly secondary causes. Competition from the Bordeaux vineyards, which had begun to export their product to the British Isles, seems to have been so great that, whatever the climate, the unfortunate English wine growers could but give in and replace their vines by cereals or pasture. The iron law of profit operated even in the fourteenth century!

Then came the little ice age, so clearly visible in the seventeenth century and affecting the years from roughly 1560 to 1850. Here Lamb is on much surer ground and his solid argument and erudition, supported by the recent discoveries of Christian Pfister, are highly impressive. It is no longer a question of mere conjecture, as in the case of fluctuations in temperature in England during the thirteenth century; in many respects, this is an area of virtual certainties. Around 1600, the Alpine glaciers crushed the most exposed hamlets around Chamonix, marking the start of the new seventeenth-century cold spells which were to continue, though with "sunny intervals," until around 1850. Temperatures during the "bad" decades of the seventeenth century (the worst occurring in the 1690s) may have fallen to an annual average of 0.9°C. below the norms for the warmer years 1920–1960. Cold winters and poor summers brought famine, killing off seeds and crops and affecting Scotland, France, and Finland especially in the fatal decade of the 1690s. But, even here, human agency and simple historical freedom do not lose their rights: English agriculture was already more technically advanced than that of France or Scotland, so the English escaped without too much harm. What was bad for Louis XIV was good for William of Orange.

It should be said at once that the little ice age was no more a single entity than the French Revolution was. There were fine, warm periods within it, for example the mild years 1710 to 1739, which coincided with the economic revival in Western Europe (admittedly also favored by the inflow of new supplies of Brazilian gold, the end of the great wars, and the political thaw which followed the death of Louis XIV). Once ended, these rises in temperature were followed by fresh disturbances, especially when accompanied by volcanic eruptions. Volcanic ash, projected into the atmosphere, intercepts the heat of the sun: in 1815, the eruption of Tamboro in the East Indies produced a famine in the cold winter and wet summer of 1816–17 and glacial advances in the northern Alps.

As I have said, the recent warm period began in the 1850s and 1860s, and culminated in the 1940s. The mild west and southwest winds became more frequent in Britain from 1860 to 1960, and after 1900 British rivers froze over completely on many fewer occasions than previously. Rainfall was heavier than before over the interior of the old continent, and all the climatic regions (Arctic, temperate, subtropical) appeared to be shifting *northward*, contracting around the North Pole. This led to a paradoxical consequence: the Antarctic region was also extending northward, so that there was a precise correlation (0.75) between the increase in mild southwest winds in London and snowfall at the South Pole.

Since the 1950s and more especially in the 1960s, the rise virtually worldwide in temperature has given way once more to some reduction, as part of a more or less irregular cycle: in twenty years, we have lost two-tenths of a degree Centigrade on overall world averages.

A word now on the short- and medium-term effects of these climatic variations (the long-term is a complete mystery). Leaving aside Greenland and even Iceland, two countries where even a minimal fall in temperature such as occurred in the Middle Ages was enough to threaten an already extremely marginal agriculture and stock breeding, it is evident that in the leading countries of Western Europe (France, Germany), or offshore (Scotland), cold winters and wet summers led to shortages. The effect of these on the price of grain could spread over several successive years, as happened during the cold period around and following 1770 in Switzerland [Christian Pfister]. Such commonsense observations throw a grim light on the history of human suffering. But, beyond that, the historian is once more in the delicate field of "it may well be that . . ." Choleras and the Black Death *may* have arisen in India and China, respectively, following the floods of 1816 and 1332 in each of those countries. But for the moment these are merely interesting speculations and the burden of proof rests on the frail shoulders of climatic historians.

More convincing is what might be called "the pinprick effect." The very cold year of 1879 (comparable to 1740) produced a bad harvest, thus inaugurating forty lean years for English agriculture, which was henceforth swamped by imports of American and Russian wheat. In the same way, among the undoubted causes of the French Revolution (though it was only one among many), was the very poor harvest of 1788, a breeding ground for the *grande peur*, which arose from the bad weather of 1787–88. In both cases, 1879 and 1789, short-term climatic and long-term human factors combined to "make history." Today, the growth in world population is both rapid and dangerous, increasing the threat from climatic events and the poor harvests they produce, as, for example, in 1972. The great natural catastrophes of our time are due to floods (40 percent), cyclones and typhoons (20 percent), and drought (15 percent).

So the climate is not without its consequences for mankind, and its variations are due in their turn to different causes, which Lamb discusses at some length. Sunspots are often invoked, with their eleven-year cycle, but, in the end, they give us only a vague indication of what is going on around them on the sun's surface or in its depths. The remarkable absence of sunspots between 1645 and 1715 might be connected with the unusual cold of the mini ice age. As for climatic variations over eleven years and multiples or sub-multiples of that figure (five and a half years, twenty-two years, etc), they might also be explained by slight fluctuations in the output of solar energy. But there is no solid evidence: according to Lamb, this is an area of pure speculation. According to Milankouvitch and his followers, even slight variations in the axis of rotation of the earth and the earth's orbit produce much more certain effects. These long-lasting cyclical changes go a long way to explain the chronology of the great ice ages during the quaternary period, with intervals of 20,000, 40,000, and 100,000 years. But the historian must here put a plain objection to the climatologist: these orbital and axial variations already existed in the long tertiary and secondary eras, when ice ages are remarkable for their complete absence.

Returning completely to earth—to the earth considered as a physico-chemical object: volcanic eruptions throw out dust which remains suspended for a long time in the upper atmosphere, is carried all round the globe by winds, and has a temporary cooling effect on the climate because of the obstacle it presents to the sun's rays. In 1783, there were two major eruptions, in Iceland and Japan: the Northern Hemisphere at once lost 1.3°C. on summer temperatures and took four to five years to return to normal, with famine in Japan as a result. On the other hand, the absence of major eruptions in the Northern Hemisphere from 1912 to 1963 favored a rise in temperature between these dates.

However, volcanoes are only circumstantial causes. More important, for Lamb, is the part played by the circumpolar vortex, that immense westward flux which in the northern latitudes sweeps across the temperature zone, encircling the globe like a ring. If this flow is in fact, or tends to be, ring-like, constricted toward the pole, then subtropical and warm anticyclones and southwest winds will predominate over France, and even over Britain. We will enter a period, lasting a season, or a decade, or a century, or even a millennium, of interglacial temperature rise. All depends chronologically on the long or short duration of the basic phenomenon of the "circumpolar vortex"; and yet, its size apart, this remains substantially the same in its physical structure on all time scales.

On the other hand, how can we explain cooler periods, whether over a few chilly years in the twentieth century, the mini ice age of the seventeenth century, or the great glacial period over the past 100,000 years? To understand this, we have only to imagine the opposite situation to the one described above: instead of contracting, the circumpolar vortex now thrusts southward, forcing the Azores anticyclone out of Western Europe, and allowing a corresponding flow of Arctic air down toward London, Paris, and Marseilles. This southward drift is accompanied by changes in the form of the vortex: it becomes sinuous, carrying huge loops of air extending across a subcontinent, and formed from anticyclone ridges and cold (cyclonic) valleys. Ironically, one of these warm ridges may lead locally to a warm dry summer, as happened in Europe during the celebrated 1976 heat wave. But over most of the Northern Hemisphere, despite such regional spells of exceptionally warm weather, this kind of movement, reaching southward and characterized by its huge loops, generally produces cold weather; and this was indeed the case during the summer of 1976, baking hot in London but chilly and even cold over most of Russia.

A further paradox is that the southward shift of the circumpolar vortex from the Northern Hemisphere, as well as the tendency toward colder weather in our latitudes, goes together with a southward shift of *all* the concentric bands which also encircle the globe like a ring from east to west, over the tropics, the Equator, and down as far as the Antarctic. This means that the cold air from the Antarctic is driven back toward the South Pole and the edges of that polar continent warm up. Consequently, during the coldest parts of our little ice age between 1670 and 1840, penguins were able to move farther and farther south, and in particular to settle on Ross Sea, which was for a time less cold.

Finally, it should not be forgotten that the oceans, those great stores of both heat and cold, act in liaison with the circumpolar vortex: when it shifts northward, bringing warmer climates to the temperate zones, the Gulf Stream follows suit; and vice versa. So the two phenomena, maritime and atmospheric, react upon each other. Herring and cod also move north or south with the colder waters. Holland was never so prosperous as in the little ice age: for her a golden age of fishing, the century of the herring.

Historians (though with little success) are always willing to make predictions. Is this the case for a great specialist in climatology like Lamb? He modestly reminds us that even peasants are capable of accurate forecasting and every American plains farmer knows, from 160 years of experience, that a serious drought will occur, on average, every twenty-two years. Historians and "cliometeorologists" know that every hundred years, in the nineties of each century (1490, 1590, 1690, 1790, 1890 . . .), very harsh winters recur. More generally, the present spell of cold weather should continue until about 2015 at the rate of $-0.15°C$. on average each decade. Then there should be a further warming up of $0.08°C$. per decade until around 2030, with an eventual stabilization which will (or should) last until a renewed slight decline in temperatures in the twenty-second century. Naturally, these forecasts do not take account of volcanic eruptions, which cannot be foreseen and might well cause temporary upsets. They also leave out the effects on the climate of pollution by man, which I shall come to in a moment.

As for the next millennium, there is an admirable study by A. Berger, of Louvain-la-Neuve, which suggests, on the basis of the major cycles of the past, that a new ice age proper will start gradually during the next one thousand years. Real glaciation, though not yet too serious, will set in over a period lasting from A.D. 5000 to 9000; then, 15,000 years from now, there will be a slight but temporary improvement, quickly followed by deep glaciation. Present temperature levels will not return for 60,000 years, or perhaps, according to another model, for 114,000 years! The previous period as warm as the interglacial era we are now enjoying lasted 11,000 years (as ours will, perhaps?), ending some 125,000 years ago and gradually giving way to a deep freeze of 100,000 years or more. This in turn finally retreated, to allow the 10,000-year warm spell of the neolithic age, protohistory, and history. But the shift to the paleo-lithic ice, 125,000 years ago, happened with terrifying abruptness, in around 125 years. A sudden event, lasting a century in this case, actually heralded one of very considerable extent (100,000 years of ice): on this subject, see the fine studies by Mme Woillard, also of Louvain-la-Neuve, who is a specialist in the peat bogs of Vosges.

These prospects and perspectives take no account of the influence of human activities: industrial dust, acting as a screen against the sun's heat, may act like volcanic dust to intensify the cooling pro-cess which has been in operation since 1960 and is likely to carry on for the next three or four decades. Much more devastating, how-ever, is liable to be the opposite, so-called greenhouse effect, due to the accumulation of carbon dioxide in the atmosphere, also from in-dustrial sources. The temperature rise caused by CO_2 might even be intensified by heat inevitably produced by peaceful uses of nu-clear energy. Both of these together—the artificial warming effect of CO_2 plus nuclear heat—might result in a rise of 2° C. (over current temperatures) in around 2100. Will we then return to the pre-historic climatic optimum, with a certain amount of coastal flood-ing due to melting glaciers? A more apocalyptic view suggests that, if "hot" industry continues in its wicked ways, we shall advance in a few centuries toward a tropical climate like that of the tertiary era, with the disappearance of the Arctic ice cap and a corresponding rise in sea levels which will submerge the great plains of Europe, and with them many towns and capital cities.

Toward the end, Lamb's book thus tends to become quite dra-matic, perhaps because he feels the need to enliven a subject which, in itself, is colder and grayer than it might at first seem. The history of climate is fascinating in itself. But, from a strictly human point of view, it explains chiefly temporary famines (that of 1694 in France) and marginal decline (the prehistoric Sahara, late medieval Greenland). Once again, in Lamb's splendid work, one must set on one side, especially for antiquity and the Middle Ages, what is purely hypothetical, doubtful, or speculative as to the influence of climate on a given civilization, whether of Ancient Greece or the Indus. Here there is a need for reliable temperature curves drawn up on the basis of series of direct observations: let us hope that this gap will one day be filled. After 1500, we are on the firm ground of the little ice age and the rises and falls in temperature which followed it. Conclusions become more solid, but must be kept in their place. Lamb has not always managed to avoid self-contradictory explanation, as when he explains the Mongol emigration, on the one hand, by the drought in Central Asia which drove them out of their homeland and, on the other, by the rainfall which increased the number of their flocks and nomadic groups, thus obliging them to emigrate.

But, on the whole, he has written a fine and important book; he has managed to turn himself into a historian, a considerable achievement for a scientist. And he has succeeded in placing himself at the strategic center of the problem. From there he has been able to undertake fascinating excursions into all sorts of fields and in all sorts of directions, without ever losing the thread. This interdisciplinary work is the crowning achievement of a splendid career.

NOTES

INTRODUCTION

[1] Utterström, 1954, 1955, and 1962.
[2] Le Roy Ladurie, 1966.
[3] Angot, 1883; Bruckner, 1890; Arakawa, 1955 and 1956.
[4] Garnier, 1955.
[5] Angot, 1883.
[6] Matthew, 1942.
[7] This bibliography was of course known to glaciologists (cf. Lliboutry, 1965, II, pp. 724–27, 765, 833), but by very reason of its historiographical nature they did not always exploit it fully. Conversely, historians rarely had occasion to go into the work of early glaciologists, though its foundations and methods were archivistic and typically documentary.

CHAPTER I

[1] Sutcliffe, 1963, p. 282.
[2] Wagner, 1940.
[3] For example, Mitchell, 1963; Pédelaborde, 1957; Lliboutry, 1956 and 1965; Flint, 1957; Godwin, 1956; Elhaï, 1963.
[4] Huntington, 1907, pp. 378–79; Le Danois, 1950 (especially p. 153 and his fanciful way of determining climatic periods).
[5] Olagüe, 1951, 1958, and 1963.
[6] This article has a very full and useful bibliography of recent articles on the history of climate. My criticism is directed not at the author's specific and perfectly relevant study of Scandinavia in the eighteenth century, but at some general ideas put forward in this article.
[7] Generally speaking, it is difficult and even dangerous to interpret recent evidence about changes in areas where crops are reared in terms of "change of climate." For example, in France, during the whole of the Little Ice Age (1550–1850), and in particular from 1550 to 1600, which were particularly cold decades, the southern limit of olive-growing moved steadily northward. (Texts in Le Roy Ladurie, 1966, Vol. 1, Part II, Chapter 1.) But is this to be interpreted as proof of a warming-up of the climate? Certainly not. We know from other evidence that the climate in Europe has never been so cold as it was then. If the olive nevertheless persisted in "drawing toward the Little Bear" (phrase of Olivier de Serres, quoted in above), it was because of the valiant efforts of the growers, who extended northward in spite of the climate in order to exploit an expanding market. Paradoxically, the northern limit of the olive in France did not move southward until the twentieth century, and that right in a warming-up period in which in theory the opposite might have been expected. But French olive-growers today disregard these new conditions, physically favorable as they might appear. They abandon their northern olive groves because irresistible competition from Italian, Spanish,

and Tunisian oil makes them no longer worth while. So in both instances—expansion toward the north in the Little Ice Age, and withdrawal toward the south in the twentieth century—changes of climate played no part in the wanderings of the olive. And I dare say the same is true of the varying northern limit of the vine during the historical period.

8 Douglass, 1919, 1928, and 1936.

9 Bibliography of their work in Guitton, 1958, p. 89, and Beveridge, 1922, pp. 412–54.

10 Brückner, 1890.

11 Tardetsky, 1961.

12 Frolow, 1958.

13 P. Jeannin.

14 Baehrel, 1961.

15 J. Favier, for example, in a stimulating book on the close of the Middle Ages, has raised with all the requisite caution the problem of this hypothetical recurrence of wet years (Favier, 1968, pp. 127–29). Other authors, including Heers, 1966, pp. 92–93; and Carpentier (E.), 1962, pp. 1078–79, etc., have done the same, with similar prudence.

16 The best account of this subject is in Lucas, 1930.

17 *Guide de la France mystérieuse*, Tchou, Paris, 1964, p. 964.

18 This decade was in fact very wet; see the chronology of Tchou, 1960.

19 Fourquin, 1964.

20 On all this see Fourquin, 1964, pp. 195–221 (graphs); and Fourquin, 1966, pp. 52 ff.

21 On these questions my basic documentation is taken from Richard, 1892, and Fossier, 1955.

22 See the excellent map given in H. H. Lamb, 1966 (*The Changing Climate*), p. 191.

23 *Archives nationales*, LL1210.

24 See Fourquin, 1964, Metrological introduction.

25 In this connection see the accounts of Widow Couet for 1552–1572 (MSS. of *Bibliothèque nationale*, new French acquisitions, MS. no. 12, 396).

26 For the ten-yearly indices see H. H. Lamb (*The Changing Climate*), 1966, p. 217.

27 Russell, 1948; Carpentier, 1962, p. 1083.

28 Huntington, 1907, pp. 378–79.

29 Brooks (C. E. P.), 1950, p. 321, who does not hesitate to write: "My rainfall graph for Asia (Fig. 33 in Brooks, 1950) is derived principally . . . from Toynbee's diagram on migrations, checked against the levels of the Caspian." Similarly Gumilev, *Cahiers du monde russe*, 1965.

30 Gumilev, 1965, is not always entirely convincing on this point.

31 Angot, 1883; Aymard, 1951.

32 Huntington, 1907, who is A. Aymard's target.

33 Pédelaborde, 1957, pp. 29–37.

34 Bloch, 1949 ed.

35 Angot, 1895.

36 Lamb, 1966.

37 Titow, 1960.

38 Angot, 1883; Mougin, 1910–1934; Richter, 1877, 1888, 1891, and 1892; Allix, 1929; on the series by Renou, see Garnier, 1955.

39 Braudel, 1949.

40 Kinzl, 1932; Mayr, 1964.

41 For example Letonnelier, 1913.

42 Angot, 1883; Garnier, 1955.

43 Mayr, 1964; see also Corbel and Le Roy Ladurie, 1962.

44 Drygalski and Machatschek, 1942, pp. 214–16; and below, Chapter 3.

[45] *Proceedings*, 1962; Le Roy Ladurie, 1963 and 1965.

[46] The period in which rigorous observations were made (the nineteenth and twentieth centuries) is really the business of meteorologists, not professional historians. I shall only deal with this period (below, p. 58) in so far as it provides models which help us to understand earlier times.

CHAPTER II

[1] C. Lévi-Strauss, *Le Cru et le cuit*, Paris, 1964.

[2] For example, Fuster, 1845.

[3] As it has been for example by Lamb, 1966.

[4] Huntington, 1915, pp. 262–70.

[5] Within the framework of this book it is not possible to give a complete account of the methods used by dendroclimatologists. But it is worth pointing out that since the average thickness of the rings diminishes from the center (which shows the years of the tree's youth and vigor) to the outside edge (which corresponds to the tree's old age), calculations are based not on the absolute thickness of each ring but on the difference between the absolute thickness and the average thickness the ring ought to have in terms of its distance from the center.

[6] Fritts, Smith, Stokes, 1965. See also Fritts's excellent graph, 1966 (*Science*), Fig. 3.

[7] H. C. Fritts, D. G. Smith, J. W. Cardis, and C. A. Budelsky, 1965.

[8] Cold year in this context is short for cold growth period.

[9] Fritts, 1961.

[10] The American school has discovered illustrious predecessors for itself. According to Studhalter, 1956, the problem of dendroclimatology was considered as a possible subject by Leonardo da Vinci, Buffon, Duhamel, Du Monceau, and Candolle.

[11] The above is taken from the *Tree-ring Bulletin*, May 1962, pp. 3 ff.; and Haury, 1962.

[12] Fritts, 1964 and 1965 a and b, and my Fig. 3.

[13] It is the large number of trees studied which is important in establishing climatic fluctuation: "A wide sampling of annual rings from the base of many semi-arid site trees appears more appropriate for evaluating past fluctuations in climatic factors than an intensive sampling of rings at several heights in only a few trees." Fritts, Smith, Cardis, Budelsky, 1965, p. 3.

[14] Garnier, 1955.

[15] H. C. Fritts, D. G. Smith, M. A. Stokes, 1965.

[16] Stockton and Fritts, 1968.

[17] On the bristlecone pine as a climatic indicator over thousands of years, see Fritts, 1969.

[18] In 1965 a slightly shorter chronology (6600 years) was published by Currey in *Ecology*, 46, No. 4564 (1965) (after Ferguson, Huber, and Suess, 1966, p. 1174).

[19] C. W. Ferguson, 1968 and other reprints.

[20] Ferguson, 1968, and also Ferguson, Huber, and Suess, 1966.

[21] *Carnegiea gigantea* (Engelm—Britt and Rose).

[22] Turner, Alcorn, Olin, and Booth, 1966.

[23] In particular Peter F. Mehringer, who explained to me in his Tucson laboratory the problems presented by his on-the-spot researches in Death Valley.

[24] Giddings, 1947.

[25] This rise should probably be related to the slight warming up demonstrated by the general withdrawal of glaciers throughout the world from 1850 on; but Giddings does not give any opinion on this.

[26] Leboutet, 1966; Pons, 1964.
Dendroclimatological work in progress in France and French-speaking countries includes J. Florence, 1962; P. de Martin, 1969, unpublished, and A. V. Munaud, 1966 (Belgium).

[27] Schove and Lowther, 1957.

[28] Dr. J. M. Fletcher of the Department of Forestry, University of Oxford, has recently developed a dendroclimatological method based on radiography and density charts. He outlined the technique, which he is applying to oak from medieval timbers used in England, to the Group for the Study of the History of the British Climate at a meeting at Oxford on April 19, 1969.

[29] Kolchin, 1962, 1963, and undated; quoted in Schove, 1964.

[30] B. Huber and W. Von Jazewitch, *Tree-ring Bulletin*, 1956.

[31] Huber and other authors, 1964 and 1969.

[32] See Huber and Siebenlist, 1963, Fig. 2 and pp. 258–59; and Huber, Siebenlist, and Niess, 1964, pp. 31–33.

[33] On the general problem of the climatic interpretation of wineharvest dates, see p. 51.

[34] See Angot, 1883, and Appendix 12.

[35] Weikinn, 1958, pp. 170–80.

[36] Huber and other authors, 1964 and 1969.

[37] See Huber and Jazewitsch, 1956; Farmer, 1969; J. Titow, 1960 and 1969; Postan and Titow, 1959, Lucas, 1930; Weikinn, 1957. See also on German dendrochronology Huber, 1964 and 1969 (master chronology of the oak in central Germany); E. Hollstein, 1965 (long series on the oak from A.D. 820 to 1960 for the area of Trier); Bauch, Liese, and Eckstein, 1967, p. 285.

[38] Siren, 1961; see also Appendix 16.

[39] Schulman, 1956, p. 478; and 1951, p. 1028.

[40] More generally, after a very careful study of the climatological results obtained by the searchers after cycles and periodicities, J. M. Mitchell comes to the conclusion that of all the cycles suggested (they vary in length from a few weeks to several centuries or thousands of years), only two stand up to criticism. "Only two cycles appear to be established at impressive levels of statistical significance. These are the cycle in precipitation, whose period is slightly less than fifteen days, and the so-called biennial oscillation, which appears with dramatically large amplitude in surface climate. Neither of these cycles evidently accounts for enough variance in surface climate to be worth incorporating into routine procedures of prediction." (Mitchell, 1964, ad fin.)

[41] Huber and Jazewitsch, 1956, p. 29.

[42] Fritts, 1964 and 1965 a and b.

[43] On phenology, see the basic article by Angot, 1883. See also Garnier, 1955 and Samson, 1954, pp. 453–56; Duchaussoy, 1934. There are convincing examples of the method in Lindzey and Newman, 1966 (a good demonstration here of the close correlation between temperature and the flowering dates of a great number of plants). On the practical applications of phenology to agriculture, see Golzov, Maximov, Iaroschevskii, 1955. See also another general account in Le Roy Ladurie, 1966, Chap. 1.

[44] Municipal archives of Montpellier, HH20.

[45] Municipal archives of Lunel, BB21.

[46] R. Laurent, 1958, Vol. 1, p. 128, and Vol. 2, pp. 170 and 246–47. Laurent says wine harvests were later during the period of the so-called "general proclamation" of the vintage date in Burgundy, from 1804 to 1842. But in fact it is only in 1853, ten years after the abolition of the "general proclamation," that the dates become definitely earlier than before. Human factors, such as the authoritarian general proclamation, though not negligible were not overriding; it was climate which governed the main chronological variations.

[47] Angot, 1883.

[48] Duchaussoy, 1934.

[49] Le Roy Ladurie, 1966, Chap. 1.

[50] G. Anes Alvarez, 1967; Bennassar, 1967, pp. 43–45; and below, Appendix 12. The Burgundy series worked out for the vineyard of Chambolle by Danguy and Aubertin, 1892, is not of much use and has not been incorporated into the averages in Appendix 11.

[51] Sévigné, 1862 edition, Vol. III, pp. 499, 506, 523.

[52] A parish pariest noted: "Since the last day of March we have seen no rain and our fountains are very low . . . the olives and grapes are almost all withered and the olives fallen from the trees." (A. C. Aniane, Hérault, AA2, fol. 67, 25-8-1718).

[53] "The heat was extreme that year and lasted all the months of July and August; it was only on the sixth of September that the rain began which refreshed the whole country." (D'Aigrefeuille, ed., 1885, Vol. II).

[54] Samson, 1955.

[55] Quoted by Roupnel, ed., 1955, p. 33.

[56] J. Garnier, a meteorologist, says of the fluctuations in the phenological curve: "The variations in wine harvests swing sometimes toward the late. . . . Earlier and later periods correspond respectively to periods that are dry and hot or those that are damp and cold." (J. Garnier, 1955, p. 299).

[57] In spite of the winter of 1709, which comes outside the vegetative period. Generally speaking, phenology throws no light on the winter, when the plant is not actually growing.

[58] Roupnel, ed., 1955, p. 33.

[59] Sanson, 1956.

[60] Goubert, 1958.

[61] Jacquart, 1960.

[62] This expression is used to denote the period from March to September which determines the dates of the wine harvests.

[63] See Le Roy Ladurie, 1966, p. 36, for exact chronology.

[64] Schove, 1966.

[65] Pepys, *Diary*, September 2, 1666; see also texts quoted by Schove, 1966.

[66] Le Roy Ladurie, 1966, p. 34, on English thermometric observations showing a warm trend in the spring-summers of the decade 1680, cf. Manley, 1965.

[67] Von Rudloff, 1967, p. 93.

[68] Baker, 1883, quoted by Easton, 1928, p. 117.

[69] Manley, 1953.

[70] Von Rudloff, 1967, pp. 125–31.

[71] Von Rudloff, 1967, p. 139.

[72] Von Rudloff, 1967, p. 154.

[73] Müller, 1953, p. 224; Von Rudloff, 1967, p. 154.

[74] Enjalbert, 1953, p. 462.

[75] It is not altogether impossible that a close and complicated analysis of wine harvest dates may some day finally yield up indications of secular climatic variation. Some progress seems to have been made in this direction as regards the sixteenth century; see Appendix 12.

[76] Angot, 1883, pp. B47 to B50, and B74.

[77] Angot, 1883, pp. B117 and B118.

[78] On the climatological problems of 1816, see Von Rudloff, 1967, p. 140 and p. 54, Fig. 17.

[79] Author's "Introduction" to *Frankenstein, or the Modern Prometheus*. See also Boujut, 1965, pp. 9–12.

[80] Quoted by Gueneau, 1929.

[81] Text published by J. M. Desbordes, about 1965.

[82] Gueneau, 1929, pp. 18 ff., 94, and passim.

[83] Baulant and Meuvret, 1960, Vol. I; Hauser, 1896.

[84] Hoskins, 1964, pp. 28, 34, and 39.
[85] Goubert, 1960, Vol. 2, graph of wheat prices at Beauvais; Deyon, 1968.
[86] Hoskins, 1964, p. 38.
[87] Hoskins, 1968, p. 20.
[88] See especially Hoskins, 1968, p. 22, and Baulant and Meuvret, 1962, Vol. 2, diagrams at the end. In Mediterranean France, on the other hand, the warm dry years of the 1680s had produced opposite results from those in the north, and the grain harvests of the Midi were poor in the 1680s (Le Roy Ladurie, 1966, Parts 1 and 5).
[89] Le Roy Ladurie, 1966, p. 750.
[90] Boislisle, p. 338.
[91] Boislisle, document 1256; Letter of the Intendant of Auvergne of November 4–6, 1693.
[92] Angot, 1883, p. B72.
[93] Boislisle, 1874, Vol. 1, p. 297.
[94] Desbordes, p. 7. For the bad harvest of 1692 in the west of France, see also Boislisle, Vol. 1, letters of October 9, 15, and 19, 1692, to the controller general.
[95] Desbordes, p. 16.
[96] Boislisle, Vol. 1, p. 297.
[97] Ibid., Vol. 1, p. 299. My italics.
[98] Müller, 1953, ad fin.; Angot, 1883.
[99] Boislisle, Vol. 1, p. 336; Desbordes, on the year 1693.
[100] Fénelon's expression.
[101] In northern Europe, where grain is very sensitive to excessive cold, the decade of 1690 also produced several major catastrophes: the famine of 1696–1697 killed over a quarter of the population of Finland (Jeannin, 1969, pp. 94–97).
[102] Rudé, 1964, p. 48, for the scarcity of 1725.
[103] Labrousse, 1944, p. 273. The data based on the period 1775–1785 are taken from an unpublished study on meteorological series and wine harvest dates by M. Desaive and myself which will be published shortly.
[104] Labrousse, 1944, p. 94.
[105] Le Roy Ladurie, 1966, Vol. 2, annex and graphs on wine harvest dates.
[106] Müller, 1953, p. 214.
[107] Labrousse, 1944, p. 94.
[108] Labrousse, pp. 286, 288.
[109] Ibid., pp. 324–26.
[110] Ibid., p. 327.
[111] Quoted by Labrousse, p. 363.
[112] Labrousse, p. 364.
[113] Ibid., p. 269.
[114] Labrousse, pp. 325 and 365 for Champagne and Burgundy.
[115] Labrousse, p. 269.
[116] Godechot, 1965, p. 117.
[117] Morineau, 1968, Fig. 7.
[118] Morineau, to be published.
[119] Desbordes, 1969.
[120] This is an early date for a Paris vineyard, where late or even normal harvests were generally in October.
[121] In 1846 too the wheat harvest in northern France was shriveled in the course of a very hot and stormy summer (Lahr, 1950, p. 236). The grain shortage that followed, combined with potato blight, was one of the many causes of the unrest which culminated in the troubles of 1848.
[122] Dupaquier, 1968.
[123] For a more complete view of the meteorological causes of the bad har-

vest of 1788, see the texts and references quoted in Appendix 12B. On the food crisis of 1789, see also Rose, 1959.

[124] A hot summer means not only good wine but also a good honey harvest: on this see the striking correlations put forward by Davis, 1968, pp. 312–13 and Fig. 2.

[125] Angot, 1883, p. B117; and, for a much more detailed study, K. Müller, 1953, pp. 234 and 240.

[126] Müller, pp. 183–233. See also Lahr, 1950, ad fin. (history of the quality of wines in Luxembourg: see Appendix 12A).

[127] Müller, 1953, p. 207.

[128] Von Rudloff, 1967.

[129] Hoinkes, 1968, ad fin.

CHAPTER III

[1] Craig and Willett, 1951; Pédelaborde, 1957, p. 403.

[2] For example, Labrijn, 1945; Lysgaard, 1949 and 1963; Liljequist, 1943; Hesselberg and Birkeland, 1940–1943; Manley, 1946 and 1953; Eythorsson, 1949; Glasspoole, 1942; supplementary references in Ahlmann, 1949.

[3] Wagner, 1940; Ahlmann, 1949; Willett, 1950; Lysgaard, 1949; Pédelaborde, 1947; Mitchell, 1963; Veryard, 1963; Lliboutry, 1966.

[4] Labrijn, 1945.

[5] Manley, 1946 and 1953; Lamb, 1966, p. 125.

[6] Glasspoole, 1942 and 1955.

[7] Eythorsson, 1949.

[8] Hesselberg and Birkeland, 1940–1943; Lysgaard, 1949 and 1963, especially p. 153.

[9] Scherhag, 1939; Lange, 1959.

[10] Ibid.

[11] Lysgaard, 1963, p. 154; Veryard, 1963; Pédelaborde, 1957.

[12] Rosenan, 1963, p. 68.

[13] Lysgaard, 1963, p. 155.

[14] Mentioned by Wagner, 1940.

[15] Mentioned by Veryard, 1963, p. 20. In European Russia the recent amelioration has also been traced by the phenological method (increasingly early growth in agricultural plants and others between 1906 and 1940). See Dolgoshov and Savina, 1969.

On the amelioration during the twentieth century at Leningrad, see also Borisov, 1967.

[16] Kincer, 1933 and 1946, quoted by Veryard, 1963.

[17] Brooks (C.), 1954; Conover, 1951, quoted by Veryard, 1963, p. 21. On city effects as an additional factor in the rise of temperature curves, see Lamb, 1965, pp. 198–99. On the amelioration in the eastern United States as between the decade 1830 and 1931–1960, see also Wahl, 1968, who gives important maps and interesting methods for the intensive and selective use of early data covering one decade (in this case the 1830s, which are particularly well documented).

[18] Currie, 1956; Veryard, 1963, p. 21.

[19] Rosenan, 1963, pp. 67 and 73; Dubief, 1959, mentioned in Dubief, 1963, p. 78.

[20] Ramdas and Rajagopalan, 1963, p. 87.

[21] Mentioned by Veryard, 1963, p. 22.

[22] Wexler, quoted in Veryard, 1963, p. 22. Quite recently Lamb has shown that the usual world trend of climate in the twentieth century—amelioration up to 1940–1950, then relative cooling—does not apply as such to the particular but enormous case of the Antarctic continent (Lamb, 1967, pp. 428–53).

23 Wagner, 1940; map in Lysgaard, 1949, reproduced in Lysgaard, 1963, p. 155.

24 Callendar's data go from 1891–1920 to 1921–1950; those of Willett and Mitchell from 1890–1919 to 1920–1950. The table is taken from Mitchell, 1963, p. 163; cf. also Callendar, 1961, mentioned by Veryard, 1963, and Willett, 1950.

25 These curves are in Mitchell, 1963, pp. 161–62. See also Fig. 8.

26 For maps see Mitchell, 1963, p. 168.

27 World map in Mitchell, 1963, p. 170. An excellent regional study of Greenland, covering the great cold of 1881–1889, in marked amelioration of 1926–1955, and then the moderate cooling of 1955–1964, can be found in Rodenwald, 1968.

28 Lysgaard, 1963, p. 153.

29 On the cooling of the 1960s as it concerns rainfall, which increased again after ten years or so in certain tropical and equatorial zones and in the low latitudes on the east coasts of continents, see Lamb, *Geog. Journal*, 1966, pp. 183–212, especially the excellent graph showing the sudden rise of Lake Victoria in the 60s; see also Kraus, 1963, particularly p. 146, Table I.

30 Lamb, 1966, p. 180. But whichever interpretation is chosen, the cooling phase, and particularly the return to hard winters, presents important problems to the planners, town planners, agronomists, and so on, who have to take this new factor into account in their calculations and projections. (On this, see Clark, pp. 145–46). The blizzards of recent years, such as that in New York from 29–31 January 1966, have raised great problems for large cities.

31 Cf. the rainfall curves for Rome and Milan in Lysgaard, 1963, p. 156.

32 Veryard, 1963, p. 26.

33 Mentioned by Veryard, 1963, p. 267.

34 In the temperate zone, on the other hand, the recent amelioration, insofar as it corresponds to an intensification of zonality and of the airflow from the west, is naturally accompanied by an increase in rainfall on the west coasts of the continents (Britain), but not, of course, on the east coasts (New England). On these problems see Lamb, 1966, p. 191 (increase in precipitation on the west coasts of the continents), and Kraus, 1955, p. 198 and pp. 430–39 (the accompanying decrease in rainfall in the tropics and on the east coasts of the continents).

35 Von Rudloff, 1967.

36 Lamb, 1966, p. 175; Von Rudloff, 1967, pp. 98, 102, 108, 114, 120.

37 Jeannin, 1969, p. 94.

38 The summer warming up of the eighteenth century, contrasting with the cooler periods at the end of the seventeenth century and the beginning of the nineteenth, seems to have been fairly general not only in Europe but also in the high latitudes of the northern hemisphere, near the Arctic. Tree rings show a strong thrust in growth as a result of summer warmth in the eighteenth century, in Scandinavia, the polar Urals, and the Canadian Northwest (Bray, 1965 and 1966; Bray and Struik, 1963; Adamenko, 1963).

39 Utterström, 1961.

40 Dupâquier, 1968, p. 180.

41 F. Lebrun, 1967, p. 190. (In my quotation the passage is abridged.)

42 Hélin, 1963, pp. 490–93.

43 Goubert, 1960.

44 Von Rudloff, 1967.

45 Rudé, 1956 and 1961.

46 Lamb, 1969.

47 Perhaps one should mention among the biological consequences of the optimum the recent introduction into Sicily of *Boerhavia repens viscosa*, orginally from the Sahara. See Trasselli, 1968, citing De Leo, 1967.

[48] For a single study of a cold year (1968), typical of the recent cooling, see Wallen, 1969.

[49] Von Rudloff, 1967, p. 204.

[50] Von Rudloff, 1967, pp. 167–85.

[51] Mougin, 1912 and 1910–1934; Bouverot, 1957 and ref. in Harrison, 1952–1956, pp. 666–68; Heusser and Marcus, 1964; Mercanton, 1916; Theakstone, 1965; see also Figs. 9, 10, 11, etc., and Grove, 1966.

[52] Lliboutry, 1965, II, p. 731.

[53] See Appendix 1, which collects together references concerning the withdrawals of the Alpine glaciers. For an overall view, cf. Lliboutry, 1962, pp. 720 and 727.

[54] R. Bonaparte, 1892, p. 28, and 1896; R. Brunet, 1956.

[55] Rogstad, 1947–1951, pp. 151 and 551. Eythorsson, 1947–1951, pp. 250 ff.; Ives and King, 1952–1956, p. 477. On the great retreat of the Tunsbergdalsbre glaciers, in the Jostedal group, between 1937 and 1961, see Kick, 1966, pp. 1–17.

[56] Jennings, 1952–1956, pp. 133–45; Jennings, 1948, pp. 167–82 (on Jan Mayen island). The glaciated area of Novaya Zemlya has been in vast retreat since 1907 (Koryakin, 1968).

[57] Wylie, 1952–1956, p. 704; cf. also, on the conclusions drawn from the ice cores from Greenland and the retreat of local glaciers, Diamond, 1958; Lliboutry, 1965, p. 728; and recently, the monumental work of Weidick, 1968, p. 55 and passim; Mock, 1966, especially the excellent curve showing the *continuous* retreat of a Greenland glacier from 1916 to 1965; and Temple, 1969 (continuous retreat of the Ruwenzori glacier from 1958 to 1966).

[58] *J. Glac.*, Vol. 2, 1952–1956, p. 640; and Vivian, 1965.

[59] Hale, 192–196, p. 22; Ward and Baird, 1954; *J. Glac.*, Vol. 2, 1952–1956, p. 653.

[60] Webb, 1948; Field, 1949; Mathews, 1951.

[61] Lawrence, 1950; Sharp, 1958; Cooper, 1937; Field, 1932; Heusser and Marcus, 1964.

[62] Butzer, 1957, pp. 21–35, and 1958, p. 126 (data and bibliography); Erinç, 1954, p. 22; Ahmad and Saxena, 1963; Kotliakov, 1967.

[63] Africa: Whitow, Shepherd, Goldthorpe, and Temple, 1963; Humphries, 1959; Spink, 1947–1951 and 1952–1956; J. de Heinzelin, 1952; Charnley, 1959. South America: Lliboutry, 1956 and 1965, pp. 729–30; Dollfus, 1959, p. 37; Broggi, 1943; White, 1952–1956. New Zealand: Odell, 1952–1956; Gage, 1947–1951; Harrington, 1952–1956.

[64] Mellor, 1959 a and b; Lliboutry, 1965, pp. 512–13.

[65] Lliboutry, 1952–1956, p. 168.

[66] Hoinkes, 1968.

[67] This sentence, written in 1966 in the French edition of this book is even more true in 1971. Articles which have appeared since 1966 emphasize the continuity of the glacial withdrawal much more than the resumption, which turns out to have been very hypothetical and very rare, of a new advance. These recent references are collected in Appendix 2B.

[68] Lliboutry, 1965, p. 731.

[69] Wallen (C. C.), 1948; Maurer, 1914; Ahlmann, 1939, p. 187 and 1940, pp. 120–23.

[70] Hoinkes, 1955, 1962 a and b, and 1963; Lliboutry, 1965, pp. 833–35.

[71] In this respect, Hoinkes' work is an answer to pertinent critical demands made by La Chapelle, 1965, p. 755.

[72] The number of days of summer snowfall, between June and September (phenomenon A), appears to be a *determining* factor in the advance or retreat of Alpine glaciers (phenomenon B). On this see the eloquent graphs of

Hoinkes, 1954, reproduced by Von Rudloff, 1967, p. 181. Phenomenon B follows phenomenon A with a time lag of three or four years.

[73] On the complexity of this problem, see for example R. P. Goldthwait, 1966, pp. 40–41.

[74] Lliboutry, 1965, pp. 833–35; Tricart and Cailleux, 1962, pp. 45–46.

[75] On this inertia, see Callendar, 1942, for an account and bibliography; Charlesworth, 1957, Vol. 1, pp. 151–52; Lliboutry, 1965, p. 833. The terminus does not show a reaction to changes in the upper masses of the glacier until several years have passed. But it reacts very fast to increased or decreased ablation.

[76] Haefeli, 1955–1956; Zingg, 1963; Callendar, 1942; Chizov and Koryakin, 1962 and 1963; Mathews, 1951; Heusser and Marcus, 1964.

[77] These correlations have yet to be mathematically and statistically treated; see the problem as presented by Lliboutry, 1965, p. 832.

[78] Or "overall complex of extreme weather conditions"—*Witterungscharakter*, the expression used by Hoinkes, 1962 and 1963.

[79] Mitchell, 1963, working on a period of half a century. Taking the century from 1800–1820, an increase of 1 degree C is recorded (Haefeli, 1955–1956).

[80] For an overall view see, in addition to reference below, Lliboutry, 1965, pp. 465–66.

[81] Mougin, 1925, pp. 199 and 219–20.

[82] Vivian, 1960, pp. 313–29.

[83] Sitzmann, 1961.

[84] Mercanton, 1954, pp. 315–16.

[85] Finsterwalder, 1953 and 1954, pp. 306–15.

[86] Vanni, Origlia, De Gemini, 1953, p. 7 (Table 1) and p. 10.

[87] Nageroni, cited by Cailleux, 1955.

[88] Mathews, 1951.

[89] Haefeli, 1955–56 (Basle series); Mougin, 1912 and 1925; pp. 103–5 (Annecy series).

[90] Mercanton, 1916. And below, Figs. 12 to 12κ and Plates 9 to 12.

[91] Rodewald, 1963.

[92] Koch, 1945.

[93] Vebaek, 1962.

[94] Morize, 1914; Derancourt, 1933; Galtier, 1958.

[95] Guilcher, 1956, p. 439; Cailleux, 1952 and 1954; Gutenberg, 1938, and Valentin, 1952, cited by Finsterwalder, 1954; Pfannenstiel, 1954, cited by Von Rudloff, 1967, pp. 190 and 342; Lamb, 1966, p. 148. One of the latest accounts (with bibliography) is that by Fairbridge, 1966, especially pp. 479–81. On the whole, despite temporary rises of as much as 5.5 millimeters per year, the overall rise in sea level from 1900 to 1950 is a mean of 1 millimeter per year—5 millimeters in fifty years. This rise corresponds with a melting of the ice which in turn corresponds to a temperature rise of about 1 degree C in those fifty years. Mr. Moynihan addressed NATO recently on the subject of a forecast rise of ten feet between 1970 and 2000. This enormous figure seems to me exaggerated (see New York *Times*, October 22, 1969).

[96] On the extremely controversial character of the causes of sea level change, and on its dimensions, see the important article by S. Jelgersma, 1966, pp. 54–55.

[97] Landsberg, 1958.

[98] Uspenskii, 1963, cited in M.G.A., 1964.

[99] Dorst, 1956, pp. 375–76.

[100] Tricart, 1958; Le Danois, 1938 and 1950; Lliboutry, 1965, p. 825.

[101] Wallen, 1920; Hustich, 1949 and 1950–1951.

[102] Smith, 1963.

[103] Slicher van Bath, 1965, pp. 10–11.

[104] In speaking of Indian famines here I refer only to their climatic aspects, determined by the monsoon. Obviously, from the human and demographic point of view there are very different explanations of how famines arise in India.
[105] The fact that man is not necessarily the object of the historian's study emerges, at least indirectly, from the more general observations of M. Foucault, *Les Mots et Les Choses*, 1966, p. 398 and passim.
[106] On all this paragraph see Flint and Brandtner, 1961, p. 458; Manley, 1965, p. 365; Godwin, 1956, p. 306; Corbel, 1962, p. 206, and 1959, p. 173; Leroi-Gourhan (A.), 1959.
[107] Lamb, 1966, pp. 52–53, 144, 172.
[108] Godwin, 1956, pp. 27 ff.
[109] Von Post, 1946.
[110] Godwin, 1966, p. 5.
[111] Godwin, 1966, p. 7. Cf. also Raikes, 1967, pp. 55 and 58, on the chronology of the birch in northern Europe.
[112] Godwin, 1956, pp. 28, and Andersson's map in Godwin, 1956, p. 39.
[113] Godwin, 1956, pp. 28 and 38; Starkel, 1966, p. 260.
[114] Ibid., pp. 28, 62, 97–98, 239–42, and 330.
[115] Godwin, Walker, and Willis, 1957; cf. also the table in Rubin, 1963, p. 1127, which gives a reliable chronology for Holland, Germany, and England. Page, 1968, pp. 694–97, shows that perhaps from 3000 B.C. and certainly from 2600 B.C. the glaciers started to advance again in Norway.
[116] Mayr, 1964, pp. 281 and 384, Tables 4 and 6; and below, Fig. 31.
[117] A recent study by Nichols, 1969, confirms this chronology for North America. More generally, Mercer, 1967, pp. 528–33 (both the text and an extensive bibliography) shows that the "hypsithermal" optimum was in fact interrupted about 2600–3000 B.C. by a cooling phase reflected in various places (Switzerland, Alaska, the Rocky Mountains, the Andes) by radiocarbon-dated glacial advances. This chronology coincides with the end of the "sunny millennium" from 4000–3000 B.C.
[118] Demougeot, 1965, p. 4.
[119] Elhaï, 1963, pp. 229–31.
[120] Cf. once more the reservations made by Raikes, 1967.
[121] Goldthwait, 1966, pp. 40–53; and especially Heusser, 1966, p. 138 (reconstructed temperature curve for Chile, Columbia, and Alaska during the last sixteen thousand years). For the lakes of Minnesota, see Wright, 1966.
[122] Butzer, 1966, pp. 72–83.
[123] Godwin, 1966, p. 7.
[124] On this point see the critical study by Walker, 1966.
[125] Emiliani, 1961.
[126] On Emiliani's conclusions, see a critical account by Eriksson, 1968, pp. 68–91.
[127] Ibid.
[128] Wiseman, 1966. Note however that he refrains from giving absolute figures for the positive or negative trends in temperature which his methods (p. 98) allow him to arrive at.
[129] Frenzel, 1966; Demougeot, 1965, and more recently her book on *Les Invasions Barbares*, Paris, 1970.
[130] Frenzel, 1966, pp. 99–123. For comparisons with the subatlantic cooling in the New World, for example in Alaska, see Heusser, 1966, pp. 128–29, 131–32, and 137. See also Lamb, 1966, pp. 172–73, on the differences between the warm periods themselves (atlantic and subboreal), and on the complexity of the subatlantic phase, which is much less "of a piece" than was once thought.
[131] Mayr, 1964, ad fin.
[132] See overall figures given by Charlesworth, 1957, II, p. 1493.
[133] Godwin, 1966, p. 8; Lamb, 1966, p. 168, concludes, in an important bibliographical note, that the many authors who have studied the optimum "have arrived at a figure of about 2 degrees C above present levels for the annual mean air temperatures, particularly in Europe."

[134] Manley, 1951.
[135] Manley, 1965.
[136] The English climate is very oceanic; the differences would perhaps be bigger in the case of a more continental one.

CHAPTER IV

[1] Pédelaborde, 1957; Dzerdzeevskii, 1963; Lliboutry, 1965, Vol. 2.
[2] Mercanton, 1916, p. 72, Fig. 6.
[3] See Mercanton, 1916, ad fin., for a series of maps on "The fronts of the glacier."
[4] The only one proposed in my 1965 article.
[5] Mercanton, 1916, plates; see also Rey, 1835 (approach on horseback possible) and, below, plates and drawings of the Rhône glacier at dates mentioned. In 1870 a whole Swiss cavalry brigade on maneuvers camped right at the foot of the Rhône glacier (see photograph preserved at the Hôtel du Glacier du Rhône at Gletsch; Seiler collection).
[6] Mercanton, 1916, maps and plates.
[7] Heim, 1885, pp. 500 ff.; see also Braudel, 1949, p. 223.
[8] Bout, Corbel, Derruau, Garavel, Peguy, 1955, pp. 516–17 (Hofsjökull); Thorarinsson, 1943, pp. 15–16 (Glama).
[9] ADHS, 10 G 308 (year 1564); 10 G 309 (year 1560), f. 68 vo, 69 ro, 73 ro and vo, 115 vo, 129 vo, XX 169; 10 G 310 (year 1557); 10 G 314 (year 1560), f. 5 vo, 52, 33 vo and passim; 10 G 316 (years 1574–1577) f. 32 vo, 37, 141 vo and passim. 10 ADHS, 10 G 246, year 1559 and year 1562; various buildings *ruinati impetu lavanchiarum*. Similarly at Grindelwald between 1565 and 1580 (Friedli, 1908). See also Poggi, 1961.
[11] The laments at Chamonix in 1560 are typical: "The said place of Chamonix is a very bad country of mountains where there are glaciers all the time and nothing grows but a little corn, oats, barley, and almost no wheat" (Ibid., f. 240). "Chamonix . . . very poor, scanty and barren mountains where no stranger comes to dwell and where all the time there are glaciers and frosts and nothing grows but a little oats and barley" (AD Haute-Savoie, 10 G 309, f. 209 vo and 240, year 1560).
[12] ADHS, 10 G 316, years 1575–1576, f. 116 vo. See also f. 108, 122 vo, 130.
[13] On the equivalence between ruin and moraine see Appendix 5. Combet confuses the glacial valley, rift, or gorge (scissura) with the debris deposited there by the glacier (ruins or moraines).
[14] This is the way I translate the Latin *alluviones*, which in sixteenth-century French still kept the Latin meaning of overflow (see Charron in Littré). I differ on this point from R. Blanchard, 1913, who, commenting on Combet, gives the word "alluvion" its modern sense of "sediment."
[15] Rabot, 1914–1915. I followed Rabot in my 1960 article, before knowing Blanchard's work.
[16] Mougin, 1912; Blanchard, 1913.
[17] ADHS, 10 G 287.
[18] Corbel, 1957, p. 41 and passim.
[19] Poissenot, 1586 (L. Febvre called attention to this text, 1912, p. 117). On the difference between a glacier and a glacière see Lliboutry, 1965, p. 429. Cf. also Boisot, 1686.
[20] For bibliography concerning the Froidère of Chaux see Appendix 6.
[21] Chronicle of Grindelwald, Hugi's version cited by Richter, 1891, p. 7.
[22] Lütschg, 1926, pp. 78, 384–85. He quotes three contemporary texts, and see also his reproduction of an 1822 engraving which gives a clear picture of the Allalin glacial barrier. (Lütschg's suggestions about a minimum for the Allalin glacier in 1400, 1403, 1495, and 1506, though they may be correct, are not

convincing. He gets his information from nineteenth-century texts dealing not with the glaciers themselves but with the population of the area, so any deductions about the position of the glaciers is necessarily indirect and vague.)

[23] Text cited by Richter, 1891, p. 7; critical study of what happened, pp. 22–23.

[24] Lütschg, 1926, pp. 422–23.

[25] Many contemporary texts and critical study in Baretti, 1880, and Sacco, 1917.

[26] See note 29.

[27] AC Cham., CC1, no. 19, text of May 2, 1605, published by G. Letonnelier, 1913. On the date of tallage reform in Savoie, see G. Peroux, pp. xxxi and xxxii. The places mentioned in the text are all in the Chamonix valley, below the glaciers of Des Bois (the Mer de Glace) and Argentière (see map, Fig. 25).

[28] AC Cham., CC1, document 81. report by François Bertier in April 1610.

[29] I have not been able to find out what the equivalent of a *journal* was at Chamonix in the seventeenth century.

[30] See the beginning of this chapter. There were plenty of local jeremiads before 1590, but they were not yet directed against the glaciers, which up till then had been fairly harmless.

[31] AC Cham. CC1, no. 81, f.4, year 1610. Text published by Letonnelier, 1913, p. 291.

[32] Rabot, 1920.

[33] Reproduced in Payot, 1950, pl. XV; and, more clearly, in Bernardi, 1965, pp. 96–97.

[34] Rabot, 1920; Kinzl, 1932; Lliboutry, 1965, p. 726 (map of moraines of Mer de Glace; cf. Fig. 16).

[35] Letter from M. Bossoney, mayor of Chamonix, to Rabot, Rabot, 1920.

[36] The hamlet of Le Châtelard is mentioned again in 1643 for the last time, though whether as inhabited or as a ruin is not clear. But from 1632 it was not among those communities which paid tallage (AC Cham., CC3, No. 84, year 1632).

[37] AC Cham., CC1, no. 81, cited by Mougin, 1912, p. 144. Bertier, who had just been reconnoitering the glacier Des Bois, then went upstream along the Arve to the northwest, "and passing beyond, past a village called La Bonneville, then another called La Rosière and another named Largentière, saw that *the glacier called of Largentière or of La Rosière had damaged much land.*"

[38] AC Cham., CC1, no. 81, year 1616.

[39] Ibid., f. 47 vo.

[40] Ibid.; "It is about thirty months since the village of La Bonneville was destroyed by the Arve." (Text of 1616.) See map, Fig. 25.

[41] AC Cham., CC1, no. 81, year 1616.

[42] Ibid., f. 57.

[43] Ibid., f. 44.

[44] Reproduced in Mougin, 1910.

[45] Kinzl confused the Arveyron of Argentière with the stream of the same name which flows out of the glacier Des Bois. In this he follows Mougin, who for once is mistaken (Mougin, 1912, p. 157). But the original document shows clearly which Arveyron is in question here (AC Cham., CC1, no. 81, year 1616).

[46] See Crans texts previously cited.

[47] As Mougin said, 1912, p. 145: "it is certain that the accumulation of large masses of ice in the valley, at a fairly low altitude, must have caused much melting during the warm season, and thus exceptionally high levels in all the water courses flowing out of the glaciers." (He is referring to the high levels of the seventeenth century.)

[48] AC Cham., CC1, no. 81, cited by Mougin, 1912, p. 183.

[49] The text reads: *Au nom de Dieu, Amen. L'an mil six centz et le jour sixiesme avril par devant moy notayre ducal soubisgné s'est p.nté* (présenté) *Jacques Cochet, du village du Boet* (Les Bois) *p.* (paroisse) *de Chamonix en Foucigny, lequel affermé p.* (par) *serment et en vérité estre envoyé et venu expres po* (pour)

soy informer sy la p. (paroisse) *de Cormayeur, p. pnt* (pour présent) *pays d'ouste, seroynt allez à Rome vers sa Saincteté po. scavoyr sy les royses* (glaciers) *des montagnes dud* (du dit) *lieu de Cormayeur sont retornées en dernière ainsy qu'ils ont entendu et s'estant addressé avec honneste Alexandre Genon dud. lieu de Cormayeur et aultres lesquels ont dict et attesté les d.* (dites) *royses icelles estre encores de p.* (présent) *à la mesme qualité ainsy qu'il a plou à Dieu et non* (aultrement) *s'estre changées ny mouvoyr et moiyns que la dite p. de Cormayeur ny aultre lieu soyt allé audit lieu de Rome p.* (pour) *ledict faict, remettant le tout à la main de Dieu. De quoy de tout ce que dessus led.* (le dit) *parant* (comparant) *a requis moy d.* (dit) *not.* (notaire) *s.* (soussigné) *fere le p.* (présent) *acte d'attesta. on po. s'en servie et valloir ainsy que de raison. Faict en la cité d'Aouste dans la ma.on* (maison) *de moy not. ssigné en p.ence* (presence) *du dit Alexandre Genon et de M.re* (maître) *Sulpis Tissieur, coutturié, habit.* (habitant) *en Aoste tesmoings requis.*

Blanc. not.

50 Text of 1755, surveys, etc. See Appendix 7 and Plate 2.
51 These accounts are preserved in ADHS, 10 G (Collegiate church of Sallanches, priory of Chamonix).
52 Mayr, 1964.
53 AC Cham., CC1, no. 23, December 21, 1610.
54 Texts in Baretti, 1880; Sacco, 1917.
55 Text cited by Lütschg, 1926, p. 385. The new path is to follow a different route.
56 The "Cronegg" is made up of early texts preserved in the papers of local families and published by Strasser in 1890. Richter, 1891, quoted long extracts from this edition, and from a fuller one by Hugi.
57 Today the Bärgel stream is 900 meters beyond the upper glacier at its nearest point; the present bridge is even farther away.
58 Text of 1603, cited by Coolidge, 1911.
59 Coolidge, 1911.
60 See Richter, 1891, p. 7.
61 Contemporary text, in Innsbruck archives, cited by Stolz, 1928.
62 Richter, 1877, 1888, 1891, and especially 1892 (publication of texts).
63 For map see Richter, 1892.
64 I verified this on the spot in 1966. In this connection, see photograph of 1846, reproduced in Von Klebelsberg, 1949, II, p. 670, which shows the glacier in a state of advance, near to and visible from the Zwerschwand.
65 I have been unable to find out the exact modern equivalent of the Tyrolean Klafter.
66 Cited by Richter, 1877, p. 168.
67 Archive of 1601 cited by Richter, quoted in Stolz, 1928; see also other contemporary texts in Richter, 1891, p. 7.
68 Hoinkes, 1968.
69 On the question of the conditions conducive to a strong glacial thrust, see Hoinkes, 1968, p. 18 ad fin.
70 AC Cham., CC4, no. 1, year 1640.
71 On the flood, see the petition of February 18, 1632, presented by the citizens of Morgex to the Council of Commissioners of the Duchy of Aosta, asking for redress (Sacco, 1917; Richter, 1891, p. 8; and particularly Baretti, 1880, p. 70 and passim).
72 For texts, see Lütschg, 1926.
73 Cited in Lütschg, 1926, p. 386. He also quotes a text on the flooding of the Mattmarksee (still called Saas-see) by Barthlome Venetsch, on July 25, 1633 (old style): *Anno 1633 in festo divi Jacobi, so gsein der 25 Juli, ist der Saserse usbrochen u die Vispen de dick mauren u gross wehri absgstossen, in die Burgschafft Visp inbrochen, die undresten höchene der heüsren, so by dem halsysen u vorab waren uffgefilt, ettliche heüsser, schür, stell u stedel wegtragen, die gewesen tieffer*

müsser mit sandt u landt uffgefilt und, usgnomen die neuwen Bynden, das gantze guot under der Burgschafft überschwembt u verlittet, die landtbruggen sambt 200 klafftern wherinen fonditus hinwegtragen; desglychen usbruch vornacher unerhert. Doher die Burgschafft mit grossem kosten in vollgenden jar die bruggen u wherinen miessen wider machen. Disen schein hab ich posteris pro memoria allhier inserieren wellen.

74 For wine harvest dates see Angot, 1883. I have given the graphs in Fig. 6, my articles of 1959 and 1960, and in *Les paysans de Languedoc*, II, graphl. The chronology here answers the shrewd objections of Taillefer, 1967.

75 Richter, 1891, pp. 8 and 23–26, also Venetz, 1820c, cited by Agassiz, 1840.

76 Corbel and Le Roy Ladurie, 1963.

77 AC Cham., CC3, no. 33.

78 Baudelaire, *Oeuvres*, Pléiades edition, p. 89.

79 Alte Pinakothek in Munich, picture no. 4963.

80 This was the hamlet of Les Rosières, between Les Praz and the priory of Chamonix (not to be confused with La Rosière, near Argentière). See text cited by Mougin, 1922, p. 175, and taken from AC Cham. CC1, no. 3: "Also in the month of June 1641 there was destroyed and flooded a little village called Les Ronzières above the village of the Priory, a globe below Les Tines, toward which came the torrent that came down from the glacier Des Bois." The place name Les Rosières still appears on present-day maps, on the left bank of the Arve, downstream from the confluence with the Arveyron des Bois.

81 AC Cham., CC4, no. 3, cited by Letonnelier, 1913, pp. 292–93.

82 In 1642 or 1643 there is also the last reference to "le Chastellard, against which the glacier advanced, as it did against Les Bois and Les Praz." AC Cham., CC43 (years 1642–1643)

83 All the above texts on the advance of the Mont Blanc glaciers in 1641–1644 are from AC Cham., CC3, CC4, and HH4, published in Letonnelier, 1913, and Mougin, 1912.

84 On Allalin glacier and 1644 map, see Lütschg, 1926, p. 387.

85 Baretti, 1880; Sacco, 1917.

86 *Historia Societatis Jesu in Vallegia* (Sitten Archives), p. 15, cited in Lütschg, 1926, p. 387.

87 See Hoinkes, 1968 for the relatively brief time lag (five, sometimes only two, years) which may occur between cool summers and glacier advances.

88 ADHS, 10 G 290, year 1660.

89 AC Cham., H H5, no. 6, cited by Mougin, 1912.

90 Ibid.

91 ADHS, 1 G 117, f. 53.

92 R. Le Pays, *Amitiés, amours et amourettes* (1671); new edition by C. E. Engel, 1965, pp. 23–24.

93 Lütschg, p. 388.

94 Richter, 1877, 1888, 1891, and especially 1892.

95 Richter, 1892, p. 384.

96 Ibid.

97 The thrust does seem to have been more marked in the eastern than in the western Alps.

98 Heuberger and Beschel, 1958, cited by Mayr, 1964, pp. 217 and 269.

99 Baretti, 1880; Sacco, 1917.

100 Text by the notary Peter Anthanmatten, July, 4, 1680; various chronicles in German and Latin cited in Lütschg, 1926, pp. 388 and 425.

101 I have measured this scale on the reproduction given in Lütschg, 1926, p. 388, taking as a base the distance between the village of Saas and the lowest point of the Mattmarksee. Cf. Lütschg, p. 41, map 10 and p. 414, map 27, for dimensions of the Mattmarksee in the Little Ice Age.

102 These limits are given for the two *mas* of the glebe of Montquart (the hamlet

of Les Bossons): the mas of Les Crans ". . . by La Crousaz stretching to the glacier," and the mas of Les Ducs, "stretching by the course of the stream of Les Bossons to under the glacier" (ADHS, 10 G, 282–1, 14 March 1679; reproduced word for word in 10 G, 282–83, year 1706).

[103] Mougin, 1912 (1st edition), p. 43.

[104] See Appendix 9 for depopulation in Chamonix in the seventeenth century and connected problems.

[105] R. Couvert, Vol. III b, p. 158.

[106] Mougin, p. 148; on the problems of the "icy air" emanating from the glaciers, see Lliboutry, 1965, p. 439.

[107] Accounts of the priory of Chamonix, collegiate church of Sallanches, ADHS 10 G.

[108] *Vie de Jean d'Arenthon*, Lyon, 1697, cited by Mougin, 1912, p. 147.

[109] "The advanced glacier stayed thirty years in the valley, and the ice had not quite gone until 1712"; cited by Richter, 1892, p. 384, and cf. p. 381.

[110] On this point I disagree with C. E. P. Brookes, 1949, who in an otherwise often excellent book seems to me to have exaggerated the scope of this withdrawal.

[111] Thorarinsson, 1943 and Eythorsson, 1935.

[112] Ibid.

[113] Eythorsson, 1935, pp. 128–29 (map, table, and text).

[114] H. W. Ahlmann, 1937, p. 195.

[115] J. Eythorsson, 1935, p. 128.

[116] Thorarinsson, 1939. See also Kinzl, ad fin., for destruction by the Alpine glaciers in the Middle Ages; also Thorarinsson, 1939, pp. 229–31, for Scandinavia.

[117] Werenskiold, 1939.

[118] Liestol, after Theakstone, 1965 (texts, maps, and graphs).

[119] See note 110.

[120] Drygalski, 1942, p. 214. I have not adopted another suggestion of Drygalski's, according to which the La Brenva glacier is supposed to have covered several meadows in the Val Veni between 1691 and 1694; in my opinion he has too hastily interpreted a text by Arnod, which I analyze below.

[121] Le Soubeyron is a piece of land belonging to the prior, downstream from the hamlet of Les Bois. Le Bouchet is on the left bank of the Arveyron des Bois, upstream from its confluence with the Arve.

[122] ADHS, 10 G 270.

[123] Gruner, 1760, Vol. III, p. 151.

[124] To be absolutely accurate, my measurements are taken from the excellent photographic reproduction in the B.N. of Herbord's engraving. But though the figures given here are not exactly those of the original engraving, their relative proportions are exactly the same.

[125] P. Arnod's description (1691) implies that in the Little Ice Age the Brenva glacier barred the Val Veni, and was visible from higher up the valley; this is corroborated by Plate XXV, and above all by the fine pre-Romantic or Romantic engraving, "Mont Blanc, the Brenva glacier, and the Val Veni," reproduced by P. Henry, 1969, p. 26 (see also ibid., p. 26, for a bibliography on the dimensions of the Brenva glacier during the Little Ice Age). Arnod's description was published by Vaccarone, 1881 and 1884.

[126] Scheuchzer, Vol. II, pp. 278–79, 1723 edition.

[127] "Beneath a cave, below and in the profoundest depths of the mountain of ice" (*sub rupe, imo monte glaciali*). *Mons glacialis* and *Eisgebirge* are periphrases employed by Swiss naturalists for "glacier" (cf. Gruner, 1760 and 1770, and the text by Le Pays.

[128] Mercanton, 1966, p. 44 and plates, especially those for 1874 and 1899.

[129] Ibid. See also the fine collection of engravings kept by Mme. Seiler at the Hôtel du Glacier du Rhône at Gletsch.

[130] Mercanton, sketch 4; cf. annual sketches up to August 23, 1898, and from August 28, 1899.

[131] Rey, 1835.

[132] AC Cham., CC5, no. 35, cited by Letonnelier, 1913, pp. 294–95.

[133] Contemporary text cited by Stoltz, 1928.

[134] Richter, 1888, pp. 162–63; 1891, p. 9; and especially 1892, pp. 386 and 410–20.

[135] Virgilio, 1883, p. 68; see also Richter, 1891, p. 9, and the text by Michel Bernard, notary of Courmayeur, preserved in the (unclassified) archives of the parish: "because of our sins, on the night of September 12, 1717, the glacier of Le Trioley collapsed, killing a hundred and twenty cows and seven men (*homines ad numerum septem*). Also cited in Vaccarone, 1884, p. 118.

[136] Text of 1719 cited by Gruner, 1760, III, p. 152; Richter, 1891, pp. 9–10, cites two texts.

[137] This engraving is reproduced in Gruner, 1760 and 1770. See Plates 3 and 5.

[138] Lütschg, 1926.

[139] Werenskiold, 1939.

[140] The parts of the "old map" which relates to Chamonix are preserved in the departmental archives of Haute-Savoie. On the great importance of the survey, see the fine study by Guichonnet, 1955.

[141] Not the extension of 1643–44, as Mougin thought (1912).

[142] Lütschg, 1926; Richter, 1891.

[143] Thorarinsson, 1943.

[144] Windham and Martel, 1741–42 (1912 edition). Extracts are also quoted in Rebuffat, 1962, pp. 21–27, which also has excellent pictures.

[145] Windham and Martel, 1741–42 (edition of 1912).

[146] No better confirmation could be required of the fact that the glacier Des Bois (Mer de Glace) in 1742, unlike today, extended as far as the valley of Chamonix.

[147] For all these texts and drawings, except the description by Martel, see Mougin, 1912.

[148] Mougin, 1912, p. 163; Ferrand, 1912, p. 40, n. 2; Payot, 1950, p. 100.

[149] Gruner, 1760, p. 152; Richter, 1891, p. 10.

[150] Altmann, 1751, pp. 1 (map), 33, and plate at the end of the book, in which the marble quarry is indicated by the letter F.

[151] Richter, 1891, p. 10.

[152] Lütschg, 1926.

[153] Baretti, 1880; Sacco, 1917.

[154] Werenskiold, 1939.

[155] J. Rekstad, 1906–7, p. 347.

[156] Werenskiold, 1939; Öyen, 1906.

[157] Eythorsson, 1935, p. 127; Thorarinsson, 1943.

[158] Heusser and Marcus, 1963; for similar datings relating to the 1750–1850 maximum, deduced from the age of the oldest trees on most recent moraines, see Lawrence, 1950, Sharp, 1958, and Cooper, 1937. For the maximum of the Mueller glacier in the Ben Chau range in New Zealand (A.D. 1745 according to the trees of an extreme moraine) see MacGregor, 1967, p. 746.

[159] Gruner, 1760, II, p. 153. According to Gruner, who visited Grindelwald in 1756, in that year the lower glacier was in almost the same position as in 1686 (see Plate 4). *Die betgefugte Kupfertafel, die beyde diese Gletscher vorstellt den untern Gletscher . . . ungefahr in dem gleichen Zustande, in welchem er sich dermalen befindt.* (Ibid., pp. 150, 151.)

[160] Lütschg, 1926.

[161] See Fig. 1. This engraving is in Gruner, 1760. By an error of the engraver or the publisher it is reversed, and gives a "mirror image."

[162] Eythorsson, 1935, pp. 124 and 133; Thorarinsson, 1943.

[163] The Saussure references hereafter are in Saussure, I, 1779, pp. 429, 466–67, and II, 1786, pp. 16–17.

[164] This is confirmed by Blaikie, 1775, cited by Bernardi, 1965, pp. 262–63.

165 Saussure, op. cit.
166 Bernardi, 1965, pp. 264–65.
167 Gruner, ed. 1770, pp. 329–30.
168 Mougin, 1912.
169 Mougin, 1910, p. 10, and 1912, p. 80, citing Coxe, Bourrit, and Saussure.
170 See Appendix 10 for text. Saussure's map is reproduced in Mougin, 1912, pl. II, pp. 140–41.
171 Richter, 1888, pp. 145–46; 1891, pp. 12–13 and 24; 1892, pp. 420 ff.; Stoltz, 1928, p. 20; Lütschg, p. 180 and passim. I have kept their spelling for place names.
172 Texts by curés Thammatter and Ritoldi, October 1774, cited by Lütschg, 1926, p. 391.
173 Richter, 1891, p. 12.
174 See Richter, 1891, p. 13.
175 Fellenberg, 1885–1886, cited by Richter, 1891, p. 13.
176 Bourrit, 1785, II, p. 9; see Plate 11.
177 Texts cited in Mercanton, 1916, pp. 44–50.
178 Eythorsson, 1935, p. 133.
179 Ferrand, 1912, p. 11. See also Appendix 10, par. 4, for texts of 1805–1814.
180 Mougin, 1912, pp. 154–55 and Plate VI.
181 For the *pecten* at this period, see also Maurer, 1914, p. 25, and the Seiler collection at the hotel at Gletsch (especially an English engraving of 1800).
182 Mougin, 1912, paragraph on Mer de Glace and Plates VII and VIII.
183 "From Saas to Montfort is three leagues. The path goes by the edge of a glacier (the Allalin) which comes down from the heights and encircles the foot of the mountain; it halts the course of a torrent and forms a lake (the Mattmarksee) about a league in circumference" (Letter by Abraham Thomas, July 15, 1795, quoted by Lütschg, 1926, p. 392). There are similar texts for 1793, 1798, 1803, 1808, 1809, 1816, etc. in Lütschg, 1926, pp. 392–95. It was only much later, from 1881 on, that the Allalin ceased to bar the Saas, Visp (ibid.), and the Mattmarksee began to dry up.
184 Text cited by Engel, 1965, p. 43. On the following, see Mougin, 1912.
185 Mougin, 1912, Plate VIII. See Plate 16.
186 Lütschg, 1926, p. 180.
187 Report by Ladoucette, prefect of the Hautes-Alpes, cited by Ch. Rabot, 1914, Vol. 29.
188 Thorarinsson, 1943; Werenskiold, 1939.
189 W. O. Field, 1932.
190 Mary Shelley, *Journal*. The sustained maxima of the Greenland glaciers in the eighteenth and especially the nineteenth centuries, before the retreat of the twentieth, have been studied in great detail by Weidick, 1968, pp. 1–202.
191 Mary Shelley, *Frankenstein*, Chapter X.
192 Not 30 meters as I said by mistake in an article in 1960. On the advance during the Restoration, see Mougin, 1912.
193 Text cited by Engel, 1965, p. 45.
194 Mougin, 1912.
195 Richter, 1891, pp. 26–29. I have followed his spelling.
196 Lliboutry, 1965, p. 727.
197 Kinzl, 1932.
198 Mougin, 1910, Figs. 2 and 6.
199 Mougin, 1912, p. 168 and Plate XV, Fig. 1.
200 Rey, 1835. See also Mercanton's comments, 1916. Rey was also quoted with reference to Scheuchzer (1705).
201 See also Forbes, 1843, p. 61, and Mougin, 1912, p. 168.
202 Reproduced in Mercanton, 1916, p. 16 and passim; George, 1866.
203 Mougin, 1912, p. 150 and Plate X, Fig. 2; see also Von Klebelsberg, 1949, II, p. 670, for a fine daguerreotype of the Vernagt glacier in 1846, barring the

Rofenthal and forming a lake. This photograph was taken from the Zwerschwand, from which the glacier is now invisible (ibid., p. 670), as I verified on the spot in 1966.

204 Kinzl, 1932.

205 Richter, 1891, pp. 33 ff.

206 Cited by Mougin, 1912, p. 169.

207 Humbert, 1934.

208 Mougin, 1912, p. 170.

209 A little Chamonix glacier above the mountain pastures of the same name.

210 Mougin, 1912; Martins, 1867 and 1875, has significant descriptions of withdrawals during 1860–1875.

211 Mercanton, 1916, p. 56 and map 11.

212 Lliboutry, 1965, p. 725.

213 Ibid., p. 727.

214 In fact, the Ruitor lake did not entirely disappear after 1864; it merely became much smaller and shallower and ceased to be dangerous (see Sacco, 1917).

215 Cited by Baretti, 1880, p. 50.

216 Cited by Baretti, 1880, p. 55.

217 ". . . *per la quale perpetuamente uscità l'acqua di detto lago, senza portare alcun damno a la detta valle, ne altrove . . .*"

218 "Rose," "roise," "rosa" are patois for "glacier" in the Val d'Aosta.

219 Kinzl, 1932. See also Lliboutry, 1965, p. 727.

220 Drygalski, etc., 1942.

221 Ibid., pp. 215–85.

222 Ibid., see also Lliboutry, 1965, pp. 724–25.

223 Drygalski, 1942, p. 215.

224 Ibid., Lliboutry, 1965, p. 727.

225 Drygalski, p. 216, and my Fig. 27.

226 R. Von Klebelsberg, 1949, II, p. 672.

227 Mayr, 1964.

228 From the name of a Tyrolean glacier with characteristic moraines.

229 Matthes, 1942, pp. 148–215.

230 Saint-Jean-de-Pertuis (or de Purtud) is supposed to have been destroyed and overrun by the Brenva glacier to punish its inhabitants for cutting their hay on Saint Margaret's day. This episode, if real, must be a very early one (twelfth or thirteenth century?), for Saint-Jean-de-Pertuis, which is said to be the oldest parish in the Brenva district, does not figure on any list of parishes of Val d'Aosta in the late Middle Ages. According to Forbes, 1842, the sound of vespers could still be heard being sung under the ice there. The present chalets at Purtud, below the right lateral moraine of the Brenva glacier, were probably built below the site of the village that was destroyed.

231 Matthes, 1942, p. 214.

232 Should we add the glaciers of Alaska and New Zealand? For New Zealand, see Burrows and Lucas, 1967, p. 467: these two authors, who use "lichenometry" techniques, consider that the maxima of the Mount Cook glaciers occur at the same times as those of the Alps in the seventeenth, eighteenth, and nineteenth centuries.

233 Flint, 1947 edition, pp. 499–500; Charlesworth, 1957, II, pp. 1495–96.

234 Pédelaborde, 1957; Mayr, 1964.

235 Lamb, 1963; Schove, 1949.

236 Agassiz, 1840.

237 Lliboutry, 1965, Vol. II, pp. 724–27.

238 Brooks, 1950, p. 301: "There was a retreat in the first half of the eighteenth century followed by a readvance in the first half of the nineteenth century." See also the sequence given by Schove, 1949, in an otherwise extremely interesting article: according to him the Little Ice Age (which begins in 1540) is interrupted

by an "interglacial" in 1681–1740, then by a "lull" in 1771–1800, and does not end until 1891. Brooks, 1951, gives another sequence of events just as inexact as the first.

[239] Lliboutry, 1965, insists on this point.

CHAPTER V

[1] Labrijn, 1945; Manley, 1946 and 1953; Liljequist, 1943; Lysgaard, 1949 and 1963. See also my Figs. 7 and 8.

[2] Mougin, 1912, pp. 206–8.

[3] Sokolov, 1955, pp. 96–98.

[4] Kassner, 1935.

[5] Arakawa, 1954, 1956, 1957.

[6] Mougin, 1912 (snow); Mougin, V, 1925, pp. 103–5 (detailed series, on which the following table is directly based).

[7] This answers the objections of Lliboutry, 1965, p. 834.

[8] For detailed figures see Appendix 14.

[9] See Charlesworth, 1957, I, p. 152.

[10] Lliboutry, 1965, pp. 833–36, believes, like Haefeli, Mercanton, etc., that temperatures, especially summer temperatures, play an essential role in past long-term glacial advance and retreat, and in the long recent withdrawal. On the other hand, he stresses the part played by precipitation and snow accumulation in the short-term paroxysms of the Alpine glaciers, e.g., 1818–20, 1850, etc. (Ibid., p. 833.)

[11] The figures and graphs put forward by Haefeli, 1955, pp. 695 and 696, show that there is a strict correlation between: (a) the progressive rise of annual temperatures after 1850, and especially after 1890, at Bâle, Zürich, St. Bernard, and Jungfraujoch, and (b) the retreat upstream ("Hebung") of the glacial tongues, which parallels the rise in altitude of the isotherm o degrees C.

[12] Hoinkes, 1955, 1962, 1963; Lliboutry, 1965, p. 835; Haefeli, 1955–56.

[13] Hoinkes' terminology.

[14] Easton, 1928.

[15] Wagner, 1940.

[16] Schove, 1949.

[17] See below, *Aspen Diagrams*, 1965, XVI, 1 (curve taken from Easton).

[18] Flohn, 1950.

[19] Le Roy Ladurie, 1966, Chapter 1.

[20] See below, *Aspen Diagrams*, XVI, 3.

[21] Van der Wee, 1963, Vol. I, p. 550.

[22] See below, *Aspen Diagrams*, XVI, 10; and Haller, 1875.

[23] As illustration, though not as proof, of the wave of cold winters from 1550 to 1700 indicated by Easton's researches, one may quote the many skating scenes painted in Flanders and Holland during that period, from Breughel to Avercamp. One example among many is *Games on the Ice*, by B. Avercamp (1612–1679), in the High Art Museum of Atlanta.

[24] Febvre, 1912, p. 194.

[25] Flohn, 1950, p. 356.

[26] See, for example, Lysgaard, 1963, p. 153 and passim.

[27] Manley, 1952, pp. 125–27.

[28] Mougin, 1912, p. 165.

[29] Garnier, 1955.

[30] Manley, 1952.

[31] Ahlmann, 1940, pp. 120–23; see also Lliboutry, 1965, II, p. 834.

[32] Lliboutry, 1965, II, p. 727.

[33] Angot, 1883; Duchaussoy, 1934; Garnier, 1955.

[84] Le Roy Ladurie, especially 1959, 1960, 1965, and 1966, Chapter I.
[85] Le Roy Ladurie, 1959 and 1966, Chapter I.
[86] Laurent, 1957–58.
[87] Le Roy Ladurie, 1966, Vol. II, Graph I, and a corrected version in Appendix 12.
[88] Huber and Siebenlist, 1963; see also my Fig. 9 (I should like to express my thanks to D. J. Schove, who pointed out this remarkable parallel to me).
[89] On the general problem of the alternation of hot and cool summers, see Davis (N. E.), 1967 (with bibliography); Miles, 1967; Huber, Siebenlist and Niess, 1964, p. 31; Schove, 1969; and Murray and Moffitt, 1969.
[40] Brehme, 1951.

CHAPTER VI

[1] Aario, 1944 (1945).
[2] Mayr, 1964.
[3] Kinzl, 1932.
[4] Aario, 1945; Firbas, 1949; Mayr, 1964, Plates 4 and 6.
[5] Heusser, 1954 and Breschel, 1961, cited in Manley, 1966, pp. 37 and 39. See also the comparable chronology of Mercer, 1965.
[6] Lamb, 1966, p. 156.
[7] Pédelaborde, 1957, p. 412.
[8] Pédelaborde, 1957, p. 413.
[9] Text cited by Richter, 1892, p. 384.
[10] Kinzl, 1932, ad fin.
[11] I was shown them myself in 1961, by M. Berchtold, a local guide.
[12] Oechsger and Rothlisberger, 1961.
[13] Radiocarbon experts fix "the present era" at A.D. 1950.
[14] Text in the *Cronegg* cited and discussed by Richter, 1891, p. 16 and passim.
[15] Gruner, 1760, I, p. 83, and 1770, pp. 63 and 330.
[16] BM 95, Schreckhorn, 680±150 (B. P.) Wood (*pinus cembro*) from the surface of the old right lateral moraine beside the path from Baregg to Schwarzegg at 1700 meters above sea level on the right bank of the lower Grindelwald glacier (47 deg. 35N, 8 deg. 05E). Collected in July 1947 by the late Pastor Nil of Grindelwald and submitted by Sir Gavin de Beer. (From the Radiocarbon Supplement of the *American Journal of Science*, Vol. 3, p. 44) Pastor Nil and Sir Gavin de Beer did not know Gruner's text.
[17] Stuiver and Suess, 1966, pp. 534–40, especially p. 537.
[18] Mayr, 1964.
[19] Mayr, 1964, p. 277.
[20] Lütschg, 1926, pp. 77 and 384.
[21] "*Item rogamus vos ut debeatis nobis censare dictam alpem a glacerio superius, secundum consuetudinem ab hominibus de valle Soxa, ut ipsi non constringant bestie nostre ut non vadant as pasculendum usque ad glacerium.*" The original Latin text is given in Gremaud, Vol. III, pp. 14–15, text 1156. German translation and discussion of topography in Lütschg, ibid.
[22] Reproduced in Lütschg, 1926
[23] Norlund, 1924, pp. 228–29, and 1936
[24] Longstaff, 1928, pp. 61–68
[25] Text published in *Meddelelser om Gronland*, 1899, XX, p. 322. See comments on this by P. Norlund, 1924, pp. 223 ff. See also Lamb, 1966, p. 149, for facts suggesting a similar chronology for the southern coasts of Scandinavia.
[26] Mayr, 1964, p. 279; see also Mercer, 1965, p. 410, for data on Yakutat Bay in Alaska and Adela in Patagonia (glacial thrusts about A.D. 1100 in the western hemisphere).

[27] Manley, 1965, p. 373; Lamb, 1963, p. 125, and 1964; Lamb, Lewis, Woodroffe, 1966, p. 185.
[28] See figures put forward by Lamb, 1965, pp. 13–37.
[29] Lamb, 1966, pp. 96, 182, 217–21.
[30] Wiseman, 1966, p. 92.
[31] Curve put forward by Koch, Lamb, and Johnson, and reproduced by Von Rudloff, 1967, p. 90. According to the ice cores made in Greenland and recently reviewed, the final culmination of the medieval optimum occurred about A.D. 1100.
[32] For a general account of recurrence surfaces, see Conway, 1948, pp. 220–38; Steensberg, 1951, pp. 672–74; Overbeck and Griez, 1954; for radiocarbon dating and chronological "bracket," see Overbeck, Munnich, Aletsee, Averdieck, 1957, especially p. 66.
[33] Boulainvilliers, 1752, VI, p. 136.
[34] Le Goff, 1964, pp. 296 and 488.
[35] Frenzel, 1966, p. 108.
[36] Bryson, Irving, Larsen, 1965, pp. 46–48; for a verification by pollen analysis, see Nichols, 1967, p. 240.
[37] Skelton, Marston, Painter, and Vietor, 1965, p. 3.
[38] Skelton, 1965, pp. 156 and 230.
[39] See Skelton, 1965, pp. 169–70 and 186.
[40] Skelton, 1965, p. 170, and *Graenlandica Saga*, 1965, p. 16.
[41] According to Skelton, 1965.
[42] From the thirteenth century on, according to Baardson. From A.D. 1140, according to the climatic datings derived from the Camp Century ice core.
[43] Dansgaard, 1969.
[44] Crone, 1966, pp. 75–78. Mr. George Kish, whom I have asked about it, considers the map authentic but reserves judgment about the far western part of it (Vinland and Greenland).
[45] Crone, 1969, p. 23.
[46] Dansgaard, Johnsen, Moller, Langway, 1969 a; see also Dansgaard and Johnsen, 1969.
[47] Research on the Camp Century ice core has in fact been carried out jointly by the researchers of the C.R.R.E.L. and those of the Physical Laboratory H.H.C. Oerstad, of Copenhagen.
[48] These episodes of temporary warming up which seem to characterize certain periods of the fourteenth, fifteenth, and eighteenth centuries, seem therefore to have occurred in Greenland during the medieval and early medieval establishment of the Little Ice Age, and even at some periods during culminations of the Little Ice Age from the sixteenth to the nineteenth century. During these warm spells of relatively short duration vegetation revived in earth that was at other times frozen. Thus the tombs and funerary remains of the last survivors of the Norman colony in Greenland, though buried in earth that has usually been frozen since the twelfth century, have at various times been bored through by the roots of plants.
[49] Dansgaard, 1969 a, p. 378.
[50] *Graenlandica Saga*, pp. 17–18 and 50.
[51] Ibid., pp. 21 and 52.
[52] Mercer, 1965.
[53] Dansgaard, etc., 1965 a, Fig. 4.
[54] Dansgaard, ibid.; Mercer, ibid., pp. 410–12.
[55] The Camp Century ice core, the lower part of which is over 100,000 years old, also corroborates, with extraordinary exactitude, the chronology of the great Quaternary Ice Ages.
[56] Dansgaard, 1969 a.
[57] Suess, 1969 a.
[58] Lamb, 1961 and 1966.
[59] Suess, 1969 a.
[60] Mayr, 1964.

[61] Hovgaard, 1925, pp. 615–16. On the fate of the Norman colony in Greenland, see Musset, 1951, pp. 218–24. Cf. also the highly critical point of view of Vebaek, 1962.

[62] In 1955, 1950, and 1963 respectively.

[63] For a slightly later period (1546), see the impressive dimensions, much more extended than today, of the Rhône glacier.

[64] Pédelaborde, 1957, p. 406.

[65] See the Orvieto series mentioned by E. Carpentier, 1962; and the unpublished Pisa series cited by D. Herlihy, in his book on Pisa.

[66] For example Baker, 1883.

[67] Titow, 1960, p. 379.

[68] When this book was in the press I learned, in Rome, of the legend about the founding of the church of Santa Maria Maggiore there. On August 4, 352, the Virgin Mary appeared in a dream to Pope Liberus I, and told him to build a church on the Esquiline Hill, where next morning he would find a carpet of freshly fallen snow. The Pope obeyed, and was able to draw the plan of the church in the snow itself.

If this story of summer cold is true, this episode may form part of the secular series of very cool summers which inhibited ablation and caused the long advance of the Alpine glaciers recorded from A.D. 400 on. Mayr lists this advance as Xb (Mayr, 1964).

[69] *Annals of the New York Academy of Sciences,* Vol. 95, art. 1, pp. 1–740, October 1961.

[70] Publication no. 58 (1962) of the *Association internationale d'Hydrologie scientifique* is entirely devoted to the work of this symposium.

[71] See *Proceedings . . . ,* 1962. See also *Annals,* July–August 1963, pp. 764 ff.

[72] The Chairman of the Conference was R. A. Bryson, and it was divided into six sections: anthropology, biology, glaciology and geography, geology, history, and meteorology.

[73] The History Section of the Congress consisted of: W. B. Watson, chairman (Massachusetts Institute of Technology); E. Giralt Raventos (University of Barcelona); K. Helleiner (University of Toronto); D. Herlihy (Bryn Mawr College); E. Le Roy Ladurie (École des Hautes Études); J. Titow (University of Nottingham); G. Utterström (University of Stockholm). Among the other researchers collaborating in the work of the History Section were: H. Arakawa (Institute of Meteorological Research, Tokyo); P. Bergthorsson (Weather Bureau, Iceland); H. Lamb (Meteorological Office, London); G. Manley (Bedford College, University of London). On the work of this committee, see, in *Proceedings,* 1962, W. B. Watson: "Summary report of the History Section" (pp. 37–43), "Census of data" (pp. 44–49), and "Bibliography" (pp. 50–58). W. B. Watson organized the history committee before the conference opened, and later directed and coordinated its work.

[74] Lindzey and Newman, 1956.

[75] Arakawa, 1955, 1956, 1957.

[76] Angot, 1883.

[77] J. Branas, 1946, pp. 56–71; G. Montarlot, n.d.; M. Godard and J. Nigond, especially their Fig. 1.

[78] For northern and central series see Angot, 1883; the southern series are by E. Le Roy Ladurie (south of France), Hyacinthe Chobaut (Comtat), and B. Bennassar (Old Castile); see also Appendix 12 and Le Roy Ladurie, 1966, Vol. II, Fig. 1.

[79] In fact, the mean includes three Swiss series (Aubonne, Lausanne, Lavaux) and one from Old Castile (Valladolid).

[80] See also Arakawa, 1954.

[81] Betin, 1957.

[82] A. Sokolov, 1955, pp. 96–98.

83 As the Catalan harvest often occurs in June, the harvest year of 1524–1525, for example, in diagram XVI–30, covers the spring-summer of 1524 and the winter-spring of 1525.

84 See the many Flemish and French texts cited in Easton, 1928, pp. 90–91.

85 It is not always possible to distinguish between the cold weather of January–February and that of December, which should be counted for different years.

86 We did not attempt gradation in the case of the eleventh century; but for the sixteenth century the texts are numerous enough, and sufficiently clear, circumstantial, and precise for cross reference to be possible within the same region.

87 Chobaut Mss., Musée Calvet, Avignon.

88 Easton, 1928, p. 200, note 1.

89 H. Lamb, 1961, pp. 152–56.

90 C. E. P. Brooks, 1949, pp. 328–30.

91 Arakawa, 1957, p. 48, and 1954, pp. 156–61.

92 J. Murray Mitchell, 1961, especially Figs. 1, 2, and 3.

93 The winter means differ by 1 degree C. See especially the series on Holland (January), Lancashire (January), Edinburgh (winter), Stockholm (January), Leningrad (January), Archangel (January), and New Haven, Connecticut (winter). They can be found in diagram form in H. W. Ahlmann, 1949, and H. C. Willett, 1950.

94 F. Braudel, 1966, I, pp. 248–49; J. Delumeau, 1959.

95 *Proceedings* . . . , 1962, pp. 18–19 (report of the anthropology section).

96 Allowing, as always in the sixteenth century, for the secular rise in prices due to economic, demographic, and monetary factors.

97 R. Shapiro, "Discussions of the Circulation Pattern, Winter 1550–1600," *Proceedings* . . . , 1962, pp. 59–74 and 91–92.

98 Pédelaborde, 1957, pp. 75–91 and 404–24.

CHAPTER VII

1 Sanson (O. N. M.), p. 3.

2 Geslin, 1954, p. 30; Ratineau, 1945, pp. 53–57.

3 The seed of winter wheat, after being exposed to low temperatures, can be sown in the spring.

4 See Wallen, 1920, pp. 332–57, for a statistical study of the yields of three varieties of winter wheat over twenty-seven years (1890–1917) in Sweden.

5 I. Hustich, 1949, pp. 90–105. See p. 92 for statistical table showing correlation between Finnish wheat harvests 1886–1939 and June-July-August temperatures.

6 Oyen, 1906; Utterström, 1955.

7 Sanson (O. N. M.), p. 34.

8 M. Garnier, 1956; a study of the influence of meteorological conditions on the yield of spring barley, based on the annual yields (1935–54) of experimental fields in the Paris basin and the west, belonging to the Society for the promotion of barley for beer.

9 Hooker, 1922.

10 Slicher van Bath, 1965.

11 Titow, 1960.

12 Ibid., p. 372.

13 Ibid., p. 374.

14 Simmons, 1959.

15 Dean, 1967, stresses the acutest phase of the great drought, between 1276 and 1299.

16 On all this episode see Woodbury, 1961, p. 708 and passim; also Fritts, 1965 a, pp. 875–77 and 1965 b, pp. 429–31.

17 Dean, 1968. This unpublished work came to my attention after the first French edition of this book in 1967.

[18] Both in northeast Arizona.

[19] This was only a general tendency, not a complete synchronism (see Lliboutry, 1965, p. 731).

[20] Schulman, 1951 and 1956.

[21] Shapiro, 1962, pp. 59 ff.; Lamb, 1962, pp. 91–92.

[22] Lamb, 1966.

[23] Pédelaborde, 1957, pp. 75 ff. and 403 ff.

[24] On the relations between general circulation at high altitude and fluctuation of the jet stream, see Reiter, 1963, p. 395 and passim.

[25] Lamb, 1966, p. 32; Von Rudloff, 1967, p. 37.

[26] Ibid., p. 207.

[27] Ibid., p. 184 (maps).

[28] Pédelaborde, 1957, p. 81.

[29] Von Rudloff, 1967, pp. 88–89.

[30] Lamb, 1966, p. 210.

[31] Ibid., pp. 150 and 163–64.

[32] Lamb, 1966, p. 163 and especially pp. 205–6 and 211.

[33] Ibid., pp. 14–17, 146, and 151.

[34] Ibid., pp. 28–30, 136, 151, 154.

[35] Ibid., p. 192.

[36] Hoinkes, 1968.

[37] Von Rudloff, 1967, pp. 37, 183, and 194.

[38] Ibid., pp. 37 and 89.

[39] Lamb, 1966, p. 201.

[40] According to Rossby, cited by Lamb, 1966, pp. 31–32, this wavelength "increases with the square root of the general speed of the westerlies." See also Lamb, ibid., pp. 207–8.

[41] Pédelaborde, 1957.

[42] 1961, p. 189.

[43] See also Kushinova, 1968, and Potapova, 1968.

[44] Dzerdzeevskii, 1963, p. 291. See also Lliboutry, 1965, II, p. 841.

[45] Dzerdzeevskii, 1961, p. 191, Fig. 1.

[46] The theory of greenhouse effect is well summarized by Callendar, 1949, p. 310: "Reduced to its simplest terms this theory depends on the fact that, whereas carbon dioxide is almost completely transparent to solar radiation, it is partially opaque to the heat which is radiated back to space from the earth. In this way it acts as a heat trap, allowing the temperature near the earth's surface to rise above the level it would attain if there were no carbon dioxide in the air." See also Plass, 1956, pp. 140–54, and Beckinsale, 1965, p. 11 (good discussion and bibliography).

[47] See also Weyl, 1968.

[48] See Fletcher, 1968, on the "Extent of polar ice pack as a sensitive climatic lever which is capable of amplifying the effects of small changes in global heating."

[49] Mitchell, 1966.

[50] Plass, 1956.

[51] In a recent article Bryson (1968) suggests a possible answer to this question: he suggests that to all the previous pollution factors there should be added that of industrial dust. This dustiness, by lessening the transparency of the atmosphere and by "scattering away more incoming sunlight" might be one of the causes of the cooling of the last twenty years or so.

[52] Mitchell's classification of causes is not the only one that has been put forward. Beckinsale, 1965, groups the theories concerning the causes of climatic variation according to whether they stress variations in solar radiation or in the transparency of the atmosphere.

[53] See Suess, 1968, and Damon, 1968 (curve showing world production of C^{14} in the twelfth and thirteenth centuries).

APPENDICES

APPENDIX 1

The Modern Retreat of the Alpine Glaciers

On this phenomenon, "plebiscitary" since 1920, and practically from 1860 to 1955, see in particular the statistical analyses of Mougin, 1910–1934; Veyret, 1952, pp. 197–99, and especially 1960, pp. 203–7; Garavel, 1955, pp. 9–26; Vanni, 1948, pp. 75–85, 1950, p. 230, and 1963; Desio, 1947–1951, pp. 421–22; Maurer, 1935, p. 22; Mercanton, 1916, pp. 52–56; Mercanton, articles in *Die Alpen* (1947–1949), according to *J. Glac.*, Vol. I, 1947–1951, pp. 139, 153, 345, 356, and ibid., Vol. II, p. 110; and the reports of the *J. Glac.*, Vol. I, 1947–1951, pp. 507, 558, and 563, and Vol. II, 1952–1956, pp. 290, 440–41, and 607; Vivian, 1960; Sitzmann, 1961; see also a general account by Thorarinsson, 1940 and 1944, and Lliboutry, 1965, especially pp. 719–21.

APPENDIX 2

The Small Glacial Advances of Recent Years (Since 1950 or 1960); Their Insignificant or Temporary Nature

For the Jan Mayen glaciers since 1954 see Lamb (H.), Probert-Jones (J.), Sheard (J.), 1962; and Kinsman and Sheard, 1963, pp. 439–47. For those of the Rocky Mountains: Harrison, 1952–1956, pp. 666–68; Rondeau, 1954, p. 193; Bengston, 1952–1956, p. 708; Hoffmann, 1958, pp. 47–60. For Spitsbergen: Kosiba, 1963; Vivian, 1965. For the Les Bossons glacier: Lliboutry, 1965, p. 720 (diagram taken from Bouverot, 1957).

But one may wonder whether these glacial "readvances" of the last one or two decades are really important. The *most recent* articles (1966–1969) suggest that they are insignificant; and what strikes me most is the slow continuity of the secular retreat up to 1970. On the insignificance of the small recent readvances, see, among the publications which have appeared since the French edition of this book: Weidick, 1968; Koryakin, 1968; Kick, 1966; Kotliakov, 1967; Loewe, 1968; Gilbert, 1969; Untersteiner, 1968; Groswal'd, 1969; Streten, 1968; Temple, 1969 (all these authors stress the *slow but sure continuity of the retreat of the glaciers even during recent years;* Kotliakov, and Loewe, and also Post, 1966, do however point to a few faint symptoms, quite recently, of readvance or "lesser retreat."

APPENDIX 3

On Certain "Intercontinental" Aspects of the Climatic Optimum

General observations on the various continents: Charlesworth, 1957, II, pp. 1484–95; Deevey and Flint, 1957, pp. 182–84. For northern Eurasia: Frenzel, 1955, pp. 40–53. For America: Butler, 1959, p. 735 (Massachusetts peat bog); Deevey (E. S.), 1951, especially p. 204 (comparison between Maine, Newfoundland, and Ireland); Darrow (R. A.), 1961, p. 41 (southwest United States); Heusser, 1952, and especially 1953, pp. 637–40 (Alaska); Zumberge (J. H.) and Potzger (J. E.), 1955, p. 1640 (Michigan); for Columbia, see Flint and Brandtner, 1961, p. 458 (Fig. 1 and caption). For New Zealand: see Deevey (E. S.), 1955, p. 324, and also Walker, 1966. On Africa, Morrisson, 1966 (pp. 142–48) considers data to be still too sparse and uncertain for comparisons to be made with the climatic chronologies put forward by European palynologists. On oceanic and marine temperatures during the optimum, see Emiliani (C.), 1955, pp. 538–79; 1958, pp. 264–76, and 1961, p. 530 (diagram); and Wiseman, 1966.

APPENDIX 4

The Rhône Glacier in 1546

(A) *Sebastian Münster's text on the glacier itself* (my italics):

"Anno Christi 1546, quarta Augusti, quando trajeci *cum equo* Furoam montem, *veni ad immensem molem glaciei* cujus *densitas,* quantum conjicere potui, fuit *dur aum aut trium phalangarum militarium*[1]; *latitudo* vero *continebat jactum fortis arcus,* longitudo sursum tendebatur, ut illius recessus et finis deprehendi nequiret, offerebat intuentibus horrendum spactaculum. *Dissilierat portio una et altera a corpore totius molis magnitudine domus,* quod horrorem magis augebat. *Procedebat et aqua canens quae secum* multas glaciei particulas *rapiebat, ut sine periculo equus illam transvadere non posset.*[2] Atque hunc fluvium putant initium esse Rhodani fluvii."

In the Museum of History and Art in Geneva, the armory room contains two Swiss infantry pikes of the fifteenth century from the arsenal at Lucerne: each is 4.6 meters long. Lapeyre (in *Charles Quint,* published in France by the *Centre National de la Recherche Scientifique,* p. 40) gives the Swiss pike as 5 meters long at the beginning of the sixteenth century.

(B) *Münster's itinerary* on August 4, 1546 (Valais, Rhonegletscher, Furka, Gothard):

[1] The text of the German edition gives: "*Zweies oder dreyes* spiess *dick,*" i.e., 10 to 15 meters. This refers to the thickness of the front, seen in cross section. Münster notes that judging by the crevasses the glaciers, higher upstream, must be 200 to 300 ells (300 to 400 meters) thick.
[2] German version: "*Es ging auch ein Bach mit Wasser und Eiss dar auch, das ich mit meinem Ross on [ohne] ein Brucken hinüber nie komme mocht.*"

"*Furca* . . . par hunc montem patet tempore aestivale *iter ad Valesiis ad Uranenses et Lepontios.* Tempore vero hyemali ob nives non licet trajicere, imo in media aestate difficulter id licet, quod ego compertum habeo qui quarta mensis Augusti tanta frigora in vertice montis hujus [Furka] passus sum, ut totus contremiscerem, cogererque tres aut quatuor nives et glacies cum equo non sine periculo trajicere; cumque eadem die *Ursellam usque ad radices montis Gothardi pervenirem,* volui et illius montis difficultates explorare."

These texts come from Münster (S.), Latin edition of 1552, pp. 332 and 342, German edition of 1567, p. 483.

APPENDIX 5

Archdeacon Combet at Chamonix (1580)

(1) *Itinerary*
"Ascendimus visitando usque ad extremum decimationis de Tines superius . . . Vidimus autem quatuor habitationes quae dicuntur de FRASSERENS, MONTRIOUD [Montroc] et LE PLANET, et hinc vero retrocedentes ascendere opportuit collem ad Aquilonem . . . ubi est locus vulgo appellatus TRELECHAMP . . . et regressum fecimus ad domum dicti Aymonis [at the priory]."

(2) *Description of the valley*
"Et est vallis inter montes posita . . . et dexteris eundo seu a meridie, continui montes intersecti habent in summitatibus albentes glacies quae etiam per diversas scissuras ipsorum montium protenduntur, et *descendunt fere usque ad planitiem, tribus saltem in locis* . . . Constat preedictas scissuras quas 'ruinas' [moraines] vocant aliquando inevitabiles causavisse alluviones tam in partibus per quas necessario decendunt quam per mediam vallem, augentibus medium illum torrentem qui *ab Alpibus de TOUR dictus est* incipere, et in flumen satis vallidum coalescit."

This text is partly unpublished, including that section concerning the itinerary (A.D. Haute-Savoie, 10 G 287); part of it was published by G. Letonnelier, 1913, pp. 288–95. The best commentary is that of R. Blanchard, 1913, pp. 443–54.

On the equivalence between "ruin" (scree) and moraine, see Fruh, 1937, I, p. 111. See also an archive of 1679 (A.D. Haute-Savoie, 10 G 282–1, folio or pp. 124–25, year 1679) relating to the *mas* or pieces of land then quite close to the Mer de Glace:

"Le mas des Gaudins, par le rey des crettes, *puis sus par la Rouyne* où est enclavé la combe du Lavanchiez et tout bonnenuict . . ."

"Les mas de Joppers et de Landrieux . . . *tirans sus par less morennes* jusqu'au gouttey et passioux (pissiou) de Bonnenuict."

In these texts the dialect word *morenne* is interchangeable with the French equivalent *rouyne.*

In the same text (A.D. Haute-Savoie, 10 G 287), Bernard Combet mentions, not far from Montquart-les-Bossons, a "ruina de Bois-David" (moraine of Bois-David), which he passed near. In 1315 Bois-David was still inhabited (see text of January 6, 1315, published by Bonnefoy, 1879–1883, I, p. 174: reference to "Michel, son of Pierre Pelarin, of Bois-David"). Was the moraine or *ruina* of Bois-David caused by a glacial advance at some unknown date between 1315 and 1580? At all events, I have found no other trace of Bois-David in the archives, or in the contemporary toponymy, written or oral, concerning Chamonix or Les Bossons.

APPENDIX 6

The Froidière of Chaux-les-Passavent (Jura)

Poissenot's visit to La Froidière (see his text in Poissenot, 1586) took place on a hot day in July: "For it was the second day of July, and the sun shone so fiercely that we sweated in drops," he wrote; this made him all the more appreciative of the "very agreeable coolness" as soon as he entered the cave.

There were many subsequent descriptions of La Froidière: quotations and bibliography are to be found in Depping, 1845, p. 438 (chapter on "Franche-Comté"), and in Fournier (E.), 1899, p. 87, and 1923, p. 85. All the texts quoted, from Poissenot and the abbé Boisot (Boisot, 1686) to Depping and Fournier, imply a contrast between the glaciation of the cave (from the end of the sixteenth to the nineteenth century) and its contemporary state of almost complete deglaciation (1910–1960). I was able to get confirmation of this on the spot from H. Humbert, present caretaker of the cave, whose father was caretaker before him about 1900–1910.

It is also interesting to note that prehistoric Bronze Age objects have been found near the entrance to the cave (Fournier, 1899).

APPENDIX 7

Habitations Destroyed by Glacial Advances Around 1600

(A) *Le Châtelard.* The first reference (May 12 and 21, 1289) is in Bonnefoy, 1879, I, p. 68 (but this may be another Châtelard, north of Vallorsine). On the toponymy, see Blanchard (R.), *Les Alpes occidentales,* Vol. VII, p. 467.

There are references which are certain to Le Châtelard and to La Rosière in the tithe accounts from 1377–1388: A.D. Haute-Savoie, 10 G 226-1, folio 1 and passim (years 1377–1388); ibid., 10 G 226-2 (year 1384) (*exitus decime de Castellari . . . de nemoribus . . . de pratis,* etc.); 10 G 227, year 1453; 10 G 228, year 1456, f.1, etc.; 10 G 229, year 1457; and Bonnefoy, 1879, I, pp. 303, 304, 305, and 308; 10 G 281, year 1458: allusion to the Côte du Piget (*costa super Castellar*); 10 G 233, year 1467: the inhabitants of Le *Castellar* and the *costa* above; 10 G 237-1, years 1521–1522: tithe of hemp at Le Châtelard (14^x) and allusion to the Côte du Piget, near Les Bois (*costae de nemoribus*); ibid., year 1523, *in molendino de chastellart*; 10 G 238, year 1530, many allusions to houses and lands at Le Châtelard; ibid., years 1533–1534–1535; similarly, 10 G 240, year 1540 (November); 10 G 241, year 1544; 10 G 242, year 1545; in the 1557 accounts, out of 255 acts of sale, two relate to Le Châtelard and three to La Rosière; ibid., 10 G 246, year 1561, three references in 94 acts; 10 G 310 and 311, year 1559, inventory of households at Le Châtelard, La Rosière, and other Chamonix hamlets (because of a withholding of "first fruits"); 10 G 312, year 1562, f. 12 vo, 13 ro, 25 and 27 ro: description of Le Châtelard in connection with an anti-tithe disturbance; 10 G 248, years 1564–1565–1566, various acquisitions of land and houses at Le Châtelard; 10 G 250, years 1570–1571, out of 201 acts of sale, five relate to Le Châtelard, in particular in October and December 1570 and July 1571; 10 G 254, year 1590: sale of lands (but not of houses) at Le Châtelard; the tithe accounts of 1590 ("receipt of tithes," 10 G 254) are the last to contain the usual mention of the "glebe of Les Bois and Le Châtelard"; from 1600 on (10 G 255) it is just "Les Bois" alone. But in 10 G 255 (year 1600) there are still purchases and exchanges of houses among people designated as inhabitants of Le Châtelard. In 1602 (10 G 256), the acme of the glacial disaster, there are only nineteen acts of sale in the whole community of Chamonix instead of the one or two hundred for a normal year in the sixteenth century. Out of ninety acts of sale from October 1621 to September 1622 (after a big gap in the records from 1602 to 1621), none makes any mention of La Rosière, Argentière, Les Bois—the hamlets buried under the ice; all that is mentioned is the acquisition of a *grangeage*, a kind of farm building or barn, at Le Châtelard. See also the texts of 1616 and 1643, cited in Chapter III. During the pastoral visitation of 1649, mention was made of the "village of Le Châtelard called Les Thines" (10 G 270); similarly in 1694 (ibid.). Thus the former hamlet becomes confused with the nearby surviving one of Les Tines. Le Châtelard itself survives not as an inhabited place but as a "mas" or purely geographical circumscription used for the assessment of dues. A text of 1755 refers to it, and to the nearby Les Tines, Le Lavancher, and Les Bois:

"Le mas du Châtelard est appelé le présent mas, *la prise* des *gaudins* et la prise none, qui se confine depuis le mas des Japers et de Landrieux par *Arve du couchant*, la pièce des terres par la chapelle *du vent* tendant par l'Enjoleiron férir au fort d'amour du Lavancher, tendant outre par le *Rey des Crètes* puis sus par la Ravine (moraine) où est enclavée la *combe du Lavancher* et tout Bonnanex sauf ce qui est des menus servis jusqu'au Passioux (cascade du Pissiou, aujourd'jui du Chapeu)." (A.D. Haute-Savoie, 10 G 282–4, March 9, 1755.)

All the land thus described is to be found on C3 and C5 of the 1945 survey (localities of Bois Gaudin, Bois de "Bonanée," and La Pendant).

Another text (10 G 282–1, year 1679, pp. 124, 126, 137, 140, and 141) shows that the "Mas of Joppers and of Landrieux containing several other unknown mas" overlaps with the land of Les Bois as far as Le Montenvers, and that the old mas of Le Gerdil includes the mas of Le Lavancher and Le Châtelard.

A final indication of how close Le Châtelard was to Les Tines is given in the records of pastoral visitations in 1693 and 1702, which still speak of "the chapel Saint-Théodule of Le Chastelard called of Les Tines" (A.D. Haute-Savoie, 1 G 121 and 122, f. 388). This chapel, which still exists, is none other than the "chapelle du vent" quoted above. On the most likely position of Le Châtelard, see Fig. 16, map of Les Bois and Les Tines.

(B) *Bonanay.* A.D. Haute-Savoie, 10 G 280, f. 53 vo., year 1458, lists the forty-five copyholders of the mas of Le Gerdil (=Châtelard+Le Lavancher); at the end one reads "pro *terra* bonenoctis, 14 denarii (f. 50 vo.); 10 G 233, year 1467, gives a similar list (f. 50 vo.). The word *terra* is the key word: it indicates land on which cereals are grown, as distinct from "pro plano et monte" in the same text (f. 50 ro.), which has no agricultural significance.

In 1523 there is another text:

"Die tercia may (1523) Johannes filius eiusdem Johannis fabri emit a guillielmo filio michaelis veytet et ab eius uxore videlicet unam peciam terre contra medietatem unius Regard, cum curtinis eiusdem et los aysous appartenentes: item portionem locum unam communem sibi pertinentem et aquam du pyssiou sitam in territorio de Bonanex precio 160 florins." (A.D. Haute-Savoie, 10 G 237–1, end of the register, no folio number.)

The sum of 160 florins was a large one: the average price for this sort of transaction was about 20.

The text of 1556–1557 says: "Michel Massat et ems (illegible) emerunt una pa terrae en chastellard a Johan Hugo de Bonanex." (A.D. Haute-Savoie, 10 G 245, accounts for June 1557).

The 1562 text is in A.D. Haute-Savoie, 10 G 312, year 1562, f. 34 ro.

The June 1571 text is in 10 G 250: "Petrus fabri bonaudy emit a perneta . . . et michaeli Leschiaz viro medietatem unius curtiils cum pertinent in loco de Bonanay, pretia, 4 flor."

The 1591 text is in 10 G 254, accounts for "1590," section on sales for January 1591.

Thereafter (after 1600), Bonanay is only a place name, referring to a locality in a moraine or *rouyne* (see above, 1679 text, Appendix 5).

(C) On *La Rosière.* Texts of 1315 and 1390 (Bonnefoy, 1879, I, pp. 174, 281, 381). See also ibid., II, pp. 28 and 54; 10 G 231, tithe accounts for 1463, etc. (the same sources as for Le Châtelard).

(D) *Sainte-Pétronille de Grindelwald.* The best study on this is Coolidge (W. B.), 1911.

But here is an additional piece of information which escaped Coolidge. I have noted (above, Chapter IV) a contemporary text which shows that the chapel of Sainte-Pétronille existed in 1520. Another text makes it possible to carry this chronology back further: in 1510 two people from Grindelwald were accused of having stolen four shillings from the collecting box of the chapel of Sainte-Pétronille (Protocols of the canton of Grindelwald, Berne archives; this text was kindly pointed out to me by M. Charles Roth, a scholar from Grindelwald).

The poem by Rebmann, who died in 1605, is cited by Friedli (E.), 1908, p. 51, by the *Echo von Grindelwald* for July 3, 1963, and especially by Coolidge, 1911. See also Mérian, 1642, p. 25; Scheuchzer, 1723; Gruner, 1760.

We may note in conclusion the curious text by Bourrit, 1785, Vol. III, p. 174: old men at Argentière told him that the silver mines which were supposed to have given the Argentière glacier its name "were not yet completely covered by the glacier two centuries ago" (i.e., about 1580). The chronology is excellent, but unfortunately it has never been proved that the mines ever existed.

APPENDIX 8

The "Events" at Chamonix About 1628–1630

A. Two texts, dated March 1640 and July 24, 1640 (A.C. Cham., CC3, nos. 33 to 99), speak of the "third of good and cultivable land lost in about the last ten years . . . through avalanches, falls of snow, and glaciers." Another text (ibid., CC 4–1, year 1640) speaks of a great overflowing of the Arve caused by the glaciers "twelve years back." Lastly, toward 1665, a document preserved in A.D.H.S., 10 G 273 b, indicates that there had been no major catastrophe due to the glaciers except in the years 1628, 1640–1644, and 1664.

B. As to the carbon datings of the morainic wood from the Taconnaz glacier (see Chapter IV), here is the exact position and aspect of the deposit as reported in 1962 by Corbel and Le Roy Ladurie:

"One of us finally located, on the edge of the Taconnaz glacier, some remains of tree trunks (*pinus cembro*) completely stripped of their bark and firmly caught under a lateral moraine. We climbed up the glacier to dig out the wood, which is sited on the right edge of the Taconnaz (or Taconna) glacier (map 1/50,000 Chamonix-Mont Blanc; map 1/10,000 Mont Blanc, I North, Auguille du Midi). The path up the glacier follows the ridge of an old moraine. According to the map, the moraine meets the rock (and the climb becomes more difficult) at 1738 meters. The wood in question is about a hundred meters below, at the base of the old moraine, whose steep slope has been exposed by the recent erosion. The deposit is at an altitude of about 1700 meters, measured by the Thommen altimeter. The wood is extremely hard, and it was difficult to extract it and take samples."

APPENDIX 9

Comparative Data on the Demography and Economy of Chamonix

1. *Demography*

Sources: Lists of taxpayers in the fifteenth century, in particular for 1458 (A.D. Haute-Savoie, 10 G 311); lists for 1679, 1706, 1733 (ibid., 10 G 282–1, 3 and 4). See also, for comparisons with other regions, Baratier, 1961; Faber, etc., 1965, p. 110; and Le Roy Ladurie, 1966, Parts 2, 4, and 5.

2. *Economy and animal production*

Deliveries of cheese from the mountain pastures looking down on the valley are recorded in the accounts of the priory of Chamonix from 1540 on (in particular in 10 G 240, 241, 244, etc.), up to the accounts for 1621–1622.

APPENDIX 10

Additional Data on the Front of the Mer de Glace (1764–1784)

1. *The Arveyron cave*

To the descriptions by Martel and Saussure cited above, Chapter III, and to the drawings by Bourrit and Bacler d'Albe reproduced in Mougin, 1912, Plates III and IV, should be added a little-known text of 1764. The anonymous author, who refers to "torrents of pieces of ice . . . petrified and immobile," writes (my italics): "From the frozen lake (the Vallée Blanche) several wide belts of pieces of ice overflow and slope down to the valley of Chamouni . . . *in the valley of Chamouni itself,* these same torrents (sic) present a novel spectacle to visitors.

The mouth of each one forms a sort of frontispiece like that of a church, with a great arcade in which is discovered a *spacious cavern* decorated and vaulted with ice, and here and there rods of ice hang down like organ pipes. Out of it flows a clear stream which goes to swell the Arve. The biggest of these streams which passes under the largest vault *is called Arbèron.*" (*Anonymous account* . . . of 1764, edited by Ferrand, 1912, pp. 46–47.)

There could be no better proof that the Arveyron cave, made by the Mer de Glace, was fully formed and visible from the valley in 1764, thus creating a situation doubly different from that of today.

2. *Position of the front of the Mer de Glace in the 1770s*

Additional texts (to be taken with the texts and engravings cited or referred to in Chapter IV): in August-September 1775, Thomas Blaikie, on a visit to Chamonix, went for a walk on the eastern flank of the glacier, which stretched almost to Les Bois; from the end of this glacier, the river Arveyron began to come out from under the ice, which formed a great arch (Blaikie, *Diary,* 1931 edition, cited by Bernardi, 1965, pp. 262–63). In connection with the glacier Des Bois "coming down almost to the plain," and the easy access to the source and cave of the Arveyron, see also: Bourrit, 1773, p. 36; 1776, pp. 18–19; and 1785, pp. 111–15. The comment above in Appendix 10, paragraph 1, applies to all these texts.

3. *Position of the front of the Mer de Glace in 1784*

In 1784 Saussure observes a certain withdrawal of the Mer de Glace in comparison with the maximum of 1777 (see Mougin, 1912, p. 164). Nonetheless the Mer de Glace remains very advanced, as is proved by the map drawn up by Saussure himself (1785), where the front of the glacier is 730 meters beyond its 1911 position (i.e., 1000 meters beyond the "ligne du chapeau"), according to Mougin, 1912, pp. 164 and 174. Another proof is an as yet uncommented text by Saussure, 1786, II, p. 19. In 1784 Saussure found, from on the spot measurements, that the (withdrawn) front of the glacier Des Bois was 500 meters as the crow flies from the old (A.D. 1600) moraine bordering the path from the priory to Les Tines[3]; measured on the map in Lliboutry, 1965, p. 726, where the old moraine is clearly represented, the glacial front of 1784 is again about 800 or 1000 meters beyond its 1958 position.

4. *Position of the front of the Mer de Glace in 1805–1814*

Evidence is sparse for this period, but H. Ferrand (1920, pp. 60–61) has brought together several references. I have been able to find some of the corresponding texts, and here are three of them:

On July 24, 1805, Henri de la Bedoyère visited Chamonix. From the valley of the Arve he saw "the glacier Des Bois curving out into the valley." And he visited, near the hamlet of Les Bois, the cave and source of the Arveyron (La Bedoyère, 1807 ed., pp. 397 and 402–3 and 1849 ed., pp. 329 and 334; on the glaciological significance of the cave, see above, Chapter IV). La Bedoyère's description (indicating the Mer de Glace in a state of expansion, curving out, and near to and *visible from the valley of the Arve*) is completely confirmed, a month later, by that of Chateaubriand (end of August 1805), which shows us the Mer de Glace well beyond Le Montenvers, and occupying the almost vertical slope *looking down on* the Chamonix valley. So once again the glacier Des Bois is visible from Chamonix: "I stopped at the village of Chamouni, and the next day went to Le Montanvert . . . From the summit, I saw what is very inaptly called the Mer de Glace. Imagine a valley the bottom of which is entirely covered by a river . . . At the head of this valley is a slope *which looks down on the valley of Chamouny. This slope, which is almost vertical, is occupied by the part of the Mer de Glace known as the Glacier Des Bois.*"[4]

In 1810, X. Leschevin (Leschevin, 1812, pp. 271–85) came to Chamonix. One of his companions said "he would not leave *the bottom of the valley,* but would just drive *to the source of the Arveyron and the foot of the glacier Des Bois,* and then he went back to sleep" (p. 251). Leschevin himself climbed the Montenvers, and from there went down to the spur of the glacier Des Bois by the wood and path of La Félia or La Filia (p. 276). He describes the glacier as *"going down to the Chamouni valley, where it takes the name of glacier Des Bois, from that of a hamlet near which it ends."* He contemplates the Arveyron cave, which may be up to 32 meters high. According to him it is only 2 kilometers from the confluence of the Arve and the Arveyron; this is probably too low an estimate, but this exaggeration is in itself striking, as the distance given on the I.G.N. map of 1949, with the glacier in its contemporary, shrunken condition, is 3.6 kilometers. The 1810 description corresponds therefore to Gabriel Lory's engraving of the same year, which also shows an advanced glacier Des Bois coming down "to the plain" (for reproduction of this engraving, see Engel, 1961).

Finally, in July 1814, Méneval (1814, pp. 78 ff.) climbed down the Montenvers by the path and wood of La Filia, which he found rather steep. By doing this he came *directly* to the glacier Des Bois, which he found "surrounded with woods of firs." His guides told him they had observed that "for some time the mountain of ice had been moving noticeably toward the plain."

All these texts are characteristic of the period of long advance of the

---

[3] On this chapel, see Appendix 7, paragraph on Le Châtelard.
[4] Chateaubriand, 1920 ed., pp. 12–13.

glacier Des Bois (cf. above, Chapter IV). Especially interesting is the repeated reference (Leschevin and Méneval) to the wood and path of La Filia. (For the position of La Filia, west of Les Mottets and La Jorasse, see the Vallot-Larminot map of the Mont-Blanc region, scale 1/50,000; or the I.G.N. map of Chamonix-Mont Blanc, scale 1/10,000.) A direct way from the Montenvers via La Filia to the glacier Des Bois is only feasible when the glacier stretches beyond the rocks of Les Mottets to the southwest. When, as is now the case, the glacier has retreated to the northwest of Les Mottets, such a route does not apply.

Lastly we may note the methodological distinction between on the one hand the long-term situation (the glacier Des Bois is "almost down to the plain" in 1805, 1810, and 1814, as in general between 1580 and 1850), and on the other hand the short and violent thrust which stands out against this background in 1814—the famous thrust which was to culminate about 1817–1820.

APPENDIX 11

Winters in the Sixteenth Century

My original intention (see *Les paysans du Languedoc*, Chapter 1) was to include in this book my whole collection of meteorological phenomena for Languedoc and the south of France from the fifteenth to the eighteenth century. But the present work could not accommodate both this material and the wine harvest dates (Appendix 12), so I give here just a summary of the most significant of my meteorological series. They relate to the winter of the sixteenth century.

The rest of the material will be published separately.

A. WINTERS IN LANGUEDOC AND THE SOUTH
OF FRANCE IN THE SIXTEENTH CENTURY
(References relate to the table and graph at the end of this book.)

1491[5] Sources: A.C. Carpentras, BB 107, f. 63, deliberation of 14.5.1491; Gaufridi, 1694, I, p. 369; Massip, 1894.
1494 Fuster, 1845, p. 289; Papon, 1786, IV, p. 18.
1495 Ibid., p. 18.
1505 (and not 1506): Fornery, 1910, I, p. 507.
1506 Ibid., and especially *Chronique consulaire de Béziers*, 1839.
1517 A.C. Valence, BB 4, 17-1-1517.
1518 *Chron. cons. Béz.*, 1839
1523 *Chron. cons. Béz.*, 1839; A.D. Tarn-et-Garonne, G 1134, f. 10; Martin, 1900, I, p. 297.
1524 (actually November 1523) Martin, 1900, I, p. 297.
1527 *Chron. cons. Béz.*, 1829, f. 102 vo.
1540 A.C.M., accounts, reg. 617, f. 24; A.D. Gironde, G 479, March 23, 1540.
1544 Fonds Chobaut.
1552 Platter (F.), 1892, p. 31.

[5] "1491" means "winter, including December 1490 and January, February, and (possibly) March 1491"; and so on.

1557 Ibid., p. 146.
1565 Devic, 1872–1892, XI, p. 465; Fonds Chobaut.
1568 Fornery, 1910, II, p. 121.
1571 Fornery, 1910, II, p. 152; Villeneuve, 1821; Papon, 1786; Devic, 1872–1892, XI, p. 542; Bouges, 1741, p. 340; A.C. Aix, BB 68, f. 64, January 15, 1571; A.C. Malaucène (Vaucluse), BB 11, f. 83.
1573 Fonds Chobaut; *Chronique Bourdeloise* for this date; Devic, 1872–1892, XI, p. 558, n. 5, and p. 563; Febvre, 1912, p. 766.
1580 Martin (A.), *Histoire de Mende*, p. 25.
1583 Fonds Chobaut.
1584 *Histoire du commerce de Marseille*, III, p. 416.
1587 Devic, 1872–1892, XI, pp. 757–62; Gaufridi, 1694, II, pp. 623–24; Massip, 1894–1895.
1590 Gaufridi, 1694, II, p. 679.
1591 A.C. Cordes, CC 147; Gaufridi, 1694, II, p. 708.
1595 (December 1594) Villeneuve, 1821; Fuster, 1845, p. 289.
1597 Platter, 1892, p. 322; d'Aigrefeuille, 1885, II, p. 20.
1598 Platter, 1892, II, p. 347.
1600 Fonds Chobaut; A.C. Narb., BB 6, April 4, 1600; Papon, 1786, IV, p. 416.
1603 Baehrel, 1961; A.C.M. (printed inventory), IX, p. 13; Villeneuve, 1821.
1608 Devic, 1872–1892, XI, p. 901.

B. WINTERS IN ANTWERP IN THE SIXTEENTH CENTURY
after Van der Wee (1963)

Decades	A	B	C	C'	C''	D	E	F	G	H
1500–1509	4	1	2		2		1	2	1	3
1510–1519	4	1	5	2	3		2		1	1
1520–1529	4					2	3	1		1
1530–1539	2	1	2		1		5	2		2
1540–1549	4	1	2	1	1		4		2	2
1550–1559	5	1	1		1		2			
1560–1569	6	2	2		2	1				
1570–1579	5		2	2			4		3	3
1580–1589	4	1	5	4	1		2		2	2
1590–1599	3	1	7	3	4	1		1		1

A: Number of winters in the decade when the documentation says nothing about temperatures.
B: Number of severe winters.
C: Number of winters mentioning frosts, ordinary or severe.
C': Number of winters mentioning ordinary frosts.
C'': Number of winters mentioning severe frosts.
D: Number of winters mentioning "a lot of snow."
E: Number of winters mentioning just "snow."
F: Number of "mild" winters.
G: Number of "very mild" winters.
H: Number of "mild" and "very mild" winters.

(Needless to say the winter of "1500" includes December 1499 and January and February 1500. No account is taken of what happened in November, March, and so on.)

This table shows:

1. Contrary to what one might expect, information about temperatures tends to decrease after 1550. Eighteen winters are unknown as to temperature between 1500 and 1549, as against 23 between 1550 and 1599. In other words 32 are known in the first half of the century, and 27 in the second.

2. The number of "severe" winters increases from 4 (out of 32) before 1550 to 5 (out of 27) after 1550.

3. The number of "frosts" during the same period increases from 11 out of 32 to 17 out of 27—from 4 out of 32 to 9 out of 27 for ordinary frosts, and from 7 out of 32 to 8 out of 27 for severe frosts.

4. The comment "a lot of snow" increases radically after 1550 (from 2 out of 32 to 8 out of 27) or more exactly after 1560. The comment "snow," on the other hand, becomes less frequent.

5. The number of "mild" and "very mild" winters falls after 1550, from 9 out of 32 to 6 out of 27. "Very mild" winters fall especially, from 4 out of 32 to 0 out of 27.

APPENDIX 12

Phenological Series: French Wine Harvest Dates, and Various Other Winegrowing Data

This appendix includes:
A. The dates of southern wine harvests (1).
B. The various means (southern, central-and-northern, and national) of the wine harvest dates for the "sixteenth century" (1484–1619), and for the seventeenth and eighteenth and nineteenth centuries.
C. Other information about wine harvest dates and other harvests.

A. DATES OF SOUTHERN WINE HARVESTS

(a) The table below deals with the following localities (the number indicates the corresponding column). Up to 75 inclusive they are all in the county of Avignon and the data are taken from the Fonds Chobaut at Avignon, from the excellent file on "Vines and wine harvests." From 76 onward the localities are in the present department of Hérault unless otherwise stated, and all data come from the communal archives, series BB mainly and also CC, FF, and HH.

1 Apt; 2 Aubignan; 3 Avignon; 4 Baume; 5 Bédarrides; 6 Bédoin; 7 Bollène; 8 Buisson; 9 Cabrières; 10 Caderousse; 11 Camaut; 12 Caromb: 13 Carpentras; 14 Caumont-sur-Durance; 15 Cavaillon; 16 Châteauneuf-du-Pape; 17 Courthezon; 18 Le Crestet; 19 Gadagne; 20 Gigondas; 21 Entraigues-sur-la-Sorgue; 22 L'Isle; 23 Jonquières; 24 Joucas; 25 Loriol; 26 Malaucène; 27 Melemort; 28 Mazan; 29 Montdragon; 30 Monteux; 31 Mormoiron; 32 Mornas; 33 Orange; 34 Calcernies; 35 Pernes; 36 Austin monastery at Pernes; 37 Piolenc; 38 Puymeras; 39 Robion; 40 Saigon; 41 Sainte-Cécile; 42 Saint-Didier; 43 Saint-Romain-en-Viennois; 44 Saint-Saturnin-d'Avignon; 45 Sarrians; 46 Sault; 47 Saumanes; 48 Sérignan; 49 Sorgues; 50 Le Thor; 51 Vaison; 52 Valréas; 53 Vedènes; 54 Villedieu; 55 Villes-sur-Auze; 56 Violé; 57 Visan; 58 Property under the protection

of Douceline de Saze; 59 Muscat de l'évêque; 60 Greyfriars of Avignon; 61 Benedictines of Sainte-Cathérine d'Avignon; 62 Greyfriars of Avignon (another vineyard); 63 Morières, belonging to Avignon (A.C. Avignon, series FF); 64 rest of the land belonging to Avignon (ibid.); 65 Avignon (series HH); 66 Avignon—day when the wine harvest was allowed to come within the gates of Avignon (ibid., series CC); 67 Avignon, decisions fixing wine harvests; 68 Chartreuse de Bonpas at Caumont; 69 Cavaillon: various parts of the district, not including Vergas; 70 Pertuis; 71 Muscat du Thor; 72 Oppède; 73 Laval; 74 Avignon (record book of a proprietor at Morières); 75, ditto at Corambaud; 76 Valence (Drôme); 77 Cordes (Tarn); 78 Saint-Jean-du-Bruel (Aveyron); 79 Castres (Tarn); 80 Saix (Tarn); 81 Gaillac (Tarn); 82 Montpeyroux; 83 Gap (Hautes-Alpes); 84 Gignac; 85 Fabrègues; 86 Frontignan; 87 Marsillargues; 88 Montpellier; 89 Lunel; 90 Béziers; 91 Aniane; 92 Lansargues; 93 Mauguio; 94 Domaine de Méric, near Montpellier; 95 Lodève; 96 Chusclan (Gard); 97 Narbonne (Aude); 98 Decisions fixing vintage dates by the courts of the county of Avignon; 99 Laudun (Gard); 100 Villefranche-de-Rouergue; 101 Cessenon-sur-Orb; 102 Bordeaux (Jurade); 103 Gironde (Abbey of Bonlieu, A.D. Gironde, H 1136); 104 Sainte-Croix de Bordeaux (A.D. Gir, series G); 106 Archbishop's vineyard at Bordeaux; 107 File of the steward of Bordeaux on authorizations of harvest dates by the chapter of Saint-Seurin (A.D. Gir., G. 1078); 108 Valladolid[6] (after B. Bennassar); 109 Montpezat-en-Quercy, effective date of wine harvest (Sources: A.D. Tarn-et-Garonne, G 846 to 848 and III E, 680 ff.; Latouche, 1923, p. 219); 110 Pichon-Longueville, Gironde (this last series is taken from Angot, 1883).

(b) Table of wine harvest dates by year and by locality. The years for which we possess data run from top to bottom in the left-hand column of the following pages and where necessary in intercalated columns. The numbers indicating localities run from left to right at the top of the pages; the columns below show the wine harvest dates, which are given in terms of number of days after September 1 (=1). Thus 28=28 September, 34=4 October, 0=31 August, −1=30 August, and so on. For all dates before 1583, when the new calendar was adopted in Languedoc, the Gregorian correction has been made.

N.B. The figure 71 in 26–1602 is doubtful and has not been taken into account in the means given in paragraph B of this appendix.

[6] On this series see also Bennassar, 1967; and Anes Alvarez, 1967.

YEARS	58	YEARS	59	YEARS	13	110	YEARS	13	19	60	62	110	YEARS	13	61	108
1330		1360		1420			1450						1480			
1331		1361		1421			1451					45	1481			
1332		1362		1422			1452						1482			
1333		1363		1423			1453						1483			
1334		1364		1424			1454						1484		6	
1335		1365	13	1425			1455						1485			
1336		1366	31	1426			1456						1486			
1337		1367	46	1427			1457						1487			
1338		1368		1428			1458	7		7			1488			
1339		1369		1429			1459			28			1489			
1340		1370		1430			1460			18			1490			
1341		1371		1431			1461			13	27		1491			
1342		1372		1432			1462			5			1492			
1343		1373		1433	7		1463			33			1493			
1344		1374		1434			1464			20			1494			
1345		1375		1435			1465			35	59		1495			
1346		1376		1436	20		1466			28			1496		27	
1347		1377		1437	16		1467				17		1497			
1348		1378		1438	14		1468						1498			
1349	19	1379		1439			1469				20		1499			49
1350	21	1380		1440	7		1470				19		1500			40
1351	21	1381		1441	15		1471						1501			37
1352	36	1382		1442	9		1472						1502			
1353	8	1383		1443	12		1473						1503			39
1354	8	1384		1444	15		1474				45		1504			36
1355	14	1385		1445			1475		8		21		1505			
1356	29	1386		1446			1476	12			4	36	1506			48
1357	25	1387		1447			1477						1507			
1358	11	1388		1448			1478				25		1508			53
1359		1389		1449		49	1479						1509			

YEARS	109	110
1480		
1481		
1482		41
1483	18	26
1484		14
1485	50	53
1486		
1487		
1488		
1489		
1490		
1491		
1492	25	
1493	29	
1494		
1495		
1496		
1497		
1498		
1499	34	
1500		
1501	16	
1502		
1503	27	
1504		
1505		
1506	36	
1507		
1508		
1509		

YEARS	14	33	50	72	76	88	90	102	108	109
1510									45	
1511									35	
1512									45	25
1513									57	
1514										
1515										
1516										
1517									45	3
1518										
1519									47	
1520										
1521										
1522										
1523										
1524							5			
1525										
1526										
1527		40							61	
1528									53	
1529									45	
1530										
1531										
1532	3			3				32		
1533					12			40		
1534			5	3				29		
1535					15					
1536						20				
1537	19					20				
1538										
1539										

YEARS	7	9	13	15
1540				
1541				10
1542				
1543				
1544				
1545				
1546				
1547				
1548				
1549				
1550	13			
1551	20		10	
1552				
1553	21			
1554				
1555.				
1556			4	
1557				
1558				
1559				
1560				
1561				
1562		5		
1563				
1564				
1565				
1566				
1567				
1568			16	
1569				

YEARS	62	76	101	106	108	109
1540		11			37	
1541					53	
1542					49	
1543					48	
1544						
1545					31	
1546					32	
1547						
1548						
1549	13					21
1550	15					10
1551	12				47	
1552	11				45	
1553	13				51	28
1554	7	30			41	
1555	12				49	
1556	6				41	
1557	18				70	
1558	15				52	
1559	2	19				
1560	12			40	54	
1561	4			28	41	
1562	3			24	47	
1563				40		16
1564				45	42	
1565				32	45	
1566				33	42	
1567				30	39	
1568				35	48	
1569				46		

YEARS	10	13	16	19	24	26	30	35	36	50	63	64
1570												
1571		2										
1572									14			
1573									19			
1574									10			
1575									8			
1576									16			
1577			11						13			
1578				13	15				5			
1579									13			
1580												
1581									14			
1582									15			
1583									5			
1584									10			
1585						22			16			
1586						23				14		
1587									22			
1588									9			
1589	14						8					
1590						19			17			
1591	16								13			
1592						14	10		11			
1593						15			10			
1594						19			9			
1595	22	15							15	15		
1596				16		22		16				
1597	22					25						
1598	21						15				9	15
1599						8			-1			

YEARS	65	71	79	83	90	97	104	105	106	107	108	109	YEARS	2	10	11	15	19	22
1570									37				1600		25				
1571									30				1601						
1572									36		52		1602						13
1573									41		49		1603	9				3	9
1574									33		45		1604		16				
1575									34		39		1605		19		12		
1576									36		55		1606				13		
1577											47		1607				1		
1578											32		1608		21		12		
1579				27							38		1609				15		
1580											43	19	1610		20		11		
1581											35		1611		19				
1582											44		1612		24				
1583				22							31		1613		23				
1584											33		1614		22		19		
1585				25							53		1615						
1586										28	35		1616						
1587										33	30		1617						
1588						14				25	33		1618						
1589		23	31								39		1619				7		
1590										28	43		1620			18			
1591									37		44		1621						
1592										27	42		1622						
1593								41			36		1623						
1594	9								50	50	37		1624						
1595									33		32	22	1625						
1596									41	36	52		1626						
1597									45		45		1627						
1598									35		35	31	1628						
1599				20					24		31		1629						

YEARS	26	28	30	33	35	36	38	50	55	63	64	66	77	83	84	87	88	89	90	91
1600			15		15	15						11		40						
1601	38					18													36	
1602	71					12						4								
1603	8					5		8						24						
1604			6			9	9	15							13					
1605	22					15	7	22	15						19					
1606	23					15		24												
1607	8			5		3		3												
1608						15		21							14			24		
1609			11			15								41						
1610						15		19								17		13		
1611						5		9												
1612				12		15		15					31		15					
1613			12			17		22						24	22					
1614				22		22		26				16		29	32		30			
1615		12				15									24		16			
1616		4	4			6									15					
1617		17	13		15	15		18				9		43	18	21	15			
1618			19			24		27				17		42	27					
1619	18				15	11						5		35	16			23		
1620			14	21		17						9								
1621			17			27	27	28												
1622	22			15		15	15	19				9		31						
1623	25		13	18		18	18							27	23					
1624	22					16	15			9				29						16
1625	26					21	22								24					25
1626		15		21	28	15	15	18	9	15		7		52	24					20
1627			30			31						27			37					
1628	44					32		36						53	39					31
1629	40					19		24				.17		40	34				47	31

YEARS	97	98	101	102	104	105	108
1600						54	32
1601		30				39	52
1602						27	35
1603				26		24	31
1604							35
1605						37	
1606					53	49	28
1607						33	33
1608	15					45	40
1609					30	24	46
1610			8	32		31	35
1611				24			37
1612				37			40
1613				39			38
1614							30
1615						31	22
1616					36	34	24
1617						28	42
1618				41			38
1619							42
1620				43			39
1621	26			50			49
1622							26
1623				30			39
1624	16						30
1625							38
1626	21				47		49
1627	31						49
1628	31						39
1629				35			52

YEARS	10	26	30	33	36	52	55	57	66	77	81
1630		29	16		16				11		
1631			15		10						
1632					22		24		11		
1633					20		23		12	38	
1634			15		20		25				
1635			15	13	17		17		10		
1636					11		15				
1637			1		4				0		
1638					6						
1639					15						
1640					13					36	
1641					23					38	
1642					32						50
1643					19						
1644					19			22			
1645					9						26
1646					13					32	
1647		23			16						
1648			22		22			30			
1649					23				20		44
1650					16						42
1651					-1						23
1652					20						44
1653					13			22	31		
1654					15				33		37
1655					15			30			
1656	18				14						31
1657	17				15			30			
1658						22			25		
1659	12								23		

YEARS	83	87	88	89	90	91	96	97	102	105	108	109	YEARS	5	16	26	36	50	57
1630		17		31									1660						
1631	29	17					19						1661				9		
1632				37	30		30						1662						
1633		29			29					40	56		1663						30
1634	40		15	27	25						39		1664			21	15		
1635	24	20			19		14				31		1665						
1636		17			18						31		1666				16		
1637		14		18	14						35		1667				26		
1638		20			20						43		1668						27
1639		19									35		1669						21
1640		24			17					38	45		1670						
1641	42	24	23			23	24				37	37	1671	11			17		22
1642	46	34									38		1672				22		
1643		28		18	28	26					40		1673				32		32
1644		26		17							40		1674				20		
1645		18		13		18	18				41		1675				41		48
1646	27	13			36	17					24		1676				22		28
1647	34	23				20					37		1677						31
1648		31				29	35		30		44		1678	19			26		
1649	44	34				30	32				38		1679	25			26		30
1650		22		19		22	26			40	36		1680	17			23		22
1651		20				15	22				32		1681				22		
1652	46	30	20			27					37		1682						30
1653		24	22	18	26	23	15			29	33		1683						27
1654		30		24			28				44	38	1684				22		
1655		22				15					38		1685					24	22
1656											41		1686				17	23	21
1657			25			28					35		1687				25		30
1658		31	31								48		1688				34	34	36
1659		18	12	15			12				38	18	1689				28	26	33

YEARS	103	105	108	110	YEARS	5	16	30	32	35	36	50	57	62	63	67	81	82	83
1660			34		1690	23						34	39					43	
1561			42		1691	30						24	22						
1662		23	39		1692							29	36						
1663		32	45		1693								31						
1664			35		1694								27						
1665		25			1695					29		33	43						
1666	24	23			1696	24							27		27				
1667					1697								29		27				
1668					1698						38		43	13					
1669			33		1699							34		29					
1670			41		1700							34		31					
1671			36		1701						33		40	36					
1672			33		1702						33		34	31				27	VERY LATE
1673			44		1703			24			27		31	31		29	40		35
1674			40		1704						22		24	25					
1675			56	60	1705								35	35				28	
1676			35		1706						18		21	20				16	
1677			38		1707						28		40	28					
1678			54		1708						28		32	30					
1679			43		1709		32				30		37	34				37	
1680			46		1710			22			22		22	18	17			29	
1681			37		1711						23		30					35	
1682			44		1712						21		26					32	
1683			37		1713			60			32		39					38	
1684			45		1714						32		31					40	42
1685			40		1715						24		31					27	
1686			35		1716						36		35					41	
1687			40		1717						30		34					32	
1688			51		1718			12			13		19					23	
1689			36		1719		14	14	15		25		25						

YEARS	62	66	73	74	75	77	79	80	81	83	84	86	87	88	89	90	91	96	98	101
1660						21			31	23			20	13	16	14				
1661		3											26		20					
1662													25	22	20		21	25		
1663													33	32	33	34				
1664		15												23	22		25	29		
1665														25		24				
1666	21			15	20					31			27	27	22	27				
1667	29			23	29				43	38			40	34	32			36		
1668	13			14	18				41				24		20					
1669	19			12									23	18	20	23				
1670	15			12	15		33						23	17		16				
1671	23												28	24	25	22				
1672	27		19			40							26	29	26		27	26		
1673	38		28				49						39	35	39					
1674	26									39			27	26			27	29		
1675					44	53	64			55			44		40		44			
1676			21		23								28		28					
1677			21		30	41		31	40				27		23					23
1678					32								33	29						
1679					30							27	32	25						
1680					28				34			25	25						17	
1681													22		22				14	
1682											29	30	38							
1683							36					28	27	22						
1684												25	28	21			19			
1685												17	29							
1686						29						21	26			27			21	
1687		24	30									30	38							
1688												38		36					31	
1689												34	35			32	29			

YEARS	84	85	86	87	88	89	90	91	92	96	98	102	105	108	YEARS	2	4	5	6
1690			39	35	25						32			52	1720				30
1691			31	26								26		38	1721			31	
1692			37	36										44	1722				
1693			42											57	1723				
1694		27						27	27					43	1724				22
1695				40		38								51	1725				
1696				31						29			38	41	1726				
1697				30			28		30				23	55	1727				
1698				45			45	43					50	52	1728			14	
1699			31	36		45		31					29	34	1729			27	30
1700			37	34		32	40							49	1730			25	
1701	39		47	47		49								54	1731			25	30
1702				39		41			40					48	1732			22	
1703			31	31		33			27					41	1733	30			
1704	25		29	26											1734	15		9	15
1705			38	36		36		31	42						1735	30			
1706	20		27	20		22		21	27						1736				
1707			35					40							1737	23			
1708			31			31		31							1738	30		24	36
1709			37					37							1739				30
1710				25		27	29								1740	40		40	
1711	36					35		36	30					43	1741				
1712			33	29		33		36						44	1742	25			31
1713	43		39	43		45	35	41						52	1743	36	37		40
1714	42		40	39		39		41	38					58	1744			35	40
1715	32		23	37		30		30	27						1745	27		21	
1716	43		42			44	35							36	1746	33		33	
1717	39	34	37	37		35								37	1747			22	32
1718	23		15		22	19		19	19					23	1748	30	30	33	
1719	20		18	25	21	21	25			28				30	1749	36		37	36

YEARS	16	22	26	28	30	32	35	36	55	57	62	79	81	82	83	84	85	86	87	88
1720						39		32								45		33		
1721						30		29	40					46	45	51			47	44
1722						28		31	28							36		31		31
1723						28		27	34							31		27	33	29
1724						19		15	25									18		12
1725						31		32								46		41		39
1726						23		30								31		19		24
1727						20		18				25				26				24
1728						22		14	23				27					9		22
1729						27		30					40			41	26	39	33	
1730						26		26					39			42	18	32		
1731					27	32		28								46		31		41
1732			30			30		26							44	44		31	44	
1733						32		32							34		23			40
1734								14									9	16	32	
1735						33		33							43		33			
1736				24				22								39		20		27
1737								23								30				27
1738						31		31							43	44		27	36	45
1739						31		28							35	43		19	28	
1740						40	40	47								55		41	44	
1741										26					35	30				
1742															46	38		33		
1743						34									44	54				
1744			35			35		36		39						43		37	41	
1745						22		22								38		33		
1746	40	33				33		33							47	48		38	40	
1747			30			29		32		26						47	30			
1748						27		30		33	34				44		34			
1749			38			33		36		37						51		40	46	

YEARS	89	90	92	94	95	96	97	99	100	101	102	108	109	YEARS	4	5	6	7	10
1720	47											45		1750		25			
1721	48	44							LATE			45		1751					
1722	33	35										36		1752	30	28			
1723	35	34										34		1753			31		
1724	22											29		1754		26			
1725	44				48				VERY LATE			49		1755		·			
1726	22				41		37					41		1756			34	35	
1727	23	30										41		1757					
1728	22											41	30	1758					
1729	33											44		1759					
1730	33						39					44		1760				29	
1731	36											43		1761	30	28		35	
1732	44			43								52		1762			24	27	
1733	30										27	43		1763					
1734	13						20					33		1764	30	31	31		
1735	35				47		40				47	53		1765		30	35		
1736	33						31				23	43		1766		29			
1737	31		30		40		30				20	51		1767	30	30			
1738			39								35	42		1768		26			
1739	26		28									55		1769	39				
1740	39	44	47											1770					
1741	23		29								19	33		1771					45
1742	33	34	45		47		32							1772	29				30
1743	45	44	51	61			47					49		1773					44
1744	44	41	45	36					38					1774	30	34			33
1745	35		42	36								38		1775		34			
1746	38		43	47		41	44				35	48		1776		33			33
1747	42	43	42	46							31	59		1777		36			36
1748	44	40	44	46								40		1778		29			21
1749		46	43	43			43				36	47		1779					34

YEARS	12	14	15	16	22	26	30	32	33	35	36	44	50	51	55	57	62	63	64	78
1750			27							25	28						29			
1751					28					29	31						35			
1752				30	30						28						27			
1753											31						32			
1754					31						32						31			
1755					30						29						30			
1756											34					45	36			60
1757										30	33					40				43
1758	32				39						32					39				52
1759											31							31		42
1760		25			25					26	32						30	29		43
1761		31			30						28						32	31		44
1762					33						29					Oct.	27	29		41
1763					33						28						35	29	33	40
1764					31					28	31						32	28		
1765			34		44						30	31					33	30	32	
1766					30						29	31					32	29	31	
1767			35		35				30		31	35					36	35		
1768		26			33					26	28	32					35	29	33	
1769		33	39		39					36	32	39					40	39		
1770		38	45		52	45					39	37					41	38		
1771		34	44			40					34	36					38	33	37	
1772			31		35	31					30	29					29	28		
1773		41	48		48	42	44				41	42					42	37		
1774		33	31		33					30	33	34	43				35	29		
1775		32	34			32				29	32	33	35	34			33	28		
1776		30	37		37	34					31	37					30	26		
1777		29	37		46					31	36	47					33	29		
1778		28	29		35	17				30	30	30					24			
1779		27	34		34	29	34				29	34	31	34				27		

YEARS	79	80	81	83	84	85	86	88	89	90	92	93	94	95	99	101	102	103	108	109
1750					43		31		37		38		33	37	35		30		44	
1751			41				36	36	38				40	37			41		44	
1752					40		26		34				39	39			39	46	37	
1753					39		25				33		33	38			38	39	33	
1754				43	38		24	31	30				37	36			46	59	40	
1755					37			33	38		38		34	39	39		35	39	45	
1756	38					41	36	38	38		38		44	48				52	54	
1757					34				34		35		40	40			38		43	
1758					47		33		40		41		46	48					49	
1759					37		25		38		39		39	33			27			
1760					36		20		37		38		36	33		37	25		48	
1761					43		29	36	39	38	39		42	43					42	
1762					42		25	34	35	38	38		41	41	37	35	24		35	
1763					42		23		34	36	34		41	42	33		38		48	
1764			27		39				32		35		38	35					29	
1765			40		45		26	34	38		38		44	45			37		44	
1766			39		44		29		33	37	36		43	45			39		42	
1767					38				32	37	33		40	40			35		35	
1768					43		32	38	33	42			41	41	43		30		38	
1769							41		48	42	43		42	50			32		36	
1770					46		42	39	46	46	47		45				47		50	
1771					40	40	34		38	45	39		40	47			28		42	
1772			28		33		31		32	31	30	36	29	36			26			28
1773			37		48				45	44	42	41	48				41		40	
1774			33		40		32		42	41	43	40	40	41					46	
1775		32	32		40				42	35	34	40	39	41					42	
1776		40			47				37	39	40	37	46							
1777		43			43			35	36	37	40	39	44							
1778		31			36		26		31	30	36	38	23	29	29					
1779		23	30		42				32	34	37		42	34	34				41	

YEARS	110	YEARS	10	12	14	16	19	22	23	26	32	33	35	36	37	39	44	50	52
1750		1780	32			34				39	29			32	32			33	
1751		1781	31			26		24		34					27			29	31
1752	33	1782	34		37	44				37		37	33	38					37
1753		1783	31		32	32				36			30	32		31			
1754	43	1784	29		24	31				30				24	30				
1755	22	1785								40				36		40			
1756	37	1786				41				39				32		34			
1757	29	1787				46				48				40					
1758	39	1788	29		38									24			22	41	
1759	24	1789												36					
1760		1790				41										37			
1761	20	1791																	
1762	17	1792				34							31						
1763	35	1793				40										37			
1764	24	1794				39	38												
1765	32	1795				35													
1766	36	1796																	
1767	35	1797																	
1768	29	1798																	
1769	29	1799																	30
1770	42	1800																	22
1771	27	1801																	26
1772	23	1802																	30
1773	37	1803																	
1774	28	1804				45			31										
1775	25	1805				50			37										
1776	33	1806							36										
1777	38	1807		28						26									
1778	29	1808								40	40								
1779	27	1809								53									

YEARS	55	63	64	68	80	81	83	84	86	89	90	92	93	94	95	99	102	108	110
1780		28	32		25			40		26	32	32	33	42	33				26
1781		24		20				39		27	34	32	34	33		31			21
1782		33	37		26			40	38	38	40	39	38		44	38		46	21
1783		32	36		25			44	36	34	38	38	37	36	40	39		45	19
1784		23						35	32	26	30	31	30	34	40			42	20
1785	40	36						43		37	35	43	41	40	37	44			15
1786		35						36		30	33	34	34	39	39	46		41	22
1787		41		42	38			46		41	39	45		38	43	45		46	34
1788		22	24	23				30		23	30	31	34	29	32		16	35	12
1789		37			36			45			38	42		42	52			43	37
1790						LATE					42			41	45			43	23
1791											29			22				43	26
1792																		39	28
1793														42				38	
1794														32				45	
1795														28				43	
1796														38				41	
1797														33				44	
1798														36				39	
1799														48				44	
1800														38					
1801																			
1802																			
1803																			
1804		31																	
1805		42																	
1806		38																	
1807		28																	
1808		36																	
1809		42																	

YEARS	7	10	12	13	16	19	20	22	23	28	29	31	33	35	45	48	52	63	64	73
1810																		31		
1811				23								19						19		
1812				35													40	35	37	
1813		40		34										35				34	36	
1814			40	40					42					40			47	43	45	
1815				25														20		
1816		51		52										51				46		
1817				32										33				31		
1818				31	39									32			36	31		
1819				28	32									30			31	28		
1820				26	28									26				21		
1821				38	42									39			40	38		
1822				10	13			11						12			12	12	16	
1823				31										36			43	30		
1824				43	45			39						44			46	42		
1825	26			22	30	28	22				21			26			24	26		
1826	32	31		32	32	39	26				32		31	32		31	32			
1827	26	24		24							28			28		24	31	26		
1828	24	15		18		22			29		22		17	22			25	16		
1829	42	37		29	42	38			30		42		38	35			50	35		
1830	27	22		22		29					27						23			
1831	26	26		22	33	26					29			23			26			39
1832	34	35			41	35											42			
1833	26	26		23	30						26						31			
1834		29		23			29				27						31			
1835	38				42		42				37				38		38			
1836		36			36	31					33						36			
1837	35	34			39	34					36					32	39			
1838	41					38					37	38	38							
1839	44			30		32					37	33	37							

YEARS	99	YEARS	5	11	13	15	16	17	19	22	25	28	29	30	31	32	33	35	37
1810		1840							35	32					35				
1811		1841							34	27		27	27	30		27		
1812		1842							33	30		26	26	33				
1813		1843			39				42	39			40	39	45		39		
1814		1844			25				30	30	28		26	30	32		24		
1815		1845			36	34			43	36			36	36	43		36		
1816		1846			22	22			24	24			19	21	28			21	
1817		1847				22			30					26	34		20		
1818		1848							32	30				32		25		
1819		1849				25			34					27	34		24	29	
1820		1850				24		37	37	39				37			33	34	
1821		1851								40				36			36	41	
1822		1852						37						34			29	34	
1823		1853					43		40					40			37	40	
1824		1854							20	25				25			25		
1825		1855							30	24							31		
1826		1856							29	30				36			36	28
1827		1857							33	28				31			35	26
1828		1858		23						25				22			24	23	27
1829		1859				28			26	24	22			22				
1830		1860					38	38	35					31			31	
1831		1861	30						25						26	31	26	
1832		1862					25		25					22		25	17	22	
1833		1863							28		24			23		23	18	22	
1834	29	1864																20	
1835	45	1865												25				
1836	40	1866																
1837	37	1867																
1838		1868																
1839	41	1869							16									

YEARS	39	41	45	46	49	50	51	52	53	69
1840										
1841							29			
1842										
1843	38									
1844										
1845										37
1846			16							
1847										
1848					40					
1849					31				31	
1850								37		
1851										
1852				29	26	39				
1853						44				
1854						25				
1855						31				
1856			36							
1857		20	29			35				
1858			22			27				
1859			22			26		21		
1860			33			38		38		
1861			30					27		
1862			22							
1863			21							
1864										
1865										
1866										
1867										
1868										
1869										

B. MEANS OF FRENCH WINE HARVEST
DATES FROM THE SIXTEENTH TO THE NINETEENTH CENTURY

(a) *Wine harvest dates for the sixteenth century.*[7] Together with Madame M. Baulant, whose researches in the winegrowing archives of the Paris region have been vital to the success of the undertaking, I have set up a new series of wine harvest dates for the sixteenth century which is more substantial than those in my previous works (L.R.L., 1966 and 1967). This new series represents the mean of six local series: Paris (the Baulant series), Dijon, Salins, Lausanne, Aubonne, and Lavau (the Angot series, 1883). In other words the column of annual figures below reflects annual and mean tendencies in spring-summer temperature in a region along the line Switzerland–Jura–Burgundy–Île de France, or northeast and central France. The dates are counted in days from September 1: thus 20 means September 20, 31 means October 1, and so on. All dates are according to the Gregorian calendar. (I should like to thank Madame Baulant for allowing me to publish here this extract from our joint work: a detailed article signed by us both will add the necessary references.)

TABLE 1

1484	31		1509	24.7		1534	20.3
1485	36.5		1510	30		1535	32
1486	20		1511	43.7		1536	8
1487	26		1512	23.7		1537	35
1488	47.5		1513	27		1538	12.7
1489	27.5		1514	28.3		1539	27.7
1490	27		1515	31		1540	11.7
1491	49		1516	10.5		1541	34.5
1492	gap[8]		1517	21.5		1542	50
1493	35.5		1518	28.3		1543	31
1494	18		1519	37		1544	29.5
1495	12		1520	23.7		1545	11.7
1496	40		1521	22.5		1546	21.5
1497	31		1522	23.3		1547	27
1498	26.5		1523	17		1548	30.5
1499	28		1524	14		1549	25
1500	14		1525	20		1550	31.3
1501	19		1526	26.7		1551	25.7
1502	25.3		1527	39		1552	18.8
1503	16.7		1528	33.5		1553	32.2
1504	17		1529	43		1554	21.3
1505	43		1530	21.5		1555	43.2
1506	28.3		1531	26		1556	0.5
1507	18.3		1532	22.7		1557	28.4
1508	32		1533	31.5		1558	25

[7] The following series is more abundant and includes more winegrowing stations than that contained in Appendix 12B of the first (French) edition of this book, in 1967.

[8] In 1492 the southern wine harvest (see above Appendix 12A, column 109, years 1492 and 1493) took place four days before those of 1493. If the same were true of the north, the figure here for 1492 would be 31.5.

1559	7.5	1579	39.5	1599	10.5
1560	32.3	1580	27.8	1600	41.9
1561	20.1	1581	38.3	1601	37.8
1562	25.5	1582	29.5	1602	21
1563	30.3	1583	14.7	1603	11
1564	37	1584	23.8	1604	27.4
1565	32	1585	36.3	1605	21
1566	25	1586	32.5	1606	39.6
1567	21	1587	38	1607	23.3
1568	33.4	1588	26.8	1608	34.5
1569	31.3	1589	26.3	1609	27.3
1570	38	1590	12.8	1610	21.5
1571	14.2	1591	30.8	1611	14.7
1572	21.4	1592	34.8	1612	32.6
1573	42.7	1593	33	1613	27.5
1574	32.3	1594	35	1614	37
1575	26.5	1595	31.9	1615	23
1576	34.1	1596	34.7	1616	7.3
1577	32.7	1597	42.6	1617	36.3
1578	24.7	1598	29.4	1618	36.3
				1619	23.3

These sixteenth-century wine harvest dates, for which the documentary basic is already very solid, clearly show the tendency toward late harvests (and cool spring-summers) which occurs between 1561 and 1600. See the simple table below (the Gregorian correction has been made throughout).

TABLE 2

Decade	No. of years in which the wine harvest date is:	
	Before Sept. 30 inc.	After Sept. 30
1491–1500 (9 years)	5	4
1501–1510 (10 years)	8	2
1511–1520	7	3
1521–1530	7	3
1531–1540	7	3
1541–1550	5	5
1551–1560	7	3
1561–1570	4	6
1570–1580	5	5
1581–1590	6	4
1591–1600	2	8
1601–1610	7	3

(b) *Wine harvest dates in the seventeenth and eighteenth centuries.* The figures which follow put forward, on the basis of several dozen scattered vineyards, two series of means for wine harvest dates in northern and southern France. Note that the previous table (Appendix 12, B, a, table 1) for the sixteenth century can easily be joined on to the column for the north below:

FRENCH WINE HARVEST DATES
(South, Center, and North)⁹

Difference in days, from the mean for 1599–1791

Year	S	N	Nat.	Year	S	N	Nat.
1599	−15	−19	−17	1639	− 8	− 8	− 8
1600	− 1	+ 8	+ 3.5	1640	− 2	− 3	− 2.5
1601	+ 2	+ 9	+ 5.5	1641	− 1	1	0
1602	− 8	−11	− 9.5	1642	6	4	5
1603	−14	−22	−18	1643	− 5	4	− 0.5
1604	−11	−11	−11	1644	− 7	−14	−10.5
1605	− 6	−14	−10	1645	−11	−14	−12.5
1606	0	6	3	1646	− 9	− 3	− 6
1607	−15	−11	−13	1647	− 6	3	− 4.5
1608	− 4	2	− 1	1648	1	2	1.5
1609	− 5	− 4	− 4.5	1649	4	8	6
1610	− 8	−19	−13.5	1650	− 3	3	0
1611	−11	−14	−12.5	1651	−12	− 8	−10
1612	− 6	1	− 2.5	1652	0	− 5	− 2.5
1613	− 5	− 6	− 5.5	1653	−11	− 9	−10
1614	− 2	7	2.5	1654	− 2	5	1.5
1615	− 9	−13	−11	1655	− 6	− 8	− 7
1616	−11	−25	−18	1656	− 6	− 3	− 4.5
1617	− 5	4	− 0.5	1657	− 5	−10	− 7.5
1618	2	6	4	1658	− 1	3	1
1619	−10	− 9	− 9.5	1659	−13	− 9	−11
1620	− 6	0	− 3	1660	−11	−15	−13
1621	7	16	11.5	1661	− 8	−11	− 9.5
1622	− 9	− 3	− 6	1662	− 6	0	− 3
1623	− 7	− 7	− 7	1663	3	0	1.5
1624	−11	−12	−11.5	1664	− 5	− 5	− 5
1625	− 5	3	− 1	1665	− 4	− 8	− 6
1626	− 5	− 3	− 4	1666	− 7	− 9	− 8
1627	10	12	11	1667	3	1	2
1628	8	15	11.5	1668	− 6	−10	− 8
1629	5	−14	− 4.5	1669	− 8	−11	− 9.5
1630	− 6	−11	− 8.5	1670	− 8	− 4	− 6
1631	− 9	− 9	− 9	1671	− 7	− 9	− 8
1632	2	6	4	1672	− 2	1	− 0.5
1633	3	0	1.5	1673	8	10	9
1634	− 4	− 4	− 4	1674	− 2	0	− 1
1635	− 8	− 4	− 6	1675	19	19	19
1636	−10	− 4	− 7	1676	− 3	−16	− 9.5
1637	−14	−22	−18	1677	− 2	0	− 1
1638	− 8	−21	−14.5	1678	2	− 8	− 3

⁹ On the methodological bases, both geographical and statistic, of this table, see Le Roy Ladurie, *Les Paysans de Languedoc*, complete edition, 1966, Vol. I, pp. 20 ff., and the beginning of Vol. II.

Year	S	N	Nat.	Year	S	N	Nat.
1679	0	− 4	− 2	1730	1	6	3.5
1680	− 4	−10	− 7	1731	4	0	2
1681	− 5	− 8	− 6.5	1732	6	1	3.5
1682	0	7	3.5	1733	1	1	1
1683	− 3	− 6	− 4.5	1734	−13	− 6	− 9.5
1684	− 4	−17	−10.5	1735	7	8	7.5
1685	− 6	− 5	− 5.5	1736	− 2	− 4	− 3
1686	− 6	−17	−11.5	1737	− 2	− 5	− 3.5
1687	3	0	1.5	1738	4	1	2.5
1688	8	− 1	3.5	1739	− 1	0	− 0.5
1689	1	3	2	1740	13	17	15
1690	5	1	3	1741	− 9	1	− 4
1691	− 3	− 5	− 4	1742	3	9	6
1692	3	15	9	1743	10	6	8
1693	9	− 1	4	1744	6	5	5.5
1694	2	− 4	− 1	1745	− 1	6	2.5
1695	9	9	9	1746	8	0	4
1696	0	0	0	1747	3	4	3.5
1697	0	0	0	1748	3	2	2.5
1698	13	17	15	1749	7	2	4.5
1699	1	− 1	0	1750	− 1	1	0
1700	6	10	8	1751	4	11	7.5
1701	11	3	7	1752	1	8	4.5
1702	4	3	3.5	1753	− 2	− 4	− 3
1703	0	6	3	1754	3	6	4.5
1704	− 6	− 8	− 7	1755	0	− 4	− 2
1705	3	6	4.5	1756	6	10	8
1706	−11	−11	−11	1757	2	− 1	0.5
1707	3	1	2	1758	7	0	3.5
1708	1	− 1	0	1759	− 1	− 5	− 3
1709	2	1	1.5	1760	− 1	− 8	− 4.5
1710	− 5	0	− 2.5	1761	1	− 8	− 3.5
1711	0	5	2.5	1762	− 2	−11	− 6.5
1712	0	1	0.5	1763	1	7	4
1713	9	12	10.5	1764	− 1	− 5	− 3
1714	7	6	6.5	1765	3	1	2
1715	− 3	4	0.5	1766	2	2	2
1716	5	11	8	1767	2	8	5
1717	2	2	2	1768	2	3	2.5
1718	−11	−17	−14	1769	6	3	4.5
1719	− 9	−12	−10.5	1770	11	14	12.5
1720	6	3	4.5	1771	5	2	3.5
1721	9	7	8	1772	− 2	0	− 1
1722	0	0	0	1773	9	9	9
1723	− 1	− 6	− 3.5	1774	3	0	1.5
1724	−11	− 7	− 9	1775	2	2	2
1725	9	17	13	1776	2	5	3.5
1726	− 3	−14	− 8.5	1777	4	8	6
1727	− 9	− 7	− 8	1778	− 4	− 3	− 3.5
1728	−11	− 6	− 8.5	1779	1	− 5	− 2
1729	3	2	2.5	1780	0	− 6	− 3

Year	S	N	Nat.	Year	S	N	Nat.
1781	− 3	−15	− 9	1787	9	7	8
1782	4	4	4	1788	− 5	−12	− 8.5
1783	1	− 8	− 3.5	1789	6	7	6.5
1784	− 4	− 9	− 6.5	1790	1	0	0.5
1785	4	3	3.5	1791	− 4	− 6	− 5
1786	1	2	1.5				

(c) *Harvest dates in the nineteenth century.* (These can follow on from the column "National mean," or if necessary "North," in the preceding table.)

—Column A: year.

—Column B: mean date of wine harvest in days counted from September 1. This date is calculated from *all* the French and Swiss wine-growing stations for which the harvest date is known. The majority of these belong to the Center and the East (Burgundy and Switzerland); but there are also some in the Loire, the Bordelais, and especially the Midi. The sources are Angot, 1883; and also above in this appendix, the table in Para. A covering 1782–1869, for the Midi.

—Column C: (every ten years): the number of stations covered for the year in question. It will be seen that this series rests on a very broad statistical basis because of the abundance of documentary sources in the nineteenth century.

A	B	C	A	B	C
1782	41		1808	34	
1783	32		1809	46	
1784	29	32	1810	37	
1785	40		1811	20	
1786	39		1812	43	
1787	45		1813	45	
1788	27		1814	42	38
1789	44		1815	35	
1790	37		1816	59	
1791	31		1817	44	
1792	38		1818	28	
1793	34		1819	32	
1794	26	20	1820	39	
1795	36		1821	49	
1796	41		1822	7.0	
1797	37		1823	45	
1798	28		1824	44	63
1799	47		1825	25	
1800	29		1826	34	
1801	31		1827	31	
1802	28		1828	32	
1803	33		1829	43	
1804	33	33	1830	33	
1805	45		1831	33	
1806	32		1832	38	
1807	26		1833	33	

A	B	C	A	B	C
1834	24	75	1857	28	
1835	40		1858	27	
1836	39		1859	27	
1837	41		1860	43	
1838	42		1861	31	
1839	35		1862	29	
1840	32		1863	31	
1841	33		1864	33	71
1842	27		1865	15	
1843	46		1866	36	
1844	30	74	1867	33	
1845	44		1868	19	
1846	22		1869	28	
1847	36		1870	22	
1848	34		1871	35	
1849	33		1872	34	
1850	41		1873	34	
1851	44		1874	28	61
1852	36		1875	31	
1853	45		1876	37	
1854	36	69	1877	34	
1855	41		1878	36	
1856	39		1879	47	

(d) *German data on quality of wine. Mean annual index of quality of wine in Rhenish vineyards, after Müller, 1953, pp. 188 ff.* (1453–1622).

Following Von Rudloff, 1967, I have used a simple numerical system which takes into account the various terms used by the chroniclers in the different regions to describe the wine of any given year. These contemporary appraisals can be found in Müller, 1953, and my numerical system therefore merely aims at showing first the annual and then the decennary trend.

The adjectives employed by the chroniclers, and after them by Müller ("extra-gut," "mittelmassig," "sauer," etc.) have been reduced to three categories: good, average, and bad. I show *good* quality as a minus quantity, *bad* as a plus, e.g.:

$$\text{Good} = -6$$
$$\text{Bad} = 6$$
$$\text{Av.} = 0$$

Figures in between result from a calculation of means, weighted in accordance with regional variations (in some years the quality of the wine is not the same in all the Rhenish vineyards), using a very simple arithmetic mean. So figures between 0 and −6 and 0 and 6 show that the quality of the wine for the year, according to the various assessments of the chroniclers for that region, has been *unequal* and/or *uncertain*. The following series results:

TABLE 1

Year						
1453	Bad					6
1454	Bad					6
1455	Bad					6
1456	Bad					6
1457			Av.			0
1458			Av.			0
1459	Bad					6
1460	Bad					6
1461				Good		−6
1462			Av.			0
1463			Av.	Good		−3
1464				Good		−6
1465	Bad			Good		0
1466	Bad					6
1467				Good		−6
1468	Bad		Av.			3
1469	Bad					6
1470			Av.	Good		−3
1471			Hail, no wine			(0)
1472	Bad		Av.	Good		0
1473				Good		−6
1474				Good		−6
1475				Good		−6
1476	Bad			Good		2
1477			Av.			0
1478				Good		−6
1479				Good		−6
1480				Good		−6
1481	Bad					6
1482				Good		−6
1483				Good	Good	−6
1484				Good		−6
1485	Bad	Bad	Av.			4
1486				Good		−6
1487			Av.			0
1488	Bad					6
1489	Bad					6
1490	Bad					6
1491	Bad					6
1492	Bad					6
1493				Good		−6
1494				Good		−6
1495				Good		−6
1496				Good		−6
1497	Bad			Good		0
1498	Bad					6
1499				Good		−6
1500				Good		−6

Year					
1501	Bad				6
1502	Bad	Av.			3
1503	Bad		Good		0
1504			Good	Good	−6
1505			Good		−6
1506			Good	Good	−6
1507	Bad		Good		0
1508		Av.	Good		−3
1509			Good		−6
1510			Good		−6
1511	Bad				6
1512	Bad		Good		0
1513	Bad		Good		0
1514			Good		−6
1515	Bad				6
1516			Good	Good	−6
1517	Bad				6
1518			Good		−6
1519			Good		−6
1520	Bad				6
1521			Good	Good	−6
1522			Good		−6
1523			Good		−6
1524	Bad				6
1525			Good		−6
1526	Bad				6
1527	Bad				6
1528			Good		−6
1529	Bad				6
1530	Bad		Good		0
1531			Good	Good	−6
1532			Good		−6
1533	Bad				6
1534			Good		−6
1535			Good		−6
1536			Good		−6
1537			Good		−6
1538	Bad				6
1539			Good		−6
1540			Good		−6
1541			Good		−6
1542	Bad				6
1543			Good		−6
1544		Av.			0
1545			Good		−6
1546			Good		−6
1547			Good		−6
1548	Bad	Av.			3
1549		Av.			0
1550			Good		−6
1551			Good		−6

Year								
1552					Good			−6
1553					Good	Good		−6
1554	Bad				Good			0
1555	Bad							6
1556				Av.	Good	Good		−4
1557	Bad	Bad			Good			2
1558					Good			−6
1559	Bad							6
1560				Av.				0
1561	Bad							6
1562					Good			−6
1563	Bad							6
1564					Good			−6
1565	Bad				Good			0
1566	Bad							6
1567					Good			−6
1568	Bad							6
1569	Bad							6
1570	Bad							6
1571	Bad							6
1572					Good			−6
1573	Bad							6
1574	Bad							6
1575					Good	Good	Good	−6
1576					Good			−6
1577	Bad							6
1578	Bad				Good			0
1579	Bad							6
1580	Bad				Good			0
1581	Bad							6
1582	Bad				Good			0
1583	Bad	Bad			Good			2
1584					Good			−6
1585	Bad	Bad						6
1586	Bad				Good			0
1587	Bad							6
1588	Bad	Bad		Av.				4
1589	Bad	Bad			Good			2
1590	Bad				Good five times			−4
1591	Bad	Bad						6
1592				Av.	Good	Good		−4
1593	Bad			Av.	Good			0
1594	Bad	Bad						6
1595	Bad	Bad			Good			2
1596					Good	Good		−6
1597	Bad	Bad						6
1598	Bad	Bad						6
1599					Good			−6
1600	Bad	Bad						6
1601	Bad	Bad	Bad					6
1602	Bad							6

Year						Good	Good	Good	
1603						Good	Good		−6
1604	Bad	Bad				Good			2
1605						Good	Good	Good	−6
1606	Bad	Bad	Bad						6
1607	Bad					Good	Good		−2
1608	Bad					Good	Good		−2
1609	Bad	Bad		Av.					4
1610						Good	Good		−6
1611	Bad	Bad				Good	Good	Good	−1.2
1612	Bad					Good	Good	Good	−4
1613	Bad	Bad				Good	Good		0
1614	Bad	Bad				Good			2
1615						Good	Good	Good	−6
1616						Good	Good	Good	−6
1617	Bad	Bad	Bad						6
1618				Av.		Good			−3
1619	Bad					Good	Good		−2
1620				Av.		Good	Good		−4
1621	Bad	Bad				Good			2
1622				Av.	Av.	Good	Good		−3

TABLE 2

Decennary indices of bad quality of wine (from Table 1)

1453–1462	30
1463–1472	−3
1473–1482	−34
1483–1492	16
1493–1502	−21
1503–1512	−27
1513–1522	−18
1523–1532	−6
1533–1542	−24
1543–1552	−39
1553–1562	−2
1563–1572	18
1573–1582	18
1583–1592	12
1593–1602	26
1603–1612	−15.2
1613–1622	−14

N.B. Note that plus figures indicate bad quality and minus figures good quality.

Of course the basic data in these tables, taken from Müller, is always rough and sometimes not very reliable. They are of interest only in so far as they demonstrate a clear trend, or give local and independent testimony on the trends toward cooling which are indicated simultaneously, and much more precisely, in the wine harvest dates and glacier series for the periods 1563–1602.

APPENDIX 12B

1787–1788: Harvest, Wine Harvests, and Climate

As regards the year 1788, it is worth citing another text which has recently come to my knowledge, on the meteorological causes of the bad harvest of 1788 which led to the food shortages and riots of 1789: "*The continual rains of October and November 1787* partly hindered the sowing of the wheat, so that the fields have not been planted. The storm of July 13, 1788, destroyed part of the crops, and the general harvest of 1788 is definitely mediocre."[10] So the factors causing the bad harvest of 1788, the historical consequences of which were literally incalculable, now appear in the following order: excessive rain in October and November 1787,[11] scorching by the sun at the beginning of summer 1788, storm and hail on July 13, 1788. On the complex connections between bad harvest and food riots, see G. Rudé, 1956; Cobb, 1945; E. P. Thompson, 1966; Rose, 1959.

APPENDIX 13

A Few Medieval Texts on the Glaciers (Cf. Chapter VI)

A curate of the nineteenth century notes in a record book in the archives of the curacy of Greissan, in the Aosta region, that in September 1284 the Ruitor lake burst through the glacial barrier ("les glaces"). The lake was swollen as the result of rain. The emptying of the lake is supposed to have swept the church of Greissan away down the valley, and to have formed the hillock or "moraine" known as that "of Gargantua." This reference is supposed to have been taken from a book by Christillin, a lawyer, about 1840. In spite of strenuous efforts I have only been able to find a horribly mutilated copy of this very rare book, and this copy did not contain the passage in question.[12]

The 1300 text on the relative advance of the Allalin is cited by Lütschg (1926).

Stolz (1928) cites a 1315 text which provides for tax and rent rebates for a whole group of farms at Otzthal where the soil was destroyed "*ex alluvionibus et inundationibus.*" But there is nothing to prove that the cause was an early thrust of the Vernagt glacier.

Du Tillier, a historian of the Val d'Aosta and native of Morgex, writes (Ms. of 1742, ed. 1880 and re-ed. 1953, p. 127): "The tower and stronghold of the nobles of Rubilly and Rovarey which must have withstood a deluge of water to judge from the remains still to be seen on the strand,

[10] Remarks presented by Messieurs Leleu to the chief minister on August 14, 1788, and cited by G. Bord, 1887, p. 24.
[11] In general, the summer and autumn of 1781 seem to have been cold and probably wet (late wine harvest date); thus the spoiled harvest of 1788 results from the paradoxical and disastrous combination of a year that was too cool (1787) and a year that was too hot (1788). Was this a very exaggerated but in a sense classic example of "biennial alternation"?
[12] It is not usual to include "personal announcements" in a book, but I should like to make an exception and ask any reader who happens to come across a *complete* edition of Christillin, 1840, to let me know at 88 rue d'Alleray, Paris XVe.

sive glair, in the neighborhood, were overthrown by the overflowing of this same lake of Ruitor, in exactly which year is not known, but it must have been before 1430, for apparently (the tower) was not furnished with a garrison at the general audiences in that year, as were all the other strongholds in the Valdigne." But we know from Duc, 1901, III, f. 403, that Rovarey still existed in 1340; also Du Tillier (ibid.) mentions Rubilly as still existing in 1371. The damage caused by Ruitor, therefore, *assuming that it actually happened*, must have occurred between 1371 and 1430.

After 1430, there is a long interval until the text of 1513, in which year a property boundary is supposed to have mentioned, in the Genderstal, under the Schwarzhorn in the Tyrol, a *Ferner* (glacier) which according to other texts is met with again in 1607, and in 1844. At present this presumed glacier seems to have been replaced by a mere névé or mass of snow. (Texts cited by Stolz, 1928.)

In 1531, in the Passeiertal (Tyrol), a vineyard is converted into a meadow "because the wine harvest will not ripen on account of the air from the glacier" (ibid.).

It is noticeable that there are no texts, not even doubtful ones, signifying glacial advance between 1430 and 1513. Does this mean that the most marked glacial minimum of the early Middle Ages occurred in this three quarters of a century (i.e., between the thrusts Mayr, 1964, lists as Xd (thirteenth century) and Xf (seventeenth century)?

As for the text asserting a marked glacial withdrawal in 1540, it is cited and criticized by Richter, 1891, pp. 5–20, with reference to the Grindelwald glaciers commented on by Strasser (Der Gletschermann). But this is a later, secondhand text taken from Altmann, 1751, p. 23, and from Gruner, 1760 (or Gruner, 1770, p. 329): its only traceable source,[13] to my knowledge, is a little-known passage from Stumpf's *Description de la Suisse* (Description of Switzerland), Vol. II (year 1548), p. 284:

"Bey etlichen heissen Summers Zeyten als im Jahr Chrysti 1540 gewesen gadt auch etwan der alt Schnee ab, doch niemermeer also gar dann das die obristen Spitsen statigs Schnee behaltend."

This text by Stumpf mentions only névés, not glaciers.

Nevertheless it is possible that Gruner got his information not from Stumpf but from early archives of the parish of Grindelwald. A short glacial withdrawal about 1540, similar to that suggested by Strasser and Gruner, is not at all inconceivable. (See the hot summer decades in the 1520s and 1530s, Annex 12 C.)

APPENDIX 14
Mean Temperature Differences at Annecy in Degrees Centigrade 1843–1913/1773–1842)

MEAN TEMPS. IN DEGS. CENTIGRADE

	Winter	Spring	Summer	Autumn	Whole year
1773–1842	0.640	9.035	18.290	9.939	9.461
1843–1913	0.619	9.766	19.114	10.702	10.050
Difference	−0.021	0.731	0.824	0.763	0.589

SOURCE: Mougin, V, 1925, p. 104, which gives decennary means.

[13] Coolidge, 1911 (see also *Jahrbuch der Schweiz*, Alp. Club, Vol. 27, p. 266, note 1), also thinks the derivation is from Stumpf, 1548.

APPENDIX 15

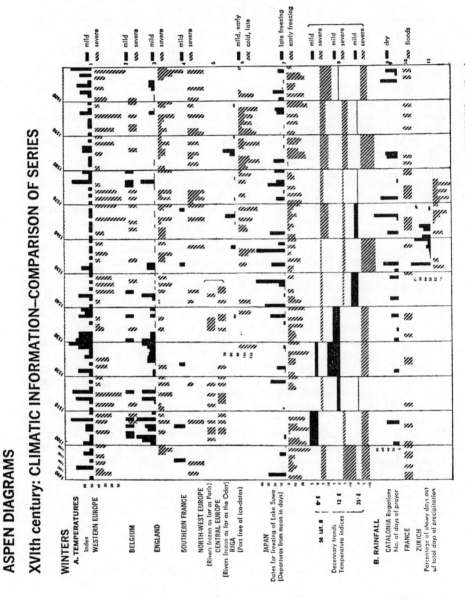

ASPEN DIAGRAMS, continued

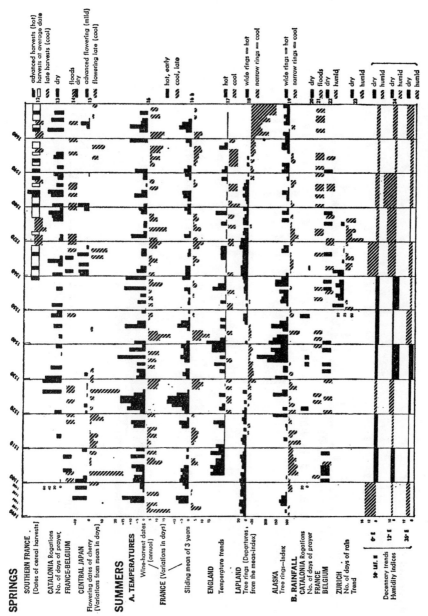

SPRINGS

SOUTHERN FRANCE.
[Dates of cereal harvests]

advanced harvests (hot)
harvests at average date
late harvests (cool)

CATALONIA Rogations
No. of days of prayer

dry

FRANCE-BELGIUM

floods
dry

CENTRAL JAPAN
Flowering dates of cherry
(Variations from mean in days)

advanced flowering (mild)
flowering late (cool)

SUMMERS

A. TEMPERATURES

Wine-harvest dates
(annual)

hot, early
cool, late

FRANCE (Variations in days)

Sliding mean of 3 years

ENGLAND
Temperature trends

hot
cool

LAPLAND
Tree rings (Departures,
from the mean-index)

wide rings = hot
narrow rings = cool

ALASKA
Tree rings—Index

wide rings = hot
narrow rings = cool

B. RAINFALL

CATALONIA Rogations
No. of days of prayer
FRANCE
BELGIUM

dry
floods
dry
humid

ZURICH
No. of days of rain
Trend

dry
humid

50 LAT. N

0-E

Decennary trends
Humidity Indices

12-E

39-E

dry
humid
dry
humid
dry
humid

ASPEN DIAGRAMS, continued

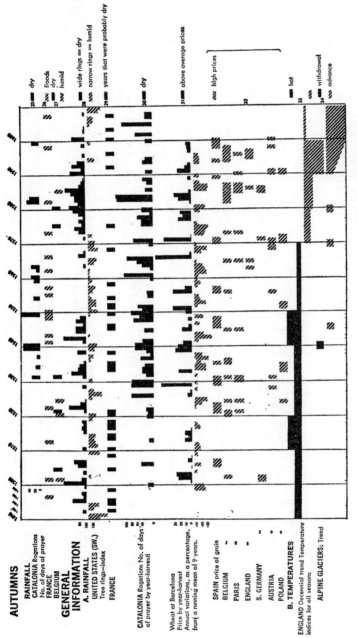

ASPEN DIAGRAMS, continued

XIth century: CLIMATIC INFORMATION–COMPARISON OF SERIES

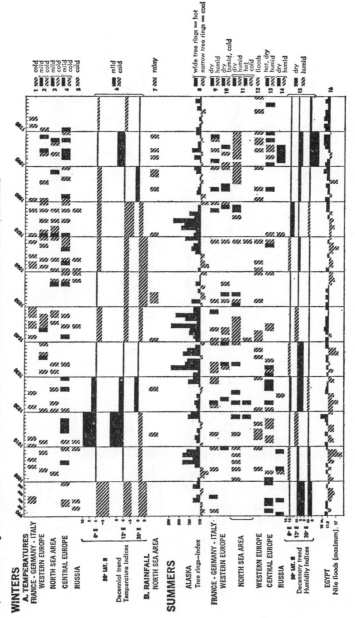

ASPEN DIAGRAMS, continued

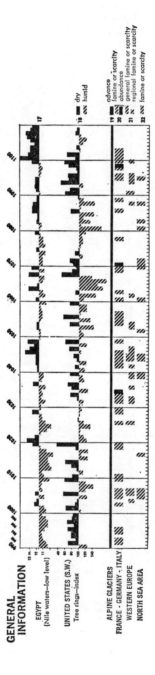

Explanatory Notes to Aspen Diagrams[14]
(with, in each case, the name of the author and his sources, etc.)

Eleventh Century

Diagram XI–1: Herlihy (D.). Source: *Monumenta Germanicae historica;* cf. Herlihy (D.), 1962.

XI–2: Helleiner (K.). Sources: Weikinn (C.), 1958, I; Curtschmann (F.), 1900; Vanderlinden (E.), 1924; Britton (C. E.), 1937.

XI–3: Manley (G.). Sources: Baker (T. H.), 1883; Britton, 1937; Vanderlinden, 1924; Short (T.), 1749; Short (T.), 1750.

XI–4: Lamb (H.). Source: see XI–6.

XI–5: Lamb. Source: Buchinsky (I. E.), 1957.

XI–6: Lamb (H.). Sources: Hennig (R.), 1904; Easton, 1928; Betin (V. V.), 1957; Vanderlinden, 1924; Buchinsky, 1957.

XI–7: Manley (G.). Source: see XI–3.

XI–8: Giddings (J. L.). Source and principles of constructing indices: Giddings (J. L.), 1948, pp. 26–32; 1952, pp. 105–10, and 1941.

XI–9: Herlihy (D.). Source: see XI–1.

XI–10: Helleiner (K.). Source: see XI–2.

XI–11: Manley (G.). Source: see XI–3.

XI–12: Lamb (H.). Source: Weikinn, 1958, I.

XI–13: Lamb (H.). Source: see XI–15.

XI–14: Lamb (H.). Source: Buchinsky, 1957.

XI–15: Lamb (H.). Sources: Hennig, Buchinsky, Vanderlinden, Betin, Britton, op. cit., and Müller (K.), 1947.

XI–16: Lamb (H.). Source: Toussoun (O.), 1925, pp. 366–410. For A.D. 822 to 1321 (ninth to thirteenth centuries, period covering the eleventh), the low-water mean is 11.64 meters, and the flood mean 17.59 meters.

XI–17: Lamb. Source: see XI–16.

XI–18: Unpublished series kindly communicated by H. C. Fritts. Index 100: mean annual growth of the trees in question during the last millennium.

XI–19: Le Roy Ladurie. Sources: for the Aletsch glacier, Œchsger and Rothlisberger, 1961, p. 191, and *Radiocarbon Supplement* of the *American Journal of Science*, Vol. I, p. 138. For the Grindelwald glacier, the fossil forest is described in Gruner, 1770, and dated in the *Radiocarbon Supplement*, ibid., Vol. 3, p. 44.

XI–20: Herlihy. Source: see XI–1.

XI–21: Helleiner. Sources: Curtschmann, 1900; Vanderlinden, 1924; Britton, 1937.

XI–22: Manley. Source: see XI–3.

Sixteenth Century

XVI–1: Le Roy Ladurie. Source: Easton, op. cit.

XVI–2: Helleiner. Source: Vanderlinden, 1924.

XVI–3: Manley. Sources: Short, Holinshed, after Baker, 1883; Vanderlinden, 1924; Mossman (R.), 1902; and various archives.

14 I should like to record my thanks to W. B. Watson, 1962 b, to whom the bibliography in these notes is largely indebted.

XVI–4: Le Roy Ladurie. Sources: Archives of Hérault and Fonds Chobaut in the Calvet Museum, Avignon, etc. (See Appendix 12.)

XVI–5: Helleiner. Source: Weikinn, 1958, I.

XVI–6: Lamb. Source: Betin, 1957. Dates counted in number of days from January 1 of year in question. The ordinate 87 (March 18) is the mean date for the opening of the port during the period 1900–1949.

XVI–7: Arakawa. Sources: his articles of 1957 and 1954. Point O corresponds to the mean date for the complete freezing of the lake in the sixteenth century (January 8 in the Gregorian calandar). The demarcation line between the white and black rectangles corresponds to the mean date for the freezing during the period 1720–1953 (i.e., January 11).

XVI–8: Lamb. Sources: see XI–6, see also Mossman, 1902, and Tycho Brahe, 1876 ed.

XVI–9: Giralt. Sources: *Arxiu Historic Municipal* of Barcelona, unpublished series "Ordinacions" and "Dietari de l'Antic conseil Barceloni."

XVI–10: Le Roy Ladurie. Source: Champion (M.), 1858–1864.

XVI–11: Lamb. Source: *Journal* of Wolfgang Haller of Zürich (1550–1576), 1875 ed.

XVI–12: Le Roy Ladurie. Sources: harvest contracts in the archives of the chapters of Narbonne, Béziers, and Agde (A. D. Aude and Héault, series G, and notaries of Agde).

XVI–13: Giralt. Sources: see XVI–9.

XVI–14: Le Roy Ladurie and Helleiner. Sources: Champion, 1858–1864; Vanderlinden, 1924.

XVI–15: Arakawa, 1955, pp. 147–50 and 1956, pp. 559–600. Point O corresponds to the mean date of flowering in the sixteenth century, in the Gregorian calendar (April 17). The demarcation line between the black and white rectangles corresponds to the mean date of flowering for the complete "cherry tree" series, which goes from the ninth to the nineteenth century. This intersecular mean date is April 15. (See Arakawa, 1957, p. 4.)

XVI–16: Le Roy Ladurie. Source: see Appendix 12.

XVI–16 b: Le Roy Ladurie (see XVI–16). If b is the departure from the mean in days for any given year, a the departure from the previous year, and c the departure from the year following, the running mean m for year b is given by the formula:

$$m = \frac{a + 2b + c}{4 \text{ shillings}}$$

XVI–17: Manley. Source: see XVI–3.

XVI–18: Siren (G.) (See complete curves and commentary in Siren, 1961.) Point O represents the mean annual growth of the trees in question for the period 1181–1960.

XVI–19: Giddings (see XI–8).

XVI–20: Giralt (see XVI–9).

XVI–21: Le Roy Ladurie (see XVI–10).

XVI–22: Helleiner. Source: Vanderlinden, 1924.

XVI–23: Lamb (see XVI–11).

XVI–24: Lamb (see XVI–8).

XVI–25: Giralt (see XVI–9).

XVI–26: Le Roy Ladurie (see XVI–10.)

XVI–27: Helleiner (see XVI–22).

XVI–28: Fritts (see XI–18).

XVI–29: Le Roy Ladurie. Source: Champion, 1858–1864 (this graph shows years with no flood recorded).

XVI–30: Giralt (see XVI–9).

XVI–31: Giralt. Source: Giralt, 1962. See also Giralt, 1958, pp. 38–61. On the displacement leftwards of one year in the chronological scale, see above, XVI–16 b.

XVI–32: Helleiner. Sources: For Austria, A. F. Pribram, 1938; for Germany, J. M. Elsass; for England, W. H. Beveridge, 1939 and T. Rogers; for Belgium, Vanderlinden, 1924; for France, Baulant (M.) and Meuvret (J.); for Poland, after the work of Rutkowsky; for Spain, after Hamilton, 1934.

XVI–33: Manley (see XVI–3).

XVI–34: (glaciers): Le Roy Ladurie. Sources: on the glacial withdrawal of 1540, see above, Appendix 13, ad fin.; for the rest, see above, Chapter IV.

The Aspen diagrams were drawn up by Hidetoshi Arakawa, Harold C. Fritts, James L. Giddings, Emili Giralt Raventos, Karl F. Helleiner, David Herlihy, H. H. Lamb, Emmanuel Le Roy Ladurie, Gordon Manley, Gustav Siren.

The diagrams were originally assembled by the Historical Committee of the Aspen Conference under the chairmanship of W. B. Watson.

The final layout is the work of Jacques Bertin and Janine Récurat.

APPENDIX 16

Dendrochronology

There are two categories of works on American dendroclimatology:
—Early publications: A. E. Douglass, 1919, 1928, 1936; Antevs, publication nos. 352 and 469; W. S. Glock, publication no. 486.
—Publications of the last twenty years, which put an entirely new complexion on the subject and on which is based the account given in this book: the whole series of *Tree-ring Bulletins* published by the University of Arizona, and Schulman, 1951 and 1956. See also Giddings, 1947; Zeuner, 1949, Chapter 1; A. Laming-Emperaire, 1952; A. Ducrocq, 1955. A last and fundamental reference is the whole series of articles by Fritts. For western Europe the basic data are in Huber, 1964 (especially the last graph), and Hollstein, 1965. See also Huber, 1969, for a very detailed diagram on the last millennium in Hesse. For France, see De Martin, an article due to appear in the *Annales*, E.S.C., 1970.

I reproduce below, because of its usefulness to researchers and to historians of Germany, Switzerland, the Benelux countries, southwestern England, and northeastern France, the annual master chronology of oak tree rings in those parts of Germany west of the Rhine, as worked out by Hollstein, 1965, for the period A.D. 820–1964.

Chronology for the oak in Germany west of the Rhine

The figures show the mean thickness of the tree rings in 1/100ths of a millimeter. The figure on the right gives the number of samples or beams consulted for each decade.

Year	0	1	2	3	4	5	6	7	8	9	n
820			152	161	115	143	186	190	175	147	1
830	144	204	185	165	138	121	174	231	148	191	4
840	181	201	216	162	199	178	187	173	172	163	4
850	168	128	130	177	153	132	148	156	137	165	4
860	133	145	138	128	135	135	156	149	127	160	4
870	141	159	143	151	80	136	134	144	147	152	7
880	128	144	138	112	150	156	163	139	149	145	7
890	119	135	154	154	159	139	139	158	110	108	7
900	141	141	117	141	118	143	132	109	117	149	9
910	95	130	126	106	94	107	78	89	79	109	10
920	82	117	88	92	91	89	82	117	81	109	10
930	83	120	118	104	117	91	115	102	131	129	12
940	131	123	153	95	100	77	100	113	129	128	13
950	144	124	134	128	107	123	111	116	112	138	13
960	115	130	107	118	99	114	114	129	124	148	13
970	119	117	124	124	91	132	118	120	104	129	13
980	100	98	119	108	101	144	119	111	108	129	13
990	86	113	102	108	97	89	105	98	95	132	14
1000	94	124	116	133	109	116	110	109	115	121	14
1010	110	115	139	138	95	122	111	114	85	114	13
1020	99	139	116	131	127	113	131	128	106	104	11
1030	148	118	119	127	116	118	105	144	114	103	11
1040	123	110	114	92	78	115	113	118	100	119	12
1050	109	110	133	98	122	123	129	127	114	93	10
1060	129	110	113	145	100	130	145	138	162	132	8
1070	107	111	118	166	122	131	121	134	108	139	7
1080	139	99	171	132	116	168	138	166	158	164	6
1090	109	163	189	180	180	164	160	187	168	170	4
1100	218	168	119	145	149	132	170	150	182	174	4
1110	174	118	193	148	196	184	256	288	216	248	4
1120	198	165	264	262	186	169	189	159	184	153	6
1130	211	188	187	142	180	178	174	135	178	180	10
1140	232	195	164	209	182	206	168	167	149	186	11
1150	147	158	121	186	169	155	184	159	175	155	11
1160	182	175	131	101	66	60	80	60	131	130	11
1170	115	149	142	154	156	148	119	81	126	124	11
1180	141	127	142	114	86	96	109	136	88	129	11
1190	130	138	102	143	128	147	133	111	140	121	11

Year	0	1	2	3	4	5	6	7	8	9	n
1200	110	160	153	160	144	120	138	89	129	128	13
1210	148	131	92	118	108	116	115	86	108	137	13
1220	115	134	106	125	124	121	118	106	103	142	15
1230	125	108	95	75	83	100	84	113	109	108	15
1240	133	120	118	132	87	135	110	138	122	140	14
1250	135	133	86	114	92	122	121	85	111	99	14
1260	102	91	73	81	119	117	138	101	128	114	13
1270	77	118	93	110	108	134	86	116	72	95	13
1280	116	98	108	141	109	116	138	75	92	81	12
1290	101	95	115	101	87	111	130	98	133	106	12
1300	111	104	118	90	91	86	88	99	79	119	12
1310	96	93	125	109	121	109	127	126	113	113	14
1320	85	117	113	94	99	89	72	107	106	108	14
1330	90	94	86	83	93	103	79	98	97	88	14
1340	74	93	99	102	77	102	101	79	90	115	15
1350	84	100	86	75	107	85	102	84	126	132	14
1360	87	90	133	131	124	110	158	158	140	175	19
1370	157	130	131	132	150	133	143	130	143	126	22
1380	134	122	107	135	132	103	142	116	137	127	29
1390	116	126	121	84	89	86	72	82	129	128	34
1400	117	135	156	161	131	141	134	122	130	144	34
1410	166	152	145	140	139	119	144	94	111	96	32
1420	80	119	72	85	104	89	128	131	130	144	36
1430	143	123	120	112	88	132	152	143	126	135	30
1440	138	120	99	120	113	119	107	119	90	137	25
1450	114	137	99	108	136	132	145	135	125	99	23
1460	129	107	91	110	78	122	105	126	154	120	28
1470	145	131	158	129	165	163	120	131	98	147	28
1480	150	149	137	138	188	139	144	179	150	135	28
1490	140	110	120	119	128	130	136	145	95	150	27
1500	117	143	138	89	73	120	84	119	109	131	27
1510	111	108	124	97	106	133	110	69	93	95	24
1520	85	112	100	103	110	98	128	111	152	137	23
1530	99	155	84	132	85	132	91	116	77	118	23
1540	100	111	124	116	136	128	108	93	109	100	23
1550	97	108	83	99	85	125	81	86	91	85	25
1560	131	124	144	120	122	125	96	87	120	101	28
1570	125	107	95	99	76	124	93	113	97	120	28
1580	118	129	99	86	136	124	136	117	94	131	26
1590	92	107	125	130	105	104	90	116	115	96	25
1600	113	88	93	69	97	96	117	148	118	118	24

Year	0	1	2	3	4	5	6	7	8	9	n
1610	98	98	120	146	125	100	101	130	115	98	22
1620	83	73	91	124	103	130	99	139	93	101	19
1630	97	101	136	118	91	86	59	85	99	76	18
1640	93	90	100	103	75	84	82	82	119	113	20
1650	109	78	63	78	100	115	100	101	106	80	21
1660	98	122	97	108	124	99	105	76	76	95	22
1670	81	102	91	129	93	100	85	105	104	117	23
1680	112	81	118	121	93	94	118	122	147	132	22
1690	137	154	132	158	137	114	92	80	115	116	20
1700	128	133	123	116	150	108	145	132	130	82	21
1710	80	86	132	141	113	137	117	135	110	93	24
1720	121	110	137	103	122	155	133	188	120	137	26
1730	147	105	133	133	140	162	130	151	162	136	24
1740	147	98	116	89	65	87	137	130	109	136	25
1750	125	133	161	108	168	123	152	119	87	96	25
1760	85	139	104	149	130	107	132	119	153	145	21
1770	154	118	101	167	166	162	137	139	114	115	18
1780	103	90	96	124	100	107	79	100	172	170	15
1790	130	128	148	107	137	135	148	147	135	108	14
1800	110	128	88	108	118	122	97	107	86	134	14
1810	110	122	95	109	104	104	129	124	107	91	14
1820	111	126	90	136	130	93	112	111	118	128	14
1830	109	117	85	95	124	99	100	113	106	108	14
1840	112	100	115	94	84	93	77	105	104	120	13
1850	121	121	105	109	70	111	95	93	72	102	13
1860	108	125	132	118	119	102	119	115	98	105	13
1870	64	115	101	102	96	127	89	121	119	118	13
1880	82	96	104	95	102	95	93	89	107	106	13
1890	106	104	89	71	130	137	123	126	154	125	14
1900	128	133	129	136	134	100	68	76	73	70	14
1910	116	111	127	125	133	98	135	124	128	100	13
1920	113	88	123	103	130	105	108	115	105	106	13
1930	107	131	130	105	97	99	112	97	114	87	12
1940	89	84	63	90	88	80	108	80	97	120	12
1950	108	117	104	124	119	123	84	67	98	97	10
1960	103	105	88	100	85	·	·	·	·	·	

APPENDIX 17

Recent Evidence on the Climate of the Thirteenth Century

The first edition of this book was already finished when Gabrielle Démians d'Archimbaud, the eminent archaeologist and expert on the lost villages of Provence, sent me the following note:

"*Rougiers* (Var).—In the course of the excavations of 1965 and 1966 we were able to study, on the site of the feudal village, the two caves C and F, which provide new evidence on the period of semiabandonment of the *castrum* during the second half of the thirteenth century. The two things that emerge are on the one hand a 'demographic fall-off' on the higher site, leading to a new implantation in the valley, and on the other hand a temporary increase in rainfall.

"The two caves are quite dissimilar. The first is like a big, roughly rectangular room (approximately 8 by 6 meters), the north part occupied by a big stalagmite. The second is a narrow corridor penetrating the rocky limestone cliff (1–2.5 meters wide by 5 meters long), with a fault running right along it. In both cases later deposits, formed on the surface since the final abandoning of the site at the beginning of the fifteenth century, are very minor: a thin layer of sand and pieces of stone broken off from the walls and ceiling does not, in cave F, even cover the whole area.

"This contrasts with the presence lower down of a layer of apparently natural accumulations, between the levels corresponding to the beginning of the thirteenth century (when the site was first used, and those corresponding to the very end of the thirteenth century or the beginning of the fourteenth, after which human occupation continued unbroken until the fifteenth century. In cave C this layer, which contains many pieces of stone [often split by frost(?)] and gravel, mingled with loose earth, is especially large in the northern part of the cave. In cave F the barren level is much clearer; it is about 30 centimeters thick and continuous throughout the cave. It consists of yellow sand, stones, and gravel resulting from the decomposition of the rock, and lies between a layer of black earth containing pottery remains of the beginning of the thirteenth century and occupation levels belonging to a much later date (the very end of the thirteenth and beginning of the fourteenth century).

"It seems therefore that after the first occupation of the site at the beginning of the thirteenth century, there was a period of semiabandonment, during which the caves were left unused probably because of serious infiltration of rainwater(?)[15]; abundant rain is the only explanation for the big natural deposits observed at lower levels.

"The thickness of these layers is surprising; there has been no comparable accumulation at surface level since the site was abandoned in the fifteenth century.

"So we appear to have here a peculiar climatic phenomenon which

[15] Summer rain: the higher temperature of the water would facilitate the dissolving of the karst or chalk.

must have occurred toward the middle of the thirteenth century: heavy summer rains and low evaporation, which one is tempted to relate to the probable phase of cooling or humidification and wet summers observed in other parts of Europe at the same period. Is this a new piece of evidence on a climatic fluctuation in western Europe?"

The parallel is remarkable: probable wet summers and cooling in southern France: and simultaneously, and perhaps correlatively, an advance of the Swiss glaciers about 1230–1250 because of lower ablation.

APPENDIX 18

A Cave in the Ardèche

As we have seen in the preceding Annex, the caves still hold further revelations for us on the history of climate. In this connection it is worth referring to the promising work of J. Labeyrie, 1967, and his colleagues. They have analyzed variations in O^{18} content in $CaCO_3$ in the various stages of formation of a white stalagmite of very pure calcite in the Aven d'Orgnac (Ardèche, France). The stalagmite is nearly seven thousand years old. As we know, variations in O^{18} indicate changes in the surrounding temperature as reflected in the isotopic composition of water; and the stalagmite was formed partly by the action of water. For the last thousand years, which particularly concern us, Labeyrie arrives at approximately the following results, based on five successive samplings:

Tenth century A.D.	:12.1° C
Circa A.D. 1150	:11.5° C
Circa A.D. 1450	:11° C
Circa A.D. 1750–1800	:12.3° C
Circa A.D. 1940	:11.7° C[16]

Labeyrie himself is on his guard against too hasty an interpretation of his results, which call for further research and especially more and numerically closer samplings. But on the whole these approximate results do not contradict other data from Europe and Greenland referred to previously: the optima of the year 1000, the eighteenth and the twentieth centuries; and the medieval deterioration, culminating in the medieval pessima of the Little Ice Age (especially around 1450). Unfortunately Labeyrie does not give any data on the seventeenth century.[17]

[16] These figures relate in theory to the mean temperature of the cave. But in fact the variations they indicate are interesting for many other reasons than that of just the absolute value of the figures themselves. The variations indicate secular warming or cooling which at the date in question affects the *whole cycle* of the water, from the time it evaporates from the sea, through its condensation into rain, up to its final incorporation into the stalactites and other formations in the cave at Aven d'Orgnac.

[17] On these questions see also Duplessy, 1967.

APPENDIX 19

Carbon 14 and Climatic Fluctuation: The Work of H. E. Suess

The eminent American physicist and expert in C^{14}, Hans E. Suess, has recently adduced further evidence about the "little optimum."[18] He makes use of multisecular wood samples taken from dendrochronological cross sections dated to almost within a year by the tree rings. He shows that the radiocarbon content of the samples varies slightly according to their age. In a curve covering almost a thousand years, he shows the rate of radiocarbon production in living organisms; this rate is determined by cosmic factors which are constant, and by solar factors which are variable, corresponding in the long term to an index of solar activity as measured by the number of sunspots. Suess's curve, which starts just before the year 1000 and goes right up to the twentieth century, shows a minimum of natural radiocarbon production between A.D. 1000 and 1200. This minimum is followed by a rise; then by a maximum between 1500 and 1700; then by a fresh decline. But in spite of the slow regression from 1700 to 1950, the rate of radiocarbon production in our own time has never quite fallen to the very low levels of the eleventh century.

However they may differ on other points, Suess and Damon[19] agree in giving a solar and climatic explanation of this. According to them, the minima of natural radiocarbon production (the eleventh and the twentieth centuries, for example) correspond on the one hand to phases of solar activity and increased numbers of sunspots, and on the other hand to a probably slight rise in solar radiation and thus of the earth's temperatures. Conversely, the maximal phase of radiocarbon production (about 1500–1700) corresponds to periods of very slightly reduced solar activity, and probably to slightly lower temperatures, such as those recorded in the Little Ice Age which culminated in the seventeenth century.

What emerges from this excellent C^{14} curve, the climatic interpretation of which is still at the hypothetical stage, is that the muse of History may sometimes be indebted to nuclear physics, and that the mild optimal phase before A.D. 1200 is once again suggested by techniques quite independent of the many other already referred to.[20]

This mild optimal phase immediately preceded the cold thrust and glacial advances which occurred in the thirteenth century both in the Alps and elsewhere.

[18] Suess, 1968, pp. 146–50.
[19] Ibid.; and Damon, 1968.
[20] On all these problems see especially Suess, 1968, p. 167. On the correlation between solar activity measured by sunspots and world temperature, see Mitchell, 1961. See also the critical article by Damon, 1968, who agrees with Suess's solar and climatic conclusions as applied to the Middle Ages and the recent era, but disagrees when it comes to the great Quaternary Ice Ages. It is worth correcting one detail in Damon's chronology (p. 152): the Alpine villages wiped out by the glaciers were destroyed at the end of the sixteenth and at the beginning of the seventeenth century, which as far as the Alps are concerned cannot therefore be regarded as a phase of "glacial retreat."

BIBLIOGRAPHY

AARIO (L.), "Ein nachwärmezeitlicher Gletschervorstoss in Oberfernau," *Acta geographica* (Finnish review) 1944 (Actual Publication, 1945).

ADAMENKO (V. N.), "On the similarity in the growth of trees in Northern Scandinavia and in the polar Ural mountains," *J. Glac.* Feb. 1963, pp. 449–451.

AGASSIZ (L.), *Etude sur les glaciers*, Neuchâtel, 1840.

AHLMANN (H. W.), "Vatnajökull," *Geog. Ann.*, 1937–1939.

AHLMANN (H. W.), "The Styggedal glacier," *Geog. Ann.*, 1940.

AHLMANN (H. W.), "The present climatic fluctuation," *Geog. Journ.*, April, 1949.

AHMAD (N.) and SAXENA (H. B.), "Glaciation of the Pindar River Valley, Southern Himalaya," *J. Glac.*, Feb. 1963.

AIGREFEUILLE (Ch. d'), *Histoire de Montpellier* (Montpellier, 1885 edition).

ALBIGNY (P. d'), "Les calamités publiques dans le Vivarais," *Revue du Vivarais*, 1912, pp. 370–381.

ALLIX (A.), *L'Oisans au Moyen Age*, Paris, 1929.

ALTMANN (J. G.), *Versuch einer historischen Beschreibung der helvetischen Eisbergen*, Zurich, 1751.

ANES ÁLVAREZ (Gonzalo), "La epoca de las vendimias: . . . climatologia retrospectiva en España," *Estudios geograficos*, May 1967.

ANGOT (A.), "Etude sur les vendanges en France," *Annales du Bureau central météorologique de France*, 1883.

ANGOT (A.), "Premier catalogue des observations météorologiques faites en France depuis l'origine jusqu'en 1850," *Annales du Bureau central météorologique de France*, 1895, I.

ANTEVS, "The big tree as a climatic measure," *Carnegie Institut. of Wash.*, public. n° 352.

ANTEVS, "Rainfall and tree growth in the great Basin," *ibid.*, public. n° 469.

ARAKAWA (H.), This author's publications are collected in *Selected papers on climatic change*, Meteorological Research Institute, Tokyo, n.d. See especially articles on:

"Climatic change as revealed by the data from the Far East," *Weather*, vol. XII, 1957; especially the graphs.

"Twelve centuries of blooming dates of the cherry blossoms at the city of Kyoto and its own vicinity," *Geofisica pura e applicata*, vol. 30, 1955: complete tables of figures.

"Climatic change as revealed by the blooming dates of the cherry blossoms at Kyoto," *Journal of Meteorology*, vol. 13, 1956: diagrams.

"Fujiwhara on five centuries of freezing dates of Lake Suwa in the Central Japan," *Archiv. für Meteorologie, Geophysik und Bioklimatologie*, series B. vol. 6, 1954.

"Dates of first or earliest snow covering for Tokyo since 1632," *Quarterly Journal of the Royal Meteorological Society*, vol. 82, 1956.

ARLERY (M.), GARNIER (M.), etc., contributions to special issue on agricultural meteorology of *Meteorologie*, Oct.-Dec. 1954.

ARNOD (Judge Philibert Amédée), *Description des Glaciers savoyards*, 1691–1694, cf. VACCARONE, 1881 and 1884.

AUBERT (Edouard), *La vallée d'Aoste*, Paris, 1861.

AYMARD (A.), Critical note in *Revue des Etudes anciennes*, 1951, pp. 126–129.

BAEHREL (R.), *Une croissance: la Basse-Provence rurale*, Paris, 1961.

BAKER (T. H.), *Records of the seasons . . . and phenomena observed in the British Isles*, London, 1883.

BANNISTER (B.), ROBINSON (W.), WARREN (R.), Tree-ring dates from Arizona (J), *Hopi Mesas Area, Lab. of tree-ring res.* University of Ariz., Tucson, 1967, and similar list for other regions, 1968.

BANNISTER (B.), "Dendrochronology" in *Science in Archaeology*, edited by BROTHWELL (D.) and HIGGS (E.), London/ New York, 1963, pp. 161–176.

BARATIER (E.), *La démographie provençale du XIIIe au XVIe siècle*, Paris, 1961.

BARETTI, "Il lago del Ruitor," *Bolletino del club Alpino italiano*, 1880, pp. 46–76.

BAULANT (M.) and MEUVRET (J.), *Prix des céréales extraits de la mercuriale de Paris*, Paris, 1962.

BECKINSALE (R.), "Climatic change, a critique of modern theories," in WHITTOW (J.) and WOOD (P.), *Essays in geography for Austin Miller*, Reading, 1965, pp. 1–38.

BENGSTON (K. B.), "Activity of the Coleman Glacier, Mount Baker, Washington, U.S.A., 1949–1955," *J. Glac.*, vol. 2, 1952–1956, p. 708.

BENNASSAR (B.), *Valladolid au siècle d'or*, Paris, 1967.

BERNARDI (A.), *Il monte Bianco* (1091–1786), Bologne, 1965.

BETIN (V. V.), "Ledovye uslivija v raione Baltiiskovo morja i na podkhodakh knemu i ih mnogoletnge izmenenija" (The Baltic ice-sheets: long-term variations,) *Moscow Gosudarstvennyi Okeanografisceskii Institut Trudy*, n° 41, 54–125, 1957.

BEVERIDGE (W.), "Weather and harvest cycles," *The economic journal*, Dec. 1921, pp. 421–453.

BEVERIDGE (W.), "Wheat prices and rainfall," *Journal of the statistical society*, vol. 85, 1922, pp. 418–454.

BEVERIDGE (W. H.), *Prices and wages in England*, London, 1939. "Bibliography on climatic changes" in *Meteorological abstracts and bibliography*, vol. I, n° 7, July 1950.

BLANCHARD (R.), "La crue glaciaire dans les Alpes de Savoie au XVII siècle," *Recueil de travaux de l'Institut de géographie alpine*, I, 1913, pp. 443–454.

BLANCHARD (R.), *Les Alpes occidentales*, vol. VII, Grenoble, 1956.

BLOCH (M.), *Apologie pour l'histoire, ou Métier d'historien* (Cahier des Annales), Paris, 1949.

BOISLISLE (A. M. de), *correspondance des contrôleurs généraux des finances avec les Intendants des provinces*, Paris, 1864 to 1897, (especially vol. I and II, for documents on 1690–1695, and the famine of 1693).

BOISOT (L'abbé), "Lettre à l'abbé Nicaise sur la Froidière de Chaux," *Journ. des Savants*, 22 July. 1686, p. 336.

BONAPARTE (Prince R.), Notices, in *Bulletin du Club alpin français*, 1892, (p. 28) and 1896.

BONNEFOY (J. A.), *Documents relatifs au prieuré de Chamonix*, Publications of the Academy of Savoie, Chambéry, 1879–1883.

BORD (J.), *Histoire du blé en France, le pacte de famine*, Paris, 1887.

BORISOV (A.), (in Russian), (*Has the Climate of Leningrad changed?*), Leningrad, 1967, from the account in M.G.A., June 1968, p. 1229.

BOUGES (Le R. P.), *Histoire ecclésiastique et civile de la ville et diocèse de Carcassonne*, Paris, 1741.

BOUJUT (M.), Introduction to the French edition of MARY SHELLEY, *Frankenstein*, Paris, 1965 ed. 10–18.

BOULAINVILLIERS (le comte de), *Etat de la France*, London, 1752, vol. VI, p. 136.

BOURRIT (T.), *Description des glacières . . . du duché de Savoye*, Geneva, 1773.

BOURRIT (T.), *Description . . . du mont Blanc*, Lausanne, 1776.

BOURRIT (T.), *Nouvelle description générale des Alpes*, Geneva, 1785.

BOUT, CORBEL, DERRUAU, GARAVEL, PÉGUY, "Géomorphologie et glaciologie en Islande centrale," *Norois*, Oct.-Dec. 1955.

BOUVEROT (M.), "Notices sur les variations des glaciers du Mont-Blanc," *Association internationale d'hydrologie scientifique* (Toronto), vol. 46, 1957, p. 331.

BRANAS (J.), *Eléments de viticulture générale*, Montpellier, 1946.

BRAUDEL (F.), *La Méditerranée et le monde méditerranéen au temps de Philippe II*, Paris, 1949; and second edition, 1966.

BRAY (J. R.), "Forest growth in N.W.—N.America," *Nature*, 205, p. 441, 30-1-1965

BRAY (J. R.), and STRUIK, in *Canad. Journ. of Bot.*, 1963, p. 1245.

BRAY (J. R.), in *J. Glac.*, 1966, p. 322.

BREHME (K.), "Jahringchronologische und-Klimatologische Untersuchungen an Hochgebirgslärchen des Berchtesgadener Landes," *Zeitschrift für Weltforstwirtschaft*, 1951, vol. 14, p. 65–80.

BRIER (G. W.), "Some statistical aspects of long term fluctuations in solar and atmospheric phenomena," *A.N.Y.A.S.*, vol. 95, art. 1, 5 Oct. 1961, pp. 173–187.

BRITTON (C. E.), *A Meteorological Chronology to A.D. 1450*, Meteorological Committee, H.M.S.O., London, 1937.

BROGGI (J. A.), "La desglaciacion actual de los Andes del Peru," *Soc. geol. del Peru, Bol.*, vol. 15-15, 1943, pp. 59–90, cited by FLINT, ed. 1947, p. 540 (see also under FLINT).

BROOKS (C.), in *Proceedings of the Toronto meteorological Conference*, Publications of the Roy. Met. Soc., London, 1954, p. 215.

BROOKS (C. E. P.), *Climate through the Ages*, London, 1949, and 1950 edition.

BROOKS (C. E. P.), "Climatic change," in *Compendium of Meteorology*, edited by T. F. Malone, Amer. Met. Soc., Boston, 1951.

BRÜCKNER (E.), "Klimaschwankungen seit 1700," *Geographische Abhandlungen* (Vienne), 4-2, 1890, pp. 261–264.

BRUNET (R.), "Un exemple de la récession des glaciers pyrénéens," Pirineos, XII, 1956, pp. 261–284 (account by J. TRICART, in *R.G.D.*, Sept.-Oct. 1958, p. 157).

BRYSON (R.), IRVING (W.), LARSEN (J.), "Radiocarbon and soil evidence of former forest in the Southern Canadian Tundra," *Science*, Jan. 1-1965, vol. 147, n° 3653, pp. 46–48.

BRYSON (R.), "A reconciliation of several theories of climatic change," *Weatherwise*, April 1968.

BRYSON (R.) and DUTTON (J.), "Variance spectra of tree-rings," *A.N.Y.A.S.*, 95, 1, 5 Oct. 1961, pp. 580–604.

BUCHINSKY (I. E.), *Oklimate proslovo Russkoi ravniny* ("Past climate of the Russian plain"), *Gidrometeoizdat*, 2 nd. ed. Leningrad, 1957.

BURROWS (C.) and LUCAS (J.), "Variations in two New Zealand glaciers since 1100 A.D.," *Nature*, 1967, p. 467.

BUTLER (M.), "Palynological studies, Cape Cod, Mass.," *Ecology*, 1959, vol. 40, n° 4, pp. 735 ff.

BUTZER (K. W.), "Late glacial and post-glacial climatic variation," *Erdkunde*, Feb. 1957, pp. 31–35.

BUTZER (K. W.), *Quaternary stratigraphy and climate in the Near East*, Section n° 24, oB. Bonner *Bonner geographische Abhandlungen*, Bonn, 1958.

BUTZER (K. W.), 1966, article in *International Symposium . . . 1966*.

CAILLEUX (A.), "Variations récentes du niveau des mers," *Bull. de la Soc. géologique de France*, 1952, pp. 135–144.

CAILLEUX, cf. NAGERONI and TRICART.

CALLENDAR (G. S.), "Air temperature and the growth of glaciers," *Q.J.R.M.S.*, 1942, pp. 57–60.

CALLENDAR (G. S.), "Can carbon dioxide influence climate?", *Weather*, 4, p. 310–314, 1949.

CARPENTIER (E.), *Une ville devant la peste: Orvieto*, Paris, 1962.

CARPENTIER (E.), "La peste noire," *Annales,* Nov.-Dec. 1962, pp. 1080 ff.
CASSEDY (J. H.), "Meteorology and medicine in colonial America," *Journal of the History of Medicine,* April 1969.
CHAMPION (M.), *Les inondations en France depuis le VIe siècle jusqu'à nos jours,* Paris, 1858–1864, 6 vol.
Changes of Climate, Proceedings of the Rome symposium, Unesco, Paris, 1963.
CHARLESWORTH (J. K.), *The quaternary Era,* London, 1957.
CHARNLEY (F. E.), "Glaciers of Mount Kenya," *J. Glac.,* vol. 3, Oct. 1959, pp. 480–493.
CHATEAUBRIAND (F. de), *Voyage au Mont-Blanc,* ed. G. Faure, Grenoble, J. Rey, 1920.
CHAUNU (P.), "Le climat et l'histoire, à propos d'un livre récent," *Rev. Historique,* Oct.-Dec. 1967.
CHIZOV (O.) and KORYAKIN (V. S.), Contribution to the Obergurgler Symposium in Publication n° 58 of the Assoc. Intern. d'hydrol. scientif. (1962); also account of this contribution in *J. Glac.,* Feb. 1963.
CHRISTILLIN (M. l'avocat), Histoire du duché d'Aoste, published about 1840. "La chronique consulaire de Béziers," *Bull. soc. arch. Béz.,* vol. 3, 1839.
CLARK (J.), "Return to hard winters," *New Scientist,* vol. 37, 18-1-1968, p. 145–146.
Climatic change, evidences, causes and effects, edited by SHAPLEY (H.), Cambridge (U.S.A.), 1953.
COBB (R.), *Terreur et subsistances,* Paris, 1965, (esp. pp. 221–383: Studies on the scarcity of the year III).
Compendium of meteorology, edited by MALONE (T.), Boston, 1951.
CONTAMINE (Ph.), *Azincourt,* Paris, 1964.
CONWAY (V.), "Von Post's work on climatic rhythms," *New Phytologist,* 1948, pp. 220–238.
COOLIDGE (W. A. B.), *Die Petronella Kapelle in Grindelwald,* Grindelwald, 1911, Jakober-Peter éd. (a rare pamphlet, consulted in the library of the S.A.C. at Grindelwald).
COOPER (W. S.), "The problem of Glacier Bay, Alaska," *Geog. Rev.,* 1937, p. 37.
CORBEL (J.), *Les Karsts du nord-ouest de l'Europe,* Lyons, 1957.
CORBEL (J.), "Nouvelles méthodes de mesure des paléotempératures," *Rev. de géog. de Lyon,* 1959, p. 168.
CORBEL (J.), *Neiges et glaciers,* Paris, 1962.
CORBEL (J.) and LE ROY LADURIE (E.), "Datation au C 14 d'une moraine du Mont-Blanc," *Revue de géographie alpine,* 1963, p. 173.
COUVERT (Roger), *Histoire de Chamonix,* unpublished, with an unpublished text on the glaciers in 1600. (See above, chronology of alpine glaciers at this date.)
CRAIG (R.) and WILLETT (H.), "Solar variations . . . and weather changes," *Compendium of meteorology* (q.v.), Boston, 1951.
CRONE (G. R.), *The Discovery of America,* London, 1969.
CRONE (G. R.), "How authentic is the Vineland Map?", *Encounter,* Feb. 1966, pp. 75–78.
CURRIE (B. W.), "Climatic trends on the Canadian prairies" (1956), account in *Meteorological abstracts and bibliography,* 551–583, 14 (712).
CURTSCHMANN (F.), *Hungersnöte im Mittelalter. Ein Beitrag zur deutschen Wirtschaftsgeschichte des 8 bis 13 Jahrhunderts,* Leipzig, 1900.
DAMON (E.), "Radiocarbon and climate," 1968, (in MITCHELL, 1968, q.v.).
DANSGAARD (W.), JOHNSEN (S. J.), MOLLER (J.), LANGWAY (C.), "One thousand centuries of climatic record from Camp Century on the Greenland ice sheet," *Science,* vol. 166, pp. 377–381, 17 Oct. 1969.
DANSGAARD (W.) and JOHNSEN (S.), "A time scale for the ice core from Camp Century," *J. Glac.,* 1969, pp. 215–223.
DARNAJOUX (H.), "Bibliographie sur les longues séries d'observations," *Mét.,* July-Sept. 1964, p. 241.

DARROW (R. A.), "Origin and development of the vegetational communities of the southwest," *New Mexico highlands University Bulletin*, n° 212, Feb. 1961, pp. 30–46.

DAVIS (N. E.), "The summers of North-West Europe," *Met. Mag.*, 1967, pp. 178–187, and pp. 319.

DAVIS (N. E.), "An optimum summer weather index," *Weather*, Aug. 1968, pp. 305–318.

DEAN (Jeffry), *Chronological analysis of the Tsegi phase site in North-East Arizona* (Unpublished thesis, University of Arizona, Tucson, 1967).

DEEVEY (E. S.), "Late-glacial and postglacial pollen diagrams from Maine," *Amer. Journ. of Science*, vol. 249, March 1951, pp. 177–207.

DEEVEY (E. S.), "Paleolimnology of the upper swamp deposit, Pyramid valley," *Records of the Canterbury Museum* (Australia), vol. 6, n° 4, pp. 291–344, 15 Feb. 1955.

DEEVEY (E. S.) and FLINT (R. F.), "Postglacial hypsithermal Interval," *Science*, Feb. 1957, vol. 125, n° 3240, pp. 182–184.

DE LEO (A.), "Una nuova avventizia nel Palermitano," *Lavori dell'Instituto botanico et del Giardino coloniale*, vol. XXII. Palermo, 1967, p. 72, cited in TRASSELI, 1968 (q.v.).

DELUMEAU (J.), *Vie économique de Rome au XVIe siècle*, Paris, 1959.

DEMOUGEOT (E.), "Variations climatiques et invasions," *Rev. hist.*, January 1965.

DEPPING, (G. B.), *Merveilles et beautés de la nature en France*, Paris, 1845.

DERANCOURT (Commandant), Contribution to "Comité des Trav. hist. et scientif.," *Bull. de la sect de géogr.*, 133, pp. LI and 89

DESBORDES (J. M.), *La Chronique villageoise de Vareddes* (Seine et Marne) . . . *aux XVIIe et XVIIIe siècles*, Paris (édition de l'Ecole, 11, rue de Sèvres), about 1969.

DESIO (A.), "Recent fluctuations of the italian glaciers," *J. Glac.*, vol. I, 1947–1951, pp. 421–422.

DEVIC (Cl.) and VAISSETTE (J.), *Histoire générale de Languedoc*, Toulouse (1872–1892 edition).

DEYON (P.), *Amiens au XVIIe siècle*, Paris, 1968.

"Diagrammes d'Aspen: Informations climatiques, séries comparées, XIe et XVIe siècles," *Annales*, Sept.-Oct. 1965, (cf. LE ROY LADURIE, 1965).

DIAMOND (M.), "Precipitation trends in Greenland during the past thirty years," *J. Glac.*, vol. 3, March 1958, pp. 177–181.

Dictionnaire historique et géographique de la Suisse, Neuchâtel, 1926.

DOLGOSHOV (V.) and SAVINA (S. S.), "Connections of phenological phenomena with variations of climate," (in Russian); analyzed in M.G.A. July 1969, p. 1828.

DOLLFUS (O.), "Formes glaciaires et périglaciaires actuelles autour du lac Humboldt," *B.A.G.F.*, 1959.

DORST (J.), *Les migrations des oiseaux*, Paris, 1956.

DOUGLASS (A. E.), *Climatic cycles and tree growth*, Carnegie Institute of Washington, Public. n° 289 (1919, 1928 and 1936).

DRYGALSKI (E. von) and MACHATSCHEK (F.), *Gletscherkunde* (vol. VIII de l'*Enzyklopädie der Erdkunde*), Vienna, 1942.

DÜBI (H.), *Saas Fee*, Berne, 1902, p. 36.

DUBIEF (J.), Contribution to *Changes of Climate*, 1963 (q.v.).

DUC (Mgr J. A.), *Histoire de l'église d'Aoste*, Aosta, 1901.

DUCHAUSSOY (H.), "Les bans de vendanges de la région parisienne," *La météorologie*, 1934.

DUCROCQ (A.), "La dendrochronologie," *Science et Avenir*, Dec. 1955.

DUCROCQ (A.), "*La science à la découverte du passé*, Paris, 1955.

DUPÂQUIER (J.), *Mercuriales du Vexin*, Paris (S.E.V.P.E.N.), 1968.

DUPLESSY (J. C.), *Etude isotopique et paléoclimatologique du concrétionnement de l'aven d'Orgnac, unpublished thesis*, University of Paris, 1967.

DZERDZEEVSKII (B. L.), "The general circulation of the atmosphere . . . ," N.Y.A.S., vol. 95, art. 1, 5 Oct. 1961, pp. 188–200.

DZERDZEEVSKII (B. L.), Contribution to *Changes of Climate*, 1963, pp. 285–296 (q.v.).

EASTON (C.), *Les hivers dans l'Europe occidentale*, Leyden, 1928.

ELHAÏ (H.), *La Normandie occidentale*, Bordeaux, 1963.

EL KORDI, *Histoire économique de Bayeux (XVIIe et XVIIIe siècles)*, Paris, 1970. This important book contains an excellent study on the variation in grain harvests in Normandy in the eighteenth century. The bad harvests of 1788 and 1789, of which we have mentioned the climatic causes and the revolutionary consequences, were, according to the (rectified) figures of the *Etats de récoltes* of Bayeux, from a half (1788) to a third (1789) below normal (1777–1787).

ELSASS (J. M.), *Umriss einer Geschichte der Preise in Deutschland, Leyden*.

EMILIANI (C.), "Pleistocene temperatures," *Journal of geology*, Nov. 1955, pp. 538–579.

EMILIANI (C.), "Paleotemperature analysis," *Journal of geology*, May 1958, pp. 264–276.

EMILIANI (C.), "Cenozoic climatic changes as indicated by . . . the chronology of deep-sea cores," A.N.Y.A.S., vol. 95, art. 1, 5 Oct. 1961, pp. 520–536.

ENGEL (C.-E.), *La littérature alpestre*, Chambéry, 1930.

ENGEL (C.-E.), *Le Mont Blanc*, s. l., Les Editions du Temps, 1961.

ENGEL (C.-E.), *Le Mont Blanc, vu par les écrivains*, Paris, 1965.

ERINC (S.), "The Pleistocene history of the Black Sea, with reference to the climatic changes," *Review of the Geographical Institute of the University of Istanbul*, 1954.

EYTHORSSON (J.), "On the variations of glaciers in Iceland," *Geog. Ann.*, 1935.

EYTHORSSON (J.), "Variations of glaciers in Iceland, 1930–1947," *J. Glac.*, vol. I (1947–1951), p. 250.

EYTHORSSON (J.), "Temperature variations in Iceland," *Geog. Ann.*, 1949.

FABER (J. A.), etc., "Population changes . . . in the Netherlands," A.A.G. *Bijdragen*, 12, 1965, pp. 46–110.

FAIRBRIDGE (R. W.), "Mean sea-level changes, long-term, eustatic, and others," in *Encyclopedia of Oceanography*, vol. I, edited by FAIRBRIDGE (R. W.), New York, 1966, pp. 479–485.

FAVIER (J.), *De Marco Polo à Christophe Colomb (1250–1492)*, Paris, 1968.

FEBVRE (L.), *Philippe II et la Franche-Comté*, Paris, 1912.

FERGUSON (C. W.), "Bristlecone pine," *Science*, Feb. 23, 1968, vol. 1519, n. 3817, pp. 839–846.

FERGUSON (C. W.), HUBER (B.), SUESS (H.), "Age of Swiss lake dwellings, . . . dendrochronologically calibrated by radiocarbon dating," *Zeitschr. f. Natforschung*, 21 a, 7, 1966, pp. 1173–1177.

FERRAND (H.), 1912, (cf. Relation anonyme . . . and WINDHAM and MARTEL).

FERRAND (H.), *Autour du Mont Blanc*, Grenoble, 1920.

FIELD (W. O.), "Glaciers of Prince William Sound," *Geog. Rev.*, 1932.

FIELD (W. O.), "Glacier observations in the Canadian Rockies," *Canadian Alpine Journal*, vol. 32, 1949, pp. 99–114, cited in *J. Glac.*, vol. 1, p. 398.

FINSTERWALDER (R.), "Photogrammetry and Glacier research," *J. Glac.*, April 1954, pp. 306–315.

FIRBAS (F.), *Spät-und nacheiszeitliche Waldgeschichte Mitteleuropas nördl. der Alpen*, Jena, 1949.

FLETCHER (J. O.), "Climatic change and ice extent on the sea," *Rand Corporation*, Santa Mon., Cal., *Paper P 3831*, April 1968 (cited in M.G.A., April 1969, p. 878).

FLINT (R. F.), *Glacial and Pleistocene geology*, New York, 1947 and new edition of 1957.

FLINT (R. F.) and BRANDTNER (F.), "Outline of climatic fluctuation since the last interglacial age," A.N.Y.A.S., 95-1, 1961, p. 458.

FLOHN (H.), "Klimaschwankungen im Mittelalter und ihre historische-geographische Bedeutung," *Berichte zur Deutschen Landeskunde*, Band 7, Heft 2, 1950, pp. 347–357.

FLORENCE (J.), *Dendrochronologie et climat en Capcir et Cerdagne*, unpublished thesis, Toulouse, 1962.

FORBES (J. D.), *Travels through the Alps of Savoy*, Edinburgh, 1843.

FORNERY (Joseph), *Histoire du Comté Venaissin et de la ville d'Avignon*, Avignon, 1910.

FOSSIER, *La crise frumentaire du XIVe siècle*, in *Recueil de travaux, offert à Clovis Brunel*, Memoirs and documents published by the Society of the Ecole des Chartes, vol. 12, Paris 1955.

FOURNIER (E.), *Gouffres et grottes du Doubs*, Besançon, 1899.

FOURNIER (E.), *Grottes et rivières souterraines*, Besançon, 1923.

FOURQUIN (G.), *Les Campagnes parisiennes à la fin du Moyen Age*, Paris, 1964; and article in *Etudes Rurales*, July-Dec. 1966.

FRAZIER (K.), "Earth's Cooling Climate," *Science News*, vol. 96, Nov. 1969. A recent article emphasizing the part played by industrial and volcanic dust in the cooling of climate since 1950.

FRENZEL (B.), "Die vegetationszonen Nord-Eurasiens, während der postglazialen Wärmezeit," *Erdkunde*, IX, 1955, pp. 40–53, reviewed by J. TRICART, R.G.D., 1955, p. 283.

FRENZEL (B.), 1966: article in *International Symposium. . . . 1966*.

FRIEDLI (E.), *Bärndütsch als Spiegel Bernischen Volkstums*, vol. 2: *Grindelwald*, Berne, 1908.

FRITTS (H. C.).
Among the many remarkable works by this author are:
"The relation of growth rings in American beech and white oak to variation in climate," *Tree-ring Bull.*, 1961–1962, vol. 25, n° 1–2, pp. 2–10.
"Dendrochronology," in *The Quaternary of the United States . . . A Review Volume for the VII Congress of the International Association for Quaternary research*, Princeton, 1965, pp. 871–879.
"Tree-ring evidence for climatic changes in western North America," *Monthly Weather Review*, vol. 93, n° 7, pp. 421–443, 1965.

FRITTS (H. C.), SMITH (D. G.), and HOLMES (R. L.), "Tree-ring evidence for climatic changes in western north America from 1500 A.D. to 1940 A.D.," *1964 Annual Report to the United States Weather Bureau*, Washington (Project: Dendroclimatic History of the United States), 31 Dec. 1964.

FRITTS (H. C.), "Tree-ring analysis . . . for water resource research," *I.H.D. Bulletin, U. S. National Committee for International Hydrological Decade*, Jan. 1969.

FRITTS (H. C.), "Growth rings of trees and climate," *Science*, 154, 25 Nov. 1966, pp. 973–979.

FRITTS (H. C.), SMITH (D. G.), CARDIS (J.), and BUDELSKY (C.), "Tree-ring characteristics along a vegetation gradient in Northern Arizona," *Ecology*, vol. 46, 1965, n° 4.

FRITTS (H. C.), "Bristlecone pine in the White Mountains of California," *Papers of the Lab. of Tree-ring Research*, n° 4, 1969 (Tucson, Ariz.).

FRITTS (H. C.), "Growth rings of trees: a physiological basis for their correlation with climate," in *Ground level climatology* (Symposium, Dec. 1965, Berkeley), *Amer. Assoc. for the advancement of Science*, Washington, D.C., 1967.

FRITTS (H.), SMITH (D.), STOKES (M.), "The biological model for paleoclimatic interpretation of tree-ring series," *Amer. Antiquity*, vol. 31, n° 2–2, Oct. 1965.

FRITTS (H. C.), SMITH (D.), BUDELSKY (C.), CARDIS (J.), "Variability of tree-rings . . ." *Tree-ring Bulletin*, Nov. 1965.

FROLOW (V.), "Aperçu sur l'évolution climatique de Paris," *B.A.G.F.*, June-July 1958.
FRUH (J.), *Géographie de la Suisse*, Lausanne, 1937, 3 vol.
FUSTER (Doctor), *Des changements dans le climat de la France*, Paris, 1845.
GAGE (M.), "The dwindling glaciers of . . . New Zealand," *J. Glac.*, vol. I, 1947–1951, pp. 504–507.
GALTIER (G.), *Bull. Soc. Lang. de géog.*, 1958, pp. 186 and 317.
GARAVEL (L.), "Lee glacier de Sarennes de 1948 à 1953," *Revue forestière française*, 1955, n° 1, pp. 9–26.
GARNIER (M.), "Contribution de la phénologie à l'étude des variations climatiques," *Met.*, Oct.-Dec. 1955.
GARNIER (M.), "Influence des conditions météorologiques sur le rendement de l'orge de printemps," *La Météorologie*, 1956, pp. 335–361.
GAUFRIDI (J.-F. de), *Histoire de Provence*, Aix, 1694.
GEORGE (M.), *The Oberland and its glaciers*, London, 1866.
GESLIN (H.), "Influence de la température sur le tallage du blé," *La Météorologie*, 1954, p. 30.
GIDDINGS (J. L.), "Mackenzie River Delta chronology," *Tree-ring Bulletin*, April 1947 (important graphs).
GIDDINGS (J. L.), "Chronology of the Kobuk-Kotzebue sites," *Tree-ring Bulletin*, 1948, vol. 14, n° 4, pp. 26–32, 1952.
GIDDINGS (J. L.), "The Arctic Woodland Culture of the Kobuk River," *Museum Monographs*, 1952, University Museum, Philadelphia, pp. 105–110.
GIDDINGS (J. L.), "Dendrochronology in Northern Alaska," *University of Arizona Bulletin* (vol. XII, n° 4), 1941.
GILBERT (O.), and other authors, "An Afghan Glacier," *J. Glac.*, 1969, pp. 51–66.
GIRALT (E.), "En torno al precio del Trigo en Barcelona durante el siglo XVI," *Hispania*, XVIII (1958), pp. 38–61.
GIRALT (E.), "A correlation of years, numbers of days of Rogation for rain at Barcelona, and the price of one *quartera* wheat in *sous* and *diners* of Barcelona." (Contribution to the Aspen conference [ronéo'd], see *Proceedings* . . . 1962.)
GLASSPOOLE (J.), "Recent seasonal climatic trends over Great Britain," *Meteorological Magazine*, vol. 86, 1957, pp. 358–362. *Der Gletschermann, Familienblatt für die Gemeinde Grindelwald*, published by G. STRASSER, curé of Grindelwald, 1890, n° 41–47, pp. 165 ff. (Contains the Chronicle or Cronegg of Grindelwald.)
GLOCK (W. S.), *Principles and methods of tree-ring analysis*, Carnegie Institute of Washington, Public. n° 486.
GODARD (M.) and NIGOND (J.), "Le climat de la vigne dans la région de Montpellier," *Vignes et vins* [Journal of the Technical Institute of Wine] n° 66.
GODECHOT, *Les Révolutions*, Paris, 1965.
GODWIN (H.), *The History of the British Flora*, Cambridge, 1956.
GODWIN (H.), WALKER (D.) and WILLIS (E. H.), "Radio-carbon dating . . . : Scaleby Moss," *Proceed. of the Royal Soc. of G.B.*, vol. 147, pp. 352–366, 1957.
GODWIN (H.), 1966, article in *International Symposium . . .*, 1966 (q.v.).
GOLDTHWAIT, 1966, article in *International Symposium . . .*, 1966, (q.v.).
GOLZOV, MAXIMOV, IAROCHEVSKII, *Praktische Agrarmeteorologie*, Berlin, 1955.
GOUBERT (P.), "Ernst Kossmann et l'énigme de la Fronde", *Annales*, 1958, p. 115.
GOUBERT (P.) *Beauvais et le Beauvaisis*, Paris, 1960.
GRAENLANDICA SAGA in *The Vinland Sagas, The Norse discovery of America*, translation and introduction by M. MAGNUSSON and H. PÁLSSON, Penguin Books, Baltimore, 1965.
GRÉMAUD (l'abbé J.), *Documents relatifs à l'historie du Valais*, vol. III (p. 14–

15, text n° 1156), Lausanne, 1878, in *Mémoires et documents publiés par la société de l'histoire de la Suisse romande*, vol. 31.

GROSVAL'D (M.) and KOTLYAKOV (V.), "Present glaciers in the USSR, and . . . their mass balance," *J. Glac.*, Feb. 1969, p. 9–23.

GROVE (J. M.), "The little ice age in the massif of Mont Blanc," *The Institute of British Geographers, Transactions and Papers*, 1966, n° 40. This is the fullest and most important article that has yet appeared in English on the fluctuations of the Savoie glaciers.

GRUNER (G. S.), *Die Eisgebirge des Schweizerlandes*, Berne, 1760.

GRUNER (G. S.), *Histoire naturelle des glacières de Suisse* (Keralio's translation of the above).

GUENEAU (C.), "La disette de 1816–1817 dans la Brie," *Rev. d'hist. mod.* vol. 4, 1929, pp. 18–95.

GUICHONNET (P.), "Le cadastre savoyard de 1730," *Rev. de géog. alp.*, 43, 1955, pp. 255–298.

GUILCHER (A.), "L'élévation du niveau de la Méditerranée," *Ann. de géog.*, 1956, p. 439.

GUITTON (H.), *Fluctuations économiques*, Paris, 1958.

GUMILEV (L. N.), "Fluctuations de la Caspienne, . . . et histoire des peuples nomades," *Cahiers du monde russe et soviétique*, VI, 3, 1965. Gumilev's fascinating articles take up the very debatable arguments of C. E. P. Brooks and Huntington on the correlations between rainfall fluctuation and nomad migration in central Asia.

GUMILEV (On humidity and migrations in Asia), cited in M.G.A., July 1968, p. 1630.

HAEFELI (R.), "Gletscherschwankung und Gletscherbewegung," *Schweizerische Bauzeitung*, 1955, pp. 626 and 693; 1956, p. 667.

HALE (M. E.), "Moraine plant succession at the edge of the ice cap (Baffin Island)," *J. Glac.*, vol. 2, 1952–1956, p. 22.

HALLER (W.), "Journal (1550–1570)," in *Schweizerische meteorologische Beobachtungen*, vol. 9, 10, and suppl., Zurich, 1875.

HAMILTON (E. J.), *American treasure and the price revolution in Spain*, Cambridge, 1934.

HARRINGTON (H. J.), "Glacier retreat in the Southern Alps of New Zealand," *J. Glac.*, vol. II, 1952–1956, pp. 133–145.

HARRISON (A. E.), "Glacial activity (Nisqually Glacier) in the Western United States," *J. Glac.*, vol. 2, 1952–1956, pp. 666–668 and 675 (graph.).

HAURY (E. W.), "A dramatic moment in South-western archaeology," *Treering Bull.*, May 1962.

HAUSER (H.), "La Rebeine de Lyon," *Rev. hist.*, 1896.

HEERS (J.), *L'Occident aux XIVe et XVe siècles*, Paris (P.U.F.), 1966.

HEIM (V.), *Handbuch der Gletscherkunde*, Stuttgart, 1885.

HEINZELIN (J. de), "Glacier recession in the Ruwenzori Range," *J. Glac.*, 1952, p. 138.

HÉLIN (E.), "Le déroulement de trois crises à Liège au XVIIIe siècle," *Actes du colloque international de démographie historique, Liège*, 18, 20 April 1963 (Problems of mortality) published by P. HARSIN and E. HÉLIN, Paris (Th. Genin), 1964.

HENNIG (A.), "Katalog bemerkenswerter Witterungsereignisse von der ältesten Zeiten bis zum Jahre 1800," *Berlin K. Preussisches Meteorologisches Institut. Abhandlungen*, 2, 4, 1904.

HENRY (P.), "Les ascensions légendaires au Mont-Blanc," *La Montagne* (Journal of the French Alpine Club), Feb. 1969.

HERLIHY (D.), "Some references to weather in eleventh century chroniclers," (contribution to Aspen conference, ronéo'd) 1962.

HESSELBERG (T.) and BIRKELAND (B.), "Säkulare Schwankungen des Klimas

von Norwegen. Teil I: die Lufttemperatur," *Goefysiske Publiskasjones*, Oslo, vol. 14, n° 4–6, 1940–43.
HEUBERGER (H.), "Gletschergeschichtliche Untersuchrengen in den Zentralalpen," *Wissenschaftliche Alpenvereinshefte*, Heft 20, Innsbrück, 1966.
HEUSSER (C. J.), Note on a radiocarbon (C14) dating in Alaska, in *Ecological monographs*, 1952.
HEUSSER (C. J.), "Radio-carbon dating of the thermal maximum in S.E. Alaska," *Ecology*, vol. 34, 1953, pp. 637–640.
HEUSSER (C. J.) and MARCUS (M. G.), "Historical variations of Lemon Creek glacier, Alaska, and their relationship to the climatic record," *J. Glac.*, Feb. 1964, p. 77.
HEUSSER, (C. J.), 1966: article in *International Symposium* . . . , 1966. (q.v.).
HOFFMANN (W.), "Der Vorstoss des Nisqually-Glestschers, 1952–1956," *Z. f. Glk.*, 1958, pp. 47–60, cited in *M.G.A.*, April 1960.
HOINKES (H.), "Ablation and heat balance on Alpine glaciers," *J. Glac.*, vol. 2, 1955, p. 497.
HOINKES (H.), and RUDOLPH (R.), Contributions on problems of "mass-balance" in Tyrolean glaciers, in *J. Glac.*, 1962 and publication n° 58 of the Association Internationale d'hydrologie scientifique (1962).
HOINKES (H.), cf. VON RUDLOFF's bibliography, 1967, n° 861.
HOINKES (H.), Contribution to *Bulletin* de l'Association internationale d'hydrologie scientifique, June 1963, pp. 85–86.
HOINKES (H.), "Glacier variation and weather," *J. Glac.*, Feb. 1968, pp. 3–21, with comments by LAMB (H. H.), *ibid.*, 1968, pp. 129–130.
HOLLSTEIN (E.), "Jahrringchronologische Datierung von Eichenhölzern ohne Wald Kante (Westdeutsche Eichenchronologie), *Bonner Jahrbücher*, 165, 1965, pp. 1–27.
HOOKER (R. H.), "The Weather and the crops in eastern England, 1885–1921," *Quart. Journ. Roy. Met. Soc.*, April 1922.
HOSKINS (W. G.), "Harvest fluctuations and English economic history. . . . (XVIth/XVIIIth centuries)," *Agricultural History Review*, 1964, pp. 28–46; and 1968, pp. 15–31.
HOVGAARD (W.), "The Norsemen in Greenland, recent discoveries at Herfoljness," *Geographical Review*, vol. 15, 1925, pp. 615–616.
HOYANAGI (Mutsumi), "Research on climatic change in Japan," in *Japanese Geography, its recent trends*, spec. publ. n° 1, published by *the Association of Japanese Geographers*, (Tokyo), 1966, pp. 57–60. This article contains an important bibliography, especially on pluviometric series in Korea from 1770 to 1907, and on the link between cold summers and bad rice-harvests.
HUBER (B.) and von JAZEWITSCH (W.), "Tree-ring studies," *Tree-ring Bulletin*, April 1956, p. 29.
HUBER (B.) and SIEBENLIST (V.), "Das Watterbacher Haus im Odenwald, ein wichtiges Brückenstück unserer tausendjährigen Eichenchronologie," *Mitteilungen der Floristischsoziologischen Arbeitsgemeinschaft*, N.F., Heft 10, 1963.
HUBER (B.), SIEBENLIST (V.), NIESS (W.), "Jahrringchronologie hessischer Eichen," *Büdinger Geschichtblätter*, Band V, 1964. An article of major importance in European dendrochronology: see especially diagram 16 which covers eight centuries, and diagram 2 (exceptionally narrow tree rings or "V-signaturen" for the years 1311 and 1326, which bring out by contrast the wide rings and great wave of humidity of the decade 1310).
HUBER (B.), Giertz-SIEBENLIST (V.), "Tausendjährige Eichenchronologie," *Sitzungsberichten der Österr-Akademie der Wiss., Mathem.-naturw-Kl.*, Abt. I, 178 Band, 1–4 Heft, Vienna, 1969. This recent article contains a master chronology for the German oak from 960 to 1960.
HUBER (B.), "Seeberg . . . Dendrochronologie," *Acta Bernensia*, 1967.
HUMBERT (P.), "Documents météorologiques anciens concernant la région du Mont-Blanc," *Mét.*, 1934, pp. 278–294.

414 *Bibliography*

HUMPHRIES (D. W.), "Glaciology of Kilimandjaro," *J. Glac.*, vol. III, Oct. 1959, pp. 475–480.

HUNTINGTON (E.), *The pulse of Asia*, Boston, 1907.

HUNTINGTON (E.), *Civilization and Climate*, New Haven, 1915 and 1927.

HUSTICH (I.), "On the correlation between growth and the recent climatic fluctuation," *Geog. Ann.*, 1949, pp. 90–105.

HUSTICH (I.), "Yields of cereals in Finland, and the present climatic fluctuation," *Fennia*, 73, 1950–1951.

INTERNATIONAL SYMPOSIUM ON WORLD CLIMATE (Imperial college, London, 1966): *World climate from 8000 to 0 B.C.*, Proceedings, London, Royal Met. Soc., 1966.

IVES (I. D.) and KING (C.), "Glaciological observations on Morsajökull, S.W. Vatnajökull," *J. Glac.*, vol. 2 (1952–56), p. 477.

JACQUART (J.), "La Fronde des princes dans la région parisienne," *Revue d'histoire moderne et contemporaine*, 1960.

JARDETSKY (W.), "Investigations of Milankovitch and the quaternary curve of solar radiation," *Annals of the New York Academy of Sciences*, vol. 95, 1961.

JEANNIN (P.), *l'Europe du Nord-ouest et du Nord aux XVIIe et XVIIIe siècles*, Paris, 1969 (p. 94).

JELGERSMA (S.), 1966, article in *International symposium . . .*, 1966.

JENNINGS (J. N.), "Glacier retreat in Jan Mayen," *J. Glac.*, Oct. 1948, pp. 167–182.

JENNINGS (J. N.), "Snaefell East Iceland," *J. Glac.*, vol. 2, 1952–1956, pp. 133–145.

JULIAN (P. R.) and FRITTS (H. C.), ". . . Extending climatic records by dendroclimatological analysis," *Proc. of the 1st Statistical Met. Cong.*, Hartford, Conn., Amer. Met. Soc., May 1968.

KASSNER (C.), "Das Zufrieren des Lake Champlain von 1816 bis 1935," *Met. Zeitschr.*, 1935, p. 333, cited by WAGNER (A.), 1940.

KICK (W.), "Long-term glacier variations as measured by photogrammetry; A re-survey of Tunsbergdalsbreen after 24 years," *J. Glac.* Feb. 1966, pp. 1–17.

KINCER (J. B.), "Is our climate changing?" *Monthly Weather Review*, vol. 61, 1933, pp. 251–259.

KINSMAN (D.) and SHEARD (J. W.), "The glaciers of Jan Mayen," *J. Glac.*, vol. 4, fév. 1963, n° 34, pp. 439–447.

KINZL (H.), "Die grössten nacheiszeitlichen Gletschervorstösse in den schweizer Alpen und in der Mont-Blanc Gruppe," *Zeitschrift für Gletscherkunde*, 1932.

KLEBELSBERG (R. von), *Handbuch der Gletscherkunde und Glazialgeologie*, Vienna, 1949.

KOCH (L.), "The East Greenland Ice," *Meddelelser om Grönland*, 1945.

KORYAKIN (V. S.), "Recent variations of the dimensions of glaciation in Novaya Zemlya" (in Russian), 1968, Reviewed by M.G.A., Jan. 1969, p. 225.

KOSIBA (A.), "Changes in the glaciers . . . in S.W. Spitsbergen," *Bull. A.I.H.S.*, April 1963.

KOTLIAKOV (V. M.), "New data on the present glaciation of the Caucasus," 1967 (in Russian), cited by M.G.A., March 1969, p. 773.

KRAUS (E. B.), "Recent changes of east coast rainfall regimes," *Q.J.R.M.S.*, 1963, pp. 145–146.

KRAUS (E. B.), "Secular changes of tropical . . . and east-coast rainfall regimes," *Q.J.R.M.S.* 1955, pp. 198–210 and 439.

KUSHINOVA (K. V.), (1968), article on climatic causes of climatic change (in Russian): cf. M.G.A., Aug. 1969, p. 2075.

LABEYRIE (J.), DUPLESSY (J. C.), DELIBRIAS (G.), and LETOLLE (R.), "Températures des climats anciens; mesures d'O 18 et C 14 dans les concrétions des cavernes," *Radioactive dating and methods of low-level counting, symposium*, Monaco, 1967 (International Atomic Energy Agency, Vienna, 1967).

LA BÉDOYÈRE (Henri de), *Journal d'un voyage en Savoie en 1804 et 1805,* Paris, 1807 and 1849.

LABRIJN (A.), "Het Klimaat van Nederland gedurende de laatste twee en een halve eeuw" (with summary in English), Koninklijk Nederlandsch Met. Inst., n° 102, *Meded. Verhandelingen,* Gravenhage, 49 (1945).

LABROUSSE (E.), *La crise de l'économie française à la fin de l'Ancien Régime,* Paris, 1944.

LA CHAPELLE (E. R.), Critical note, in *J. Glac.,* June 1965, p. 755.

LAHR (E.), *Un siècle d'observations météorologiques en Luxembourg,* published by the *Min. de l'Agr., Serv. mét.* Luxembourg, 1950.

LAMB (H. H.), "Climatic change within historical time," *A.N.Y.A.S.,* vol. 95, art. I, 1961, pp. 124–161.

LAMB (H. H.), Contribution to the Aspen Proceedings (mimeographed) 1962, pp. 85 sq. (q.v.).

LAMB (H. H.), PROBERT-JONES (J. R.), SHEARD (J. W.), "A new advance of the Jan Mayen glaciers," *J. Glac.,* Oct. 1962.

LAMB, LEWIS, WOODROFFE, article in *International Symposium . . . ,* 1966 (q.v.).

LAMB (H. H.), "Trees and climatic history in Scotland," *Q.J.R.M.S.,* Oct. 1964.

LAMB (H. H.), "The early medieval warm epoch and its sequel," *Palaeogeography, palaeoclimatology, palaeoecology,* I, 1965, pp. 13–37.

LAMB, LEWIS, WOODROFFE, article in *International Symposium . . . ,* 1966 (q.v.).

LAMB (H. H.), *The Changing Climate,* London, 1966.

LAMB (H. H.) "On climatic variations affecting the far south," (*World Meteorological Org., Techn.* note n° 87, 1967, pp. 428–453: reviewed in *M.G.A.,* Dec. 1968, p. 3052)

LAMB (H. H.), "Climate in the 1960's," *Geogr. Journal,* June 1966, pp. 183–212.

LAMB (H. H.), 1968, see HOINKES, 1968.

LAMB (H. H.), Contribution on "climatic fluctuations," *8ème congrès de l'INQUA* (*Union internationale pour le quaternaire*), Paris, Aug.–Sept. 1969.

LAMING-EMPERAIRE (A.), *Découverte du passé,* Paris, 1952.

LANDSBERG (H. H.), "Trends in climatology," *Science,* 1958, vol. II, p. 755.

LANCE (R.), "Zur Erwärmung Grönlands und der Atlantischen Arktis," *Ann. der Met.,* 1959, pp. 265–277.

LAPEYRE (H.), Contribution to *Charles-Quint et son temps,* Paris (C.N.R.S.), 1959, p. 40.

LATOUCHE (R.), *La vie en Bas-Quercy* (XIVe–XVIIIe *siècles*), Toulouse, 1923.

LAURENT (R.), *Les vignerons de la "Côte-d'Or" au XIXe siècle,* Dijon, 1957–1958.

LAWRENCE (D. B.), "Glacial fluctuations for six centuries in Southern Alaska," *Geog. Rev.,* April 1950.

LEBOUTET (L.), "Techniques et méthodes de la dendrochronologie . . . ," *Annales de Normandie,* Dec. 1966.

LEBRUN (F.), see LEHOREAU (R.), published 1967.

LE DANOIS (E.), *L'Atlantique,* Paris, 1938.

LE DANOIS (E.), *Le rythme des climats dans l'histoire de la terre et de l'humanité,* Paris, 1950.

LE GOFF (J.), *La civilisation de l'occident médiéval,* Paris, 1964.

LEHOREAU (R.), *Cérémonial de l'Eglise d'Angers* (1692–1721) published by LEBRUN (F.), Paris (C. Klinsieck), 1967.

LEROI-GOURHAN (Mme A.), Contribution to the 1959 Congress of the *Société préhistorique française.*

LE ROY LADURIE (E.), "Fluctuations météorologiques et bans de vendanges au XVIIIe siècle," *Féd. hist. du Lang. médit. et du Rouss.,* 30th and 31st Congresses. Sète-Beaucaire (1956–1957), Montpellier, n.d., p. 189.

LE ROY LADURIE (E.), "Historie et climat," *Annales,* 1959.

LE ROY LADURIE (E.), Climat et récoltes aux XVIIe et XVIIIe siècles," *Annales,* 1960.

LE ROY LADURIE (E.), "Aspects historiques de la nouvelle climatologie," *Rev. hist.*, 1961.

LE ROY LADURIE (E.), "La conférence d'Aspen . . . ," *Annales*, 1963.

LE ROY LADURIE (E.), "Le climat des XIe et XVIe siècles, séries comparées," *Annales*, 1965.

LE ROY LADURIE (E.), *Les Paysans de Languedoc*, Paris, 1966.

LESCHEVIN (P.-X.), *Voyage à Genève et dans la vallée de Chamouni*, Paris and Geneva, 1812.

LETONNELIER (G.), "Documents relatifs aux variations des glaciers dans les Alpes françaises," *Comité des travaux historiques et scientifiques, Bulletin de la section de géographie*, vol. 28, 1913.

LILJEQUIST (G. H.), "The severity of the winters at Stockholm, 1757–1942," *Geog. Ann.*, 1943, pp. 81–97.

LINDZEY (A. A.) and NEWMANN (J. E.), "Use of official datas in spring time temperature analysis of Indiana phenological record," *Ecology*, 37-4, Oct. 1956.

LLIBOUTRY (L.), "More about advancing and retreating glaciers in Patagonia," *J. Glac.*, vol. 2, 1952–1956, pp. 168 ff.

LLIBOUTRY (L.), *Nieves y Glaciares de Chile, Fundamentos de Glaciologia*, Santiago, 1956.

LLIBOUTRY (L.), *Traité de glaciologie*, Paris, vol. I, 1964 and vol. II, 1965.

LOEWE (F.), "Variations of the Qaumaruyuk glacier, Western Greenland 1930–1967," *Beiträge zur Geophysik*, 77 (3), pp. 232–234, 1968.

LONGSTAFF (T. G.), "The Oxford University Expedition to Greenland, 1928," *The Geographical Journal*, July 1929, pp. 61–68.

LUCAS (H. S.), "The famine of 1315–1317," *Speculum*, 5, 1930, pp. 341–377.

LUDLUM (D. M.), *Early American Winters*, Boston, (American meteorological society), 1966.

LUNN (A.), *The Bernese Oberland*, London, 1958.

LÜTSCHG (O.) (1), *Uber Niederschlag und Abfluss im Hochgebirge, Sonderdarstellung des Mattmarkgebietes*, Schweizerische Wasserwirtschaftverband, Verbandschrift C n° 14, Veröffentlichung der Schweizerischen meteorologischen Zentralanstalt in Zürich, Zürich, Sekretariat des Schweizerischen Wasserwirtschaftverbandes, 1926.

> This book is very difficult to come by: I was only able to read it at Zürich at the headquarters of the Wasserwirtschaft Verband, and more recently (1969) in the Science Library of the University of Michigan (Ann Arbor).

LYSGAARD (L.), "Recent climatic fluctuations," *Folia geographica danica* (Copenhague), 1949.

LYSGAARD (L.), "On the climatic variation" in *Changes of Climate*, 1963, pp. 151–160 (q.v.).

MACGREGOR (V. R.), "Holocene Moraines in the Ben Ohau Range, N. Zeal.," *J. Glac.*, vol. 6, n° 47, 1967, p. 746.

MAGNUSSON (M.), see *Graenlandica Saga*.

MALAURIE, (J.), *Thèmes de recherche géomorphologique . . . en Groenland*, éd. du C.N.R.S., Paris 1968 (especially pp. 443–447).

MANLEY (G.), "Temperature trends in Lancashire," *Q.J.R.M.S.*, 1946.

MANLEY (G.), "The range of variation of the British climate," *Geog. Journ.*, March 1951, pp. 43–68.

MANLEY (G.), "Variation in the mean temperature of Britain since glacial times," *Geologische Rundschau*, 1952, pp. 125–127.

MANLEY (G.), "The mean temperature of central England, 1698–1952," *Q.J.R.M.S.*, 1953, pp. 242–262 and 558.

MANLEY (G.), "Temperature trends in England," *Archiv. für Met., Geophys. und Bioklimatol.*, 1959, reprint with an additional manuscript series, kindly provided by Mr. MANLEY, and containing a table of temperatures in England from 1670 to 1697.

MANLEY (G.), "Possible climatic agencies in the development of post-glacial habitats," *Proceedings of the Royal Society*, B, vol. 161, pp. 363–375, 1965.
MANLEY (G.), article in *International Symposium* . . . , 1966.
MARTEL (P.), *Relation d'un voyage aux glacières du Faucigny*, 1743. In the library at Geneva, but not in the B.N.; was republished by H. FERRAND in 1912, but without the illustrations; cf. WINDHAM and MARTEL.
MARTIN (E.), *Histoire de Lodève*, Montpellier, 1900.
MARTIN (P. de), "Datation de poutres anciennes dans les maisons rurales du Livradois (Auvergne, France), unpublished, about 1969. See also DE MARTIN's article (to be published in *Annales* E.S.C., 1970) on seventeenth-century house-beams from Lorraine, dated by comparing their tree rings with HOLLSTEIN's chronology for western Germany, 1965 (q.v.).
MARTINS (Ch.), Articles on contemporary glaciers in *Revue des Deux Mondes*, Jan. 15, Feb. 1, Mar. 1, 1867 and April 15, 1875.
MASSIP (M.), "Variations du climat de Toulouse," *Mémoires de l'Académie de Toulouse*, 1894–1895.
MATHEWS (W. H.), "Fluctuations of glaciers . . . in S.W. British Columbia," *Journal of Geology*, 1951, pp. 357–380.
MATTHES (F. E.), "Glaciers," in *Hydrology*, a collection of papers edited by MEINZER (O. E.), New York, 1942.
MAUNDER (E.), Paper on the prolonged minimum of sun spots, 1645–1715, in *Journal of the British Astronomers Association*, 1922.
MAURER (J.), "Uber Gletscherschwund und Sonnenstrahlung," *Met. Zeitschr.*, 31, 1914, p. 23.
MAURER (J.), "Neuer Rückzug der Schweizer Gletschern," *Met Zeitschr.*, 1935, p. 22.
MAYR (F.), "Untersuchungen über Ausmass und Folgen der Klima und Gletscherschwankungen seit dem Beginn der postglazialen Wärmezeit. Ausgewählte Beispiele aus den Stubaier Alpen in Tirol," *Zeitschrift für Geomorphologie*, 1964. (See also F. MAYR's contributions to the 7th Congress of INQUA, 1966.)
MELLOR (M.), "Variations of the ice margins in East Antarctica," *Geogr. Journ.*, June 1959.
MELLOR (M.), "Mass balance studies in Antarctica," *J. Glac.*, vol. 3, Oct. 1959, n° 26, pp. 522–533.
MENEVAL (Baron de), *Récit d'une excursion l'Impératrice Marie-Louise aux glaciers de Savoie, en juillet 1814*, Paris, 1847.
MERCANTON (P.-L.), "Mensurations au glacier du Rhône," *Neue Denkschriften der Schweiz. Naturforschenden Gesellschaft*, vol. 52, 1916.
MERCANTON (P.-L.), Detailed notes on contemporary fluctuations in Alpine glaciers in *Die Alpen*, 23 1947, pp. 313–320; 24, 1948, p. 387 and 1949, p. 267; and in *J. Glac.*, vol. II, 1952–1956, p. 110.
MERCANTON (P.-L.), "Glacierized areas in the Swiss Alps," *J. Glac.*, April, 1954, n° 15, pp. 315–316.
MERCER (J.), "Glacier variations in Patagonia," *Geog. Rev.*, 1965, pp. 390–413. In this article see especially pp. 410–413 and notes 47–53 for the substantial bibliography on glacier advances in the subatlantic period about 300 B.C., in Europe, America and New Zealand.
MERCER (J. H.), "Glacier resurgence at the Atlantic sub-Boreal transition," *Q.J.R.- M.S.*, 1967, pp. 528–534.
MERIAN (M.), *Topographia helvetiae*, n.p., 1642.
MILANKOVITCH (M.), *Canon of insolation and the ice age problem*, (translation), U. S. Department of Commerce, 1969.
MILES (M. K.), "The summers of North-west Europe," *Met. Mag.*, 1967, p. 318.
Ministère de l'Agriculture, Direction de l'Hydraulique et des Améliorations agricoles. Service d'étude des grandes forces hydrauliques (Région des Alpes) (à partir du t. III, Direction générale des Eaux et Forêts). *Etudes glaciologiques:*

Volume I: *Tyrol autrichien. Massif des Grandes-Rousses*, by MM. FLUSIN, JACOB and OFFNER, 1909.

Volume II: *Etudes glaciologiques en Savoie*, by M. MOUGIN (P.) and *Programme et méthodes applicables à l'étude d'un grand glacier*, by M. BERNARD (C.-J.-M.), 1910.

Volume III: *Etudes glaciologiques, Savoie et Pyrénées.—I. Etudes glaciologiques en Savoie*, by M. MOUGIN (P.); II. *Observations glaciaires dans les Pyrénées*, by M. GAURIER (Ludovic), 1912.

Volume IV: I. *Etude sur le glacier de Tête-Rousse*, by MM. MOUGIN (P.) and BERNARD (C.); II. *Les avalanches en Savoie*, by M. MOUGIN (P.), 1922.

Volume V: *Etudes glaciologiques en Savoie*, by M. MOUGIN (P.), 1925.

Volume VI: *Observations glaciologiques faites en Dauphiné jusqu'en 1924*, recapitulated and partly edited by M. ALLIX (André), and *Variations historiques des glaciers des Grandes Rousses*, by M. MOUGIN (P.), 1934. N. *Etudes glaciologiques*, 1912, exists in two editions with the pages differently numbered. I used one edition for my 1960 article and the other for this book.

MITCHELL (J.), "Recent secular changes of global temperature," A.N.Y.A.S., vol. 95, art. 1, 1961, pp. 235–250.

MITCHELL (J.), "Contribution to *Changes of Climate*, 1963, pp. 161–182 (q.v.).

MITCHELL (J.) "Stochastic models of air-sea interaction and climatic fluctuation," *Proceedings of the symposium on the artic heat budget and atmospheric circulation*, edited by J. O. FLETCHER, Jan. Feb. 1966, (The Rand Corporation Memorandum, R.M. 5233, N.S.F. Dec. 1966).

MITCHELL (J.), and others, *Causes of climatic change, in Meteorological monographs*, vol. 8, n° 30, Feb. 1968 (publ. by the Amer. Met. Soc.).

MITCHELL (J. M.), "A critical appraisal of periodicities in climate," reprint from *Weather and our food supply*, C.A.E.D. report 20, Iowa State Univ., Amer. (Iowa), 1964.

MITCHELL (J. M.), "Theoretical paleo-climatology," in *Quaternary of the United States*, 1965.

MITCHELL (J.) and KISS (E.), "Bibliography on climatic changes in historical times," M.G.A., 15 December 1964.

MOCK (S. J.), "Fluctuations of the terminus of the Harald Moltke Brae glacier, Greenland," *J. Glac.*, Oct. 1966, pp. 369–373.

MONTARLOT (G.), "Facteurs météorologiques et végétation de la vigne," *Annales de l'Ecole nationale d'agriculture de Montpellier*, vol. XXII, fasc. 5, p. 236.

MONTERIN (U.), "Il clima sulle Alpi ha mutato in epoca storica" *Bolletino del Comitato glaciologico italiano*, 16, 57, 1936–1937. This is the best synthesis in Italian glaciological literature on "recent" (i. e., historic) fluctuations in the Alpine glaciers. But on one point this exemplary study needs correction: travel backwards and forwards over the passes of the Alps is not to be accounted for in terms of climate and glaciers, except in those very rare cases when a pass is physically blocked by an advancing glacier. In all other cases the traffic simply varies according to the commercial factors involved.

Monumenta Germaniae historica, Hanovre, Published 1937–1962.

MORINEAU (M.), "d'Amsterdam à Seville . . . ," *Annales* Jan. 1968.

MORINEAU (M.), Study on the agricultural non-revolution in French in the eighteenth century, to be published in *Cahiers des Annales* (and also, see *Rev. Hist.*, April 1968, p. 299 ff.)

MORIZE, "Aiguesmortes au XIIIe siècle," *Ann. du Midi*, 26, 1914, p. 313.

MORRISSON, 1966, article in *International Symposium . . .*, 1966.

MOSSMAN (R.), "Reduction of the Edinburgh meteorological observations to 1900," *Royal Society of Edinburgh, Transactions*, 40, 1902.

MOUGIN (P.). Cf. Ministère de l'Agriculture.

MÜLLER (K.), "Weinjahre und Klimaschwankungen der letzten 1000 jahre," *Weinbau, Wissenschaftliche Beiheft* (Mainz), 1, 83, 123 (1947).

MULLER (K.), *Geschichte des Badischen weinbaus*, (*mit einer Weinchronik und einer Darstellung der Klimaschwankungen in letzen Jahrtausend*), Lahr in Baden, 1953.

MUNAUD (A. V.), Dendrochronological studies in Belgium *Agricultura* (Belgian) vol. 14, 1966, n° 2.

MUNSTER (S.) *Cosmographiae universalis lib.* VI, Bâle, 1552; *Cosmographey oder Beschreibung aller Länder*, Bâle, 1567.

MURRAY (R.) and MOFFITT (B. J.), "Monthly patterns of the quasi-biennial pressure oscillation," *Weather*, Oct. 1969, pp. 382–390.

MUSSET (L.), *Les peuples scandinaves au Moyen Age*, Paris, P.U.F., 1951.

NAGERONI (G.), "Appunti per una revisione del catalogo dei ghiacciai lombardi" (*Att. soc. Ital. Sc. Nat.*, XXXXIII, pp. 373–407), quoted by CAILLEUX, *Revue de Geomorphologie Dyn.*, 1955, card 528.

NICHOLS (H.), "Central Canadian palynology and its relevance to North-western Europe in the late quaternary period," *Rev. of paleobotany and palynology* (Amsterdam), 2, 1967, pp. 231–243.

NICHOLS (H.), "Late quatern. history of vegetation and climate . . . in Manitoba," *Arctic and Alpine Research*, 1, 3, 1969, pp. 155–167.

NORLUND (P.), "Buried Norsemen at Herfoljness," *Meddelelser om Gronland*, vol. 67, 1924, pp. 228–259.

NORLUND (P.), *Viking settlers in Greenland*, London, 1936.

ODELL (N. E.), "Mount Ruapehu, N.Z., observations on glaciers," *J. Glac.*, vol. 2, 1952–1956, pp. 601 ff.

ŒCHSGER (H.) and ROTHLISBERGER (H.), "Datierung eines ehemaligen Standes der Aletschgletschers," *Zeitschrift für Gletscherkunde*, IV, 3, 1961, pp. 191–205.

OLAGÜE (I.), *La decadencia española*, Madrid, 1951 (esp. volume IV, pp. 247–306).

OLAGÜE (I.), *Histoire d'Espagne*, Paris, 1958.

OLAGÜE (I.), "Les changements de climat l'histoire," *Cahiers d'histoire mondiale*, VII, 3/1963, p. 637.

OLIVER (J.): Professor Oliver has made a systematic study of early records and diaries for information on weather in the eighteenth century. His articles, which are models of their kind, have appeared in *Weather and Agriculture*, ed. J. A. Taylor, Pergamon, Oxford, 1967; *Q.J.R.M.S.*, vol. 84, n° 360, April 1958, pp. 126–133; *Weather*, vol. 13, n° 8, Aug. 1958; vol. 16, n° 10, Oct. 1961; vol. 20, n° 12, Dec. 1965; *The National Library of Wales Journal*, X, 3, summer 1958; *Transactions of Anglesey Antiquarian Society and Field Club*, 1958.

OSBORNE (D.), and NICHOLS (R.), "Dendrochronology of the Wetherhill Mesa," *Tree-Ring Bull.*, May 1967.

OVERBECK (F.) and GRIEZ (I.), "Mooruntersuchungen zur Rekurrenzflächen-frage . . . in der Rhön," *Flora*, 141 (1954), pp. 51 sq.

OVERBECK (F.), MUNNICH (K.), ALETSEE (L.), AVERDIECK (F.), "Das Alter des Grenzhorizonts nord-deutscher Hochmoore nach Radiocarbon-Datierungen," *Flora*, 145, (1957), p. 37.

OYEN (P. A.), "Klima und Gletscherschwankungen in Norwegen," *Zeitschrift für Gletscherkunde*, May 1906, pp. 46–61 and 173–174.

PAGE (N.), "Atlantic/early sub.Boreal glaciation in Norway" *Nature*, 219, Aug. 17, 1968, pp. 694–697.

PAPON (L'abbé), *Histoire générale de Provence*, Paris, 1786.

PAYOT (P.), *Au Royaume du Mont Blanc*, Bonneville, 1950.

PÉDELABORDE (P.), "La circulation sur l'Europe occidentale," *Annales de géog.*, 1953.

PÉDELABORDE (P.), *Le climat du Bassin parisien*, Paris, 1957.

PÉDELABORDE (P.), Reviews of recent climatological publications, *Annales de géog.*, 1967, pp. 79–92 and especially pp. 203–220.

PEROUX (G.), Inventaire des Archives départementales de la Savoie, Series E, vol. I, introduction.

PFANNENSTIEL (M.), "Die Schwankungen des Mittelmeerspiegels als Folge der

420 Bibliography

Eiszeiten," Freiburger Universitätsreden, Neue Folge, H. 18, Freiburg, 1954, quoted by H. V. RUDLOFF, 1967, p. 342.

PILLEWIZER (W.), account of D.D.R. expedition to Spitzberg, *Petermanns geogr. Mitt.*, 1962, p. 286.

PLASS (G. N.), "The carbon dioxide theory of climatic change," *Tellus*, vol. 8, May 1956, pp. 140–154.

PLATTER, *Félix et Thomas Platter à Montpellier* (1552–1559, 1595–1599). *Notes de voyage de deux étudiants bâlois, Montpellier, 1892.*

POGGI (A.), "La neige et les avalanches," *Mét.*, April–June 1961.

POISSENOT (Bénigne), *Nouvelles histoires tragiques . . . ensemble une lettre à un ami contenant la description d'une merveille appelée la Froidière veue par l'autheur en la Franche Comté de Bourgogne*, Paris, Richou, 1586, pp. 440–451 Copy in Bibl. nationale.

POLGE (H.) and KELLER (R.), "La Xylochronologie perfectionnement logique de la dendrochronologie" *Annales des Sciences forest.*, 1969, 26 (2), pp. 225–256. This article proposes a new and sophisticated approach to dendroclimatology.

POST (A.), "The recent surge of Walsh glacier (Yukon and Alaska)," *J. Glac.*, vol. 6, n° 45, Oct. 1966.

POTAPOVA (L. S.), 1968, article on climatological causes of climatic change (in Russian) cf. M.G.A., August 1969, p. 2075.

POWELL (W.), *Silver, Soldiers and Indians*, University of California, 1955.

PRIBRAM (A. F.), *Materialen zur Geschichte der Preise . . .*, Vienna, 1938.

"Proceedings of the Conference on the Climate of the Eleventh and Sixteenth Centuries, Aspen, June 16–24, 1962," National Center for Atmospheric Research, Boulder, Colorado (N.C.A.R. Technical Notes, 63–1).

RABOT (Ch.), "Les glaciers du Pelvoux au début du XIXe siècle," *La Géographie*, 1914, vol. 29.

RABOT (Ch.), "Récents travaux glaciaires dans les Alpes françaises," *La Géographie*, vol. 30, 1914–1915, pp. 257–268.

RABOT (Ch.), "Les catastrophes glaciaires dans la vallée de Chamonix au XVIIe siècle," *La Nature*, 28 August 1920.

RAIKES (Robert), *Water, weather, and prehistory*, London, 1967.

RAMDAS (L. A.), and RAJAGOPALAN (N.), Contribution to *Changes of Climate*, 1963, pp. 87–91 (q.v.).

RATINEAU (J.), *Les céréales*, Paris, 1945, pp. 53–57.

REBUFFAT (G.), *Mont-Blanc . . .*, Paris, 1962.

REITER (E. R.) *Jet-stream meteorology*, Univ. of Chicago Press, 1963.

REKSTAD (J.), "Gletscherschwankungen in Norwegen," *Zeitschrift für Gletscherkunde*, 1906–07.

Relation anonyme d'une visite à Chamouni en 1764, edited and annotated by H. FERRAND, Lyon, 1912 (B.N. LK⁷, 39 030).

RENOU (E.), "Etudes sur le climat de Paris," *Annales du Bureau central météorologique de France*, 1887, vol. I, B 195 à B 226.

REY (M.), "La source et le glacier du Rhône en 1834," *Nouvelles Annales des voyages*, Paris (Pihan), 1835.

RICHARD (J. M.), "Thierry d'Hireçon," *Bibliothèque de l'Ecole des Chartes*, vol. 53, 1892.

RICHTER (E.), "Zur Geschichte des Vernagtgletschers," Z.D.O.A., 1877.

RICHTER (E.) *Die Gletscher der Ostalpen*, Stuttgart, 1888.

RICHTER (E.), "Geschichte der Schwankungen der Alpengletscher," Z.D.O.A., 1891.

RICHTER (E.), "Urkunden über die Ausbrüche des Vernagt—und Gurglergletschers im 17 und 18 Jahrhundert, aus den Innsbrucker Archiven herausgegeben," *Forschungen zur deutschen Landes—und Volkskunde*, 6, 1892.

RODEWALD (M.) Contribution to *Changes of Climate*, 1963, pp. 97–108 (q.v.)

RODEWALD (M.) "Die Klimaschwankung in Westgrönland," *Wetterlotse* (Hamburg), Feb. 1968.

ROGERS (Th.), A *history of agriculture and prices in England*, London, 1886–1887.
ROGSTAD (O.), Articles on variations of Norwegian glaciers in *Norsk geografisk Tidskrift*, 1941, p. 273; and 1942, p. 129, cited by J. *Glac.* vol. 1 (1947–1951, p. 151).
ROGSTAD (O.), "Variations in the glacier mass of Jostedalsbreen," J. *Glac.*, vol. 1, 1947–51, p. 551.
RONDEAU (A.), "Recherches morphologiques dans l'État de Washington," *B.A.G.F.*, Nov. Dec. 1954, pp. 183–195.
ROSE (B.), "18th century price riots," *International review of Social History*, III, 1959, pp. 432–445.
ROSEMAN (N.), Contribution to *Changes of Climate*, 1963, pp. 67–74 (q.v.)
RUBIN (M.), Contribution to *Changes of Climate*, 1963, pp. 223–229 (q.v.)
RUDÉ (G.), "La taxation populaire de Mai 1775 (la guerre des farines)," *Annales d'hist. de la Révol. franç.*, 1956, pp. 139–179; and 1961, pp. 305–326.
RUDÉ (G.), *The Crowd in History*, New York, 1964.
RUDLOFF (H. VON): See VON RUDLOFF (H.)
RUSSELL (J. C.), *British Medieval Population*, Albuquerque, 1948.
SACCO (F.), "Il ghiacciaio ed i laghi del Ruitor," *Bol. della Soc. geol. Ital.*, 36, 1917, pp. 1–36.
SANSON (J.), *Relations entre le caractère météorologique des saisons et le rendement du blé*, Publications of O.N.M., Paris, n.d.
SANSON (J.), "Températures de la biosphère et dates de floraison des végétaux," *La Météorologie*, Oct.–Dec. 1954, pp. 453–456.
SANSON (J.), "Y a-t-il une périodicité dans la météorologie?" *La Météorologie*, 1955.
SANSON (J.), "En marge météorologique de l'histoire," *La Météorologie*, 1956.
SAUSSURE (H.-B. de), *Voyages dans les Alpes*, Neufchâtel, 1779 and 1786.
SCHERHAG (R.), "Die Erwärmung des Polargebiets," *Ann. hydrogr.* (Berlin), 1939, pp. 57–67 and 292–303.
SCHEUCHZER (J.), ουρεσιφοιτης helveticus, sive itinera et Helvetiae alpinas regiones, Lugduni Batavorum, 1723.
SCHMUCK (A.), "Frequency of dry spells in Wroclaw, 1883–1960, and max. of droughts, 1946–1960," *Wroclawskie Towarzystwo Naukowe (Annales Silesiae)*, Wroclaw, 1951.
SCHOVE (K. J.), Contribution to Post-glacial climatic change," *Q.J.R.M.S.*, 1949, pp. 175–179 and 181.
SCHOVE (D. J.), *Climatic fluctuations in Europe in the late historical period*, M. Sc. Thesis, University of London, 1953 (unpublished).
SCHOVE (D. J.), "Medieval Chronology in the USSR," *Medieval Archaeology*, vol. 8, 1964, pp. 216–217.
SCHOVE (D. J.), "Fire and drought, 1600–1700," *Weather*, Sept. 1966. An important article giving many texts on the drought which preceded the Great Fire of London in 1666, and drawing attention to the link between dry years and narrow tree rings in the English oak.
SCHOVE (D. J.), "The sunspot cycle (a historical record)," *Journ. of geophys. res.*, 60, 2 1955, pp. 127–146. This article contains an important bibliography on the history of the sunspot cycle *before* the time of rigorous observations, i.e., during the Middle Ages and the early modern period. The bibliography is particularly interesting on early references to the occurrence of the northern lights.
SCHOVE (D. J.), "The biennial oscillation," *Weather*, Oct. 1969, pp. 390–396.
SCHULMAN (E.), "Tree-ring Indices of Rainfall, Temperature and River Flow," *Compendium of Meteorology*, The American Meteorological Society, Boston, 1951.
SCHULMAN (E.), "Tree-ring and History in the Western United States," *Smithsonian Report for 1955*, 459–473, Smithsonian Institute of Washington, 1956.
SÉVIGNÉ, *Lettres*, Hachette, Paris, 1862.
SHAPIRO (R.), "Circulation pattern," contribution to Aspen *Proceedings*, 1962, pp. 59 ff. (q.v.)

422 Bibliography

SHAPLEY (H.), Cf. *Climatic change.*
SHARP (R.), "The latest major advance of Malaspina glacier, Alaska," *Geog. Rev.*, 1958.
SHORT (T.), *General Chronological History*, 1749.
SHORT (T.), *New observations . . . with an appendix on the weather and meteors,* London, 1750.
SIMMONS (L.), *Soleil hopi*, translation, Paris, 1959.
SIREN (G.), "Skogsgränstallen som indikator för klimatfluktuationerna I norra fennoskandien under historisk tid," *Metsäntutkimuslaitoksen Julkaisuja 54.2 (Communicationes Instituti forestalis fenniae 54.2)*, Helsingfors, 1961.
SITZMANN (P.), "Les variations des glaciers du bassin de la Romanche," *Revue de géographie alpine*, Jan. 1961.
SKELTON, 1965, article in SKELTON, MARSTON, and PAINTER, 1965.
SKELTON (R. A.), MARSTON (T. E.), PAINTER (G. D.), *The Vinland Map and the Tartar Relation*, with a foreword by A. O. VIETOR, New Haven, Yale Univ. Press, 1965.
SLICHER VAN BATH (B. H.), "Les problèmes fondamentaux de la société pré-industrielle en Europe," *A.A.G. Bijdragen*, 12, 1965.
SMILEY (T. L.), "A Summary of tree-ring dates from South-western archaeological sites," University of Arizona *Bulletin*, vol. 22, *Laboratory of tree-ring Research, Bulletin* n° 5, (Tucson, Ariz.)
SMITH (D. G.) and NICHOLS (R. F.), "A tree-ring chronology for climatic analysis (Mesa Verde)" *Tree-ring Bull.*, May 1967.
SMITH (L. P.), Contribution to *Changes of Climate*, 1963, pp. 457–469 (q.v.).
SOKOLOV (A.), "Diminution de la durée du gel des rivières en liaison avec le réchauffement du climat" (in Russian) *Priroda*, 1955, pp. 96–98.
"Solar variation, climatic change . . . ," a collection of articles by different hands (740 pages) published in A.N.Y.A.S., vol. 95, art. 1, 5 Oct. 1961.
SPINK (P. C.), "Glaciers of East Africa," *J. Glac.*, vol. 1, 1947–1951, p. 277.
SPINK (P. C.), "Recession of the African glaciers," *J. Glac.*, vol. 2, 1952–1956, p. 149.
STARKEL, 1966, article in *International Symposium . . .* , 1966, (q.v.).
STEENSBERG (A.), "Archeological dating of the climatic change in Northern Europe, about 1300," *Nature*, 20 Oct. 1951, pp. 672–674.
STOCKTON (C. W.) and FRITTS (H. C.), "Probability . . . for variation in climate, based on widths of tree-rings," *Annual report, Grant E. 88–67 (G.) Environmental Science Services Administr.* Jan. 1968.
STOLTZ (O.), "Anschauung und Kenntnis der Hochgebirge Tyrols," *Z.D.O.A.* 1928.
STRASSER (G.), Cf. *Der Gletschermann.*
STRETEN (N.) and WENDLER (G.), "An Alaskan Glacier," *J. Glac.*, Oct. 1968.
STUMPF (J.), *Gemeiner löblicher Eydgenossenschaft Landen . . . Beschreibung*, Zurich, 1547–1548 and 1586, 1606, etc.
STUDHALTER (R.), "Early History of crossdating," *Tree-ring Bulletin*, April 1956.
STUIVER (M.) and SUESS (H. E.), "On relationship between radiocarbon dates and true sample ages," *Radiocarbon*, (published by *Amer. Journ. of Science*), 1966, vol. 8, pp. 534–540 (cf. especially p. 537, table of corrections; and p. 538, graph).
SUESS (H.), see STUIVER, 1966.
SUESS (H.), "Climatic changes, solar activity, and the cosmic-ray production rate of natural radiocarbon," 1968, article in MITCHELL, 1968 (q.v.).
SUESS (H.), "The three causes of the secular C 14 fluctuations," *unpublished paper*, Aug. 1969.
SUESS (H.), "Bristlecone pine calibration of the radiocarbon time scale, 5300 B.C. to the present," *unpublished paper*, Aug. 1969.
SUTCLIFFE (R.), Contribution to *Changes of Climate*, 1963, pp. 277–285 (q.v.)

TAILLEFER (F.), "Une histoire du climat . . . ," *Rev.* (de) *géog. des Pyrénées et du Sud-Ouest*, vol. 38, 1967, p. 373.

TEMPLE (P. H.), "Survey of the Ruwenzori glaciers for recent changes," *Nature and Resources* (*Unesco*), 5, 1, March 1969, pp. 13–14.

THEAKSTONE, "Recent changes in the glaciers of Svartisen," *J. Glac.*, Feb. 1965, pp. 411 sq.

THOMSON (E. P.), "The moral economy of the poor (the food riot)," *unpublished paper*, Buffalo, 1966.

THORARINSSON (S.), "The ice dammed lakes of Iceland," *Geog. Annaler*, 1939–1940.

THORARINSSON (S.) "Vatnajökull," *Geog. Ann.*, 1943.

THORARINSSON (S.), "Present glacier shrinkage," *Geog. Ann.*, 1944.

TILLIER (J.-B. de), *Histoire de la vallée d'Aoste*, ms. de 1742, Aost, 1880, republished 1953.

TITOW (J.), "Evidence of weather in the account rolls of the bishopric of Winchester, 1209–1350," *Economic History Review*, 1960.

TOUSSOUN (O.), "Mémoires sur l'histoire du Nil," in *Mémoires de l'Institut d'Egypte*, vol. IX, 1925, pp. 366–410.

TRASSELLI (C.) "Studi sul clima," *Rivista di storia dell'agricoltura*, March 1968.

TREVELYAN (G. M.) *English Social History*, London, 1942.

TRICART (J.), "Géomorphologie quaternaire," *R.G.D.*, Nov. Dec. 1958, p. 184.

TRICART (J.), and CAILLEUX (A.), *Le modelé glaciaire et nival* (vol. II du *Traité de géomorphologie*), Paris, 1962.

TURNER (R. M.), ALCORN (S. M.), OLIN (G.), and BOOTH (J. A.), "The influence of shade, soil and water on Saguaro seedling establishment," *The botanical gazette*, vol. 127, n° 2–3, June–Sept. 1966.

TYCHO-BRAHE, *Meteor. Dagbog* (1582–1597) edited by P. LACOUR, Copenhagen, 1876.

UNTERSTEINER (N.) and NYE (J.), ". . . Berendon glacier, Canada," *J. glac.*, June 1968.

USPENSKII (S. M.), "Poteplanie Arktiki i faună vysokikh shirot," *Priroda*, Feb. 1963 (cited in M.G.A., April 1964, p. 833).

UTTERSTRÖM (G.), "Climatic fluctuations and population problems in early modern history," *The Scandinavian Economic History Review*, 1955.

UTTERSTRÖM (G.), "Population problems in pre-industrial Sweden, *Scandinavian Economic History Review*, 1954, and 1962.

VACCARONE (Luigi), "I Valichi nel Ducato d'Aosta nel secolo 17," *Bolletino del club alp. Ital.* vol. 15, 1881, pp. 181–193 (contains F. ARNOD's account, 1691–1694, of the Savoie glaciers).

VACCARONE (L.), *Le vie delle Alpi occidentali negli antichi tempi*, Torino, (Candeletti), 1884, (also contains ARNOD's account).

VALENTIN, in *Pet. geog. Mitt.*, 1952, cited by FINSTERWALDER, 1954 (q.v.).

VANDERLINDEN (E.), *Chronique des événements météorologiques en Belgique jusqu'en 1834*, Brussels, 1924.

VAN DER WEE (H.) *The growth of the Antwerp market and the European economy*, The Hague, 1963.

VANNI (M.), "Le variationi frontali dei ghiacciai italiani," *Boll. del Comitato glaciologico italiano* (Turin), 1948, pp. 75–85.

VANNI (M.), "Le variationi recenti dei ghiacciai italiana," *Geofisica pura e aplicata*, 1950, p. 230.

VANNI (M.), ORIGLIA (C.), de GEMINI (F.), "I Ghiacciai della valle d'Aosta," *Bolletino del Comitato glaciologico italiano*, 1953.

VANNI (M.), "Variations of the Italian glaciers in 1961," *J. Glac.*, Feb. 1963.

VEBRAEK (C. L.), "The climate of Greenland in the 11th and 16th centuries"; Roneo'd contribution to Aspen Conference (1962) on the climate of the XIe and XVIe centuries; see *Proceedings . . . 1962* (q.v.).

VERYARD (R. G.), Contribution to *Changes of Climate*, 1963, pp. 3–36 (q.v.).

424 Bibliography

VEYRET (P.), "Trois glaciers du Pelvoux en 1951, *Rev. de geog. alp.*, 1952, pp. 197–199 (see also *J. Glac.* vol. 2, 1952–56, p. 154).

VEYRET (P.), "La tournée glaciologique de 1959," *Rev. de géog. alp.*, Jan. 1960, pp. 203–207.

VEYRET (P.), "Les variations des glaciers du Mont-Blanc," *La Montagne*, April 1966.

VILLENEUVE (Comte de), *Statistique des Bouches-du Rhône*, Marseilles, 1821.

VIRGILIO, "Note sur la catastrophe du Pré-de-Bar en 1717," *Bolletino del Club alpino italiano*, 1883.

VIVIAN (R.), "Le recul récent des glaciers du Haut-Arc et de la Haute-Isère," *Rev. de géog. alp.*, 1960 (2), pp. 313–329. See also VIVIAN's many single studies on Alpine glaciers in recent volumes of *Rev. de Géog. alp.*

VIVIAN (R.) "Glace morte et morphologie glaciaire," *Rev. de géog. alp.*, 1965.

VON POST (L.), "Pollen Analysis . . . in earth's climatic history," *New Phytol.*, 1946, pp. 193–218.

VON RUDLOFF (H.), "Die Schwankungen der Grosszirkulation. . . . immerhalb der letzten Jahrhunderte," *Annalen der Meteorologie*, 1967.

VON RUDLOFF (H.), *Die Schwankungen und Pendelungen des Klimas in Europa seit dem Beginn der regelmässigen Instrumenten-Beobachtungen*, Braunschweig (Vieweg), 1967.

WAGNER (A.), *Klimaänderungen und Klimaschwankungen*, Braunschweig, 1940.

WAGRET (P.), "Le climat se réchauffe-t-il?" *La Nature*, Jan. and Aug. 1958, 3273 and 3280, pp. 19 and 331.

WAHL (E. W.), "A comparison of the climate of the Eastern United States, during the 1830's with the current normals," *Monthly Weather Review*, Feb. 1968, pp. 73–82.

WALKER (D.), article on climatic fluctuations in New Zealand and Australia, in *International Symposium . . .* , 1966 (q.v.).

WALLEN (A.), "Température, pluie et récoltes," *Géog. Ann.*, 1920, pp. 332–357.

WALLEN (C. C.), "Glacial meteorological investigations of the Karsa glacier in Swedish Lapland," *Geog. Ann.*, 1948.

WALLEN (C. C.), "European weather in 1968," *Weather*, Oct. 1969.

WARD (W. H.) and BAIRD (P. D.), "A description of the Penny ice cap (Baffin Island," *J. Glac.* April 1954).

WATSON (W.) Contribution to *Proceedings . . .* 1962, p. 37 (q.v.); ibid. "Census of data" and "Bibliography."

WEBB (C. E.), "Investigations of glaciers in British Columbia," *Canadian Alpine Journal*, vol. 31, 1948, pp. 107–117, cited by *J. Glac.*, vol. 1, p. 275.

WEIDICK (A.), "Observations on some holocene glacier fluctuations in West Greenland" *Meddelelser on Grönland*, vol. 165, n° 6, 1968, 202 pages.

WEIKINN (C.), *Quellentexte zur Witterungsgeschichte Europas von der Zeitwende biz zum 1850*, Akademie-Verlag, Berlin, 1958–1967, 4 vol. published.

WEYL (P. K.), "Role of the ocean in climatic change," *Meteorological Monographs*, Boston, Feb. 1968.

WERENSKIOLD (W.), "Glaciers in Jotunheim," *Norsk geografisk Tidsskrift*, 1939.

WHITE (S. E.), "The firn field of the Popocatepelt," *J. Glac.*, 1952–1956, p. 389.

WHITOW (J.-B.), etc., "Glaciers of the Ruwenzori," *J. Glac.*, June 1963, pp. 581–617.

WILLETT (H. C.), "Climatic change; temperature trends of the past century," *Centenary proceedings of the Royal Meteorological Society*, 1950, p. 195.

WINDHAM and MARTEL, *Premiers voyages à Chamouni, Lettres de Windham et de Martel* (1741–1742) published with notes by H. FERRAND, Lyons, 1912, (B. N., LK7 39 030).

WISEMAN, article on ocean thermic fluctuations, in *International Symposium* . . . , 1966.

WOODBURY (R.), "Climatic changes and prehistoric agriculture in the southwestern United States," *N.Y.A.S.*, vol. 95, art. I, 5 Oct. 1961, pp. 705–709.

World Climate (8000–0 B.C.), see *International Symposium* . . . , 1966.

WRIGHT, 1966, article in . . . *International Symposium* . . . , 1966.

WYLIE (P. J.), "Ice recession in N.E. Greenland," *J. Glac.*, vol. 2, 1952–1956, p. 704.

ZEUNER (F. E.), *Dating the past, an introduction to geochronology*, London, 1949 (chap. I).

ZINGG (Th.), Paper and graphs, in *Bull. of the Assoc. of Scientific Hydrology*, June 1963, pp. 84–85.

ZUMBERGE (J. H.) and POTZGER (J. E.), "Late Wisconsin chronology of the Lake Michigan Basin, correlated with pollen studies," *Bull. of the Geol. Soc. of Amer.*, vol. 66, 1955-2 p. 1640.

ZURLAUBEN, *Tableaux topographiques de la Suisse*, Paris (Clousier), 1780.

INDEX

INDEX

Aario, Leo, 244, 245
Abkühlung, 98
Ablation, 227–43, 254–64. *See also* Amelioration; specific aspects, centuries, glaciers, locations
Abrekke valley glaciers, 183
Abyssinia, 280
"Account of the Passes and Cols of the Alps" (Arnod), 187
Africa, 25, 85, 280 (*see also* specific locations); shrinking of glaciers in, 103–4
Agassiz, L., 223
Agriculture, climate and, 8–17, 23ff., 50ff., 56ff., 65–70, 73–77, 92–95, 118–19, 123 (*see also* Harvests; specific aspects, crops, locations); eighteenth century, 70ff., 73–77; eleventh and sixteenth centuries, 270–72, 275–76, 285; glacial fluctuations and, 88, 90–95, 118–19, 123, 166, 179, 237–43; and rainfall and famine in fourteenth century, 45–48
Ahlmann, H. W., 82, 83, 105, 223, 239
Alaska, 26, 40–41, 101, 103, 106, 124, 207, 208, 254, 262, 272–73, 280, 296
Aletsch Glacier, 172–73, 174, 249–50, 251, 254, 264, 267, 274–75
Allalin Glacier, 140–41, 166, 172, 177, 184, 192, 203, 205, 206–7, 210, 252–53, 320–21, 326, 376
Alleé Blanche Glacier, 203, 205, 210
Allix, A., 19, 267
Alpine glaciers, 4, 9–11, 58, 60, 61, 64, 68, 78–79, 93, 94–95, 96, 97–98 (*see also* specific aspects, centuries, countries, glaciers, locations); advance (1215–1350) of, 248–64; climate in eleventh and sixteenth centuries and, 281–87; climate in Middle Ages and, 244–87 *passim*; recent small advance of, 334; retreat (1855–1955) of, 99–110ff., 227–43; secular oscillations of, 129–226 *passim*, 296–97
Amelioration, 80–128 (*see also* Ablation; specific aspects, centuries, locations); causes of, 305–6; climatic fluctuations in nineteenth and twentieth centuries, 80–128; effect on life of, 115ff.; effect on oceans of, 110–15; glaciation and, 93–110 *passim*

Ander Eggen (village), 141
Anes Alvarez, G., 53, 313
Angot, André, xiv, 3, 17, 19, 52–54, 77, 239–40, 270
Animal life, effect of climatic fluctuations on, 8, 115–19, 122–23. *See also* Ecology; specific aspects
Annecy, temperature data at, 228–31, 377
Antarctic region, 85, 104, 105
Anticyclones, 62, 76, 96, 301, 302, 304
Antwerp, 234, 344–45
Arakawa, H., 228, 272
Arc basin glaciers, 108, 207
Archaeology, 20, 26–42, 294–96. *See also* specific aspects, individuals, places
Arctic Ocean region, 26, 40–41, 85, 86, 88, 96, 99, 103, 116, 261, 304
Ardèche, Aven d' Orgnac cave in, 390
Arenthon, Jean d', 174, 180–81
Argentière Glacier, 104, 135, 136, 143, 147–51, 153, 155, 157, 169, 170, 171, 174, 194, 202, 205, 209, 210, 212, 215, 216, 239
Arizona, 3, 25, 26–37, 37–40, 49–50, 272–73, 294–96, 297; University of (Tucson School), 26ff., 37–40, 49–50, 272–73, 295
Arnod, P. A., 187–88
Arve, 136, 137, 138, 143, 144, 146–51, 155, 166, 170, 171, 174, 177–80, 195–98, 199–200ff., 214, 342
Arveyron cave, 195–98, 201, 205, 209, 216, 341, 342
Asau Glacier, 103
Asia, 3, 7, 17, 88, 103, 124. *See also* specific countries, locations
Aspen (Colorado) Symposium, 268–87; and diagrams of climate in eleventh and sixteenth centuries, 378–85
Atlantic (hypsithermal) climatic phase, 121–28, 254–64, 297
Atlantic Ocean, 85ff., 125, 255, 256. *See also* North Atlantic; Oceans (seas)
Atmosphere, warming trends and pollution of, 306
Austria, 83, 102, 108, 228, 256
Aven d' Orgnac, 390

Baardson, I., 253, 257–58
Baffin Island, 103